T0254608

Distributed Machine Learning with PySpark

Migrating Effortlessly from Pandas and Scikit-Learn

Abdelaziz Testas

Apress®

Distributed Machine Learning with PySpark: Migrating Effortlessly from Pandas and Scikit-Learn

Abdelaziz Testas
Fremont, CA, USA

ISBN-13 (pbk): 978-1-4842-9750-6
ISBN-13 (electronic): 978-1-4842-9751-3
https://doi.org/10.1007/978-1-4842-9751-3

Managing Director, Apress Media LLC: Welmoed Spahr
Acquisitions Editor: Celestin Suresh John
Development Editor: Laura Berendson
Coordinating Editor: Gryffin Winkler

Cover designed by eStudioCalamar

Cover image by Aakash Dhage on Unsplash (www.unsplash.com)

Distributed to the book trade worldwide by Apress Media, LLC, 1 New York Plaza, New York, NY 10004, U.S.A. Phone 1-800-SPRINGER, fax (201) 348-4505, e-mail orders-ny@springer-sbm.com, or visit www.springeronline.com. Apress Media, LLC is a California LLC and the sole member (owner) is Springer Science + Business Media Finance Inc (SSBM Finance Inc). SSBM Finance Inc is a **Delaware** corporation.

For information on translations, please e-mail booktranslations@springernature.com; for reprint, paperback, or audio rights, please e-mail bookpermissions@springernature.com.

Apress titles may be purchased in bulk for academic, corporate, or promotional use. eBook versions and licenses are also available for most titles. For more information, reference our Print and eBook Bulk Sales web page at http://www.apress.com/bulk-sales.

Any source code or other supplementary material referenced by the author in this book is available to readers on GitHub (https://github.com/Apress). For more detailed information, please visit https://www.apress.com/gp/services/source-code.

Paper in this product is recyclable

I dedicate this book to my family and colleagues for support and encouragement.

Table of Contents

About the Author

Abdelaziz Testas, PhD, is a data scientist with over a decade of experience in data analysis and machine learning, specializing in the use of standard Python libraries and Spark distributed computing. He holds a PhD in Economics from the University of Leeds and a master's degree in Finance from the University of Glasgow. He has completed several certificates in computer science and data science.

In the last ten years, the author worked for Nielsen in Fremont, California, as a lead data scientist, focusing on improving the company's audience measurement by planning, initiating, and executing end-to-end data science projects and methodology work, and drove advanced solutions into Nielsen's digital ad and content rating products by leveraging subject matter expertise in media measurement and data science. The author is passionate about helping others improve their machine learning skills and workflows and is excited to share his knowledge and experience with a wider audience through this book.

About the Technical Reviewer

Bharath Kumar Bolla has over 12 years of experience and is a senior data scientist at Salesforce, Hyderabad. Bharath obtained an MS in Data Science from the University of Arizona, USA. He also had a master's in Life Sciences from Mississippi State University, USA. Bharath worked as a research scientist for around seven years at the University of Georgia, Emory University, and Eurofins LLC. At Verizon, Bharath led a team to build a "Smart Pricing" solution, and at Happiest Minds, he worked on AI-based digital marketing products. Along with his day-to-day responsibilities, he is a mentor and an active researcher with more than 20 publications in conferences and journals. Bharath received the "40 Under 40 Data Scientists 2021" award from *Analytics India Magazine* for his accomplishments.

Acknowledgments

I would like to express my gratitude to all those who have directly or indirectly contributed to the creation and publication of this book. First and foremost, I am deeply thankful to my family for their unwavering support and encouragement throughout this journey.

I would like to acknowledge the invaluable assistance of Apress in bringing this book to fruition. Specifically, I wish to extend my heartfelt thanks to Celestin John, the Acquisitions Editor for AI and Machine Learning, for his guidance in shaping the main themes of this work and providing continuous feedback during the initial stages of the book. I am also grateful to Nirmal Selvaraj for his dedication and support as my main point of contact throughout the development cycle of the book. Additionally, I would like to express my appreciation to Laura Berendson for her advisory role and invaluable contributions as our Development Editor.

Lastly, I extend my gratitude to the technical reviewer Bharath Kumar Bolla who diligently reviewed the manuscript and provided valuable suggestions for improvement. Your meticulousness and expertise have significantly enhanced the quality and clarity of the final product.

To all those who have contributed, whether mentioned individually or not, your contributions have played an integral part in making this book a reality. Thank you for your support, feedback, and commitment to this project.

Introduction

In recent years, the amount of data generated and collected by companies and organizations has grown exponentially. As a result, data scientists have been pushed to process and analyze large amounts of data, and traditional single-node computing tools such as Pandas and Scikit-Learn have become inadequate. In response, many data scientists have turned to distributed computing frameworks such as Apache Spark, with its Python-based interface, PySpark.

PySpark has several advantages over single-node computing, including the ability to handle large volumes of data and the potential for significantly faster data processing times. Furthermore, because PySpark is built on top of Spark, a widely used distributed computing framework, it also offers a broader set of tools for data processing and machine learning.

While transitioning from Pandas and Scikit-Learn to PySpark may seem daunting, the transition can be relatively straightforward. Pandas/Scikit-Learn and PySpark offer similar APIs, which means that many data scientists can easily transition from one to the other.

In this context, this book will explore the benefits of using PySpark over traditional single-node computing tools and provide guidance for data scientists who are considering transitioning to PySpark.

In this book, we aim to provide a comprehensive overview of the main machine learning algorithms with a particular focus on regression and classification. These are fundamental techniques that form the backbone of many practical applications of machine learning. We will cover popular methods such as linear and logistic regression, decision trees, random forests, gradient-boosted trees, support vector machines, Naive Bayes, and neural networks. We will also discuss how these algorithms can be applied to real-world problems, such as predicting house prices, and the likelihood of diabetes as well as classifying handwritten digits or the species of an Iris flower and predicting whether a tumor is benign or malignant. Whether you are a beginner or an experienced practitioner, this book is designed to help you understand the core concepts of machine learning and develop the skills needed to apply these methods in practice.

This book spans 18 chapters and covers multiple topics. The first two chapters examine why migration from Pandas and Scikit-Learn to PySpark can be a seamless process, and address the challenges of selecting an algorithm. Chapters 3–6 build, train, and evaluate some popular regression models, namely, multiple linear regression, decision trees, random forests, and gradient-boosted trees, and use them to deal with some real-world tasks such as predicting house prices. Chapters 7–12 deal with classification issues by building, training, and evaluating widely used algorithms such as logistic regression, decision trees, random forests, support vector machines, Naive Bayes, and neural networks. In Chapters 13–15, we examine three additional types of algorithms, namely, recommender systems, natural language processing, and clustering with k-means. In the final three chapters, we deal with hyperparameter tuning, pipelines, and deploying models into production.

CHAPTER 1

An Easy Transition

One of the key factors in making the transition from Pandas and Scikit-Learn to PySpark relatively easy is the similarity in functionality. This similarity will become evident after reading this chapter and executing the code described herein.

One of the easiest ways to test the code is by signing up for an online Databricks Community Edition account and creating a workspace. Databricks provides detailed documentation on how to create a cluster, upload data, and create a notebook. Additionally, Spark can also be installed locally through the pip install pyspark command.

Another option is Google Colab. PySpark is preinstalled on Colab by default; otherwise, it can be installed using the !pip install pyspark command in a Colab notebook. This command will install PySpark and its dependencies in the Colab environment. While both provide Jupyter-like notebooks, one advantage of Databricks is that Colab is a single-core instance, whereas Databricks provides multi-node clusters for parallel processing. This feature makes Databricks better suited for handling larger datasets and more complex computational tasks in a collaborative team environment.

Although Pandas and Scikit-Learn are tools primarily designed for small data processing and analysis, while PySpark is a big data processing framework, PySpark offers functionality similar to Pandas and Scikit-Learn. This includes DataFrame operations and machine learning algorithms. The presence of these familiar functionalities in PySpark facilitates a smoother transition for data scientists accustomed to working with Pandas and Scikit-Learn.

In this chapter, we examine in greater depth the factors that contribute to the ease of transition from these small data tools (Pandas and Scikit-Learn) to PySpark. More specifically, we focus on PySpark and Pandas integration and the similarity in syntax between PySpark, on the one hand, and Pandas and Scikit-Learn, on the other.

A. Testas, *Distributed Machine Learning with PySpark*, https://doi.org/10.1007/978-1-4842-9751-3_1

PySpark and Pandas Integration

PySpark is well integrated with Pandas. Both the DataFrame and MLlib (Machine Learning Library) designs in PySpark were inspired by Pandas (and Scikit-Learn) concepts. The DataFrame concept in PySpark is similar to that of Pandas in that the data is stored in rows and columns similar to relational database tables and excel sheets. Consequently, it is easy to convert a PySpark DataFrame to a Pandas DataFrame and vice-versa, making the integration between the two packages straightforward.

A PySpark DataFrame can be converted to a Pandas DataFrame by using `toPandas()` method, while a Pandas DataFrame can be converted to a PySpark DataFrame using the `createDataFrame()` method.

Let's start with an example of how to convert a Pandas DataFrame to a PySpark DataFrame using the PySpark `createDataFrame()` method:

Step 1: Import Pandas

```
[In]: import pandas as pd
```

Step 2: Create a Pandas DataFrame

```
[In]: data = {'Country': ['United States', 'Brazil', 'Russia'], 'River':
      ['Mississippi', 'Amazon', 'Volga']}
[In]: pandas_df = pd.DataFrame(data)
```

Step 3: Convert Pandas DataFrame to PySpark DataFrame

```
[In]: from pyspark.sql import SparkSession
[In]: spark = SparkSession.builder.appName("BigRivers").getOrCreate()
[In]: pyspark_df = spark.createDataFrame(pandas_df)
```

Step 4: Display the content of the PySpark DataFrame

```
[In]: pyspark_df.show()
[Out]:
```

Country	River
United States	Mississippi
Brazil	Amazon
Russia	Volga

Notice that before converting the Pandas DataFrame to the PySpark DataFrame using the PySpark createDataFrame() method in step 3, we needed to create a Spark Session named spark. There are two lines of code to achieve this. In the first line (from pyspark.sql import SparkSession), we imported the SparkSession class from the pyspark.sql module. In the second line, we created a new instance of the Spark Session named spark (spark = SparkSession.builder.appName("BigRivers").getOrCreate()). The following is what each part of this line of code does:

- SparkSession.builder: This starts the process of configuring and building a new Spark Session.
- appName("BigRivers"): The appName method sets a name for the Spark application.
- getOrCreate(): This method either retrieves an existing Spark Session if one exists or creates a new one if none is found. It ensures that only one Spark Session is active at a time.

This Spark Session concept is fundamental in Apache Spark as it is an entry point to interacting with its features and APIs in a structured manner. It provides an environment for managing configurations, creating Spark DataFrame and Dataset objects, executing Spark operations, and coordinating tasks across a cluster.

This is particularly important if we are running the code in an environment like Google Colab. When using Databricks notebooks, however, the Spark Session is already created and initialized, so we can directly use the spark object to interact with Spark features, run queries, create DataFrames, and use various Spark libraries. We can now do the reverse: Create a PySpark DataFrame and then convert it to a Pandas DataFrame using the `toPandas()` method:

Step 1: Import Spark Session

```
[In]: from pyspark.sql import SparkSession
```

Step 2: Create a Spark Session

```
[In]: spark = SparkSession.builder.appName("BigRivers").getOrCreate()
```

Step 3: Create a PySpark DataFrame

```
[In]: data = [('United States', 'Mississippi'), ('Brazil', 'Amazon'),
      ('Russia', 'Volga')]
[In]: spark_df = spark.createDataFrame(data, ["Country", "River"])
```

Step 4: Convert Spark DataFrame to a Pandas DataFrame

```
[In]: pandas_df = spark_df.toPandas()
```

Step 5: Display the data

```
[In]: print(pandas_df)
[Out]:
```

	Country	River
0	United States	Mississippi
1	Brazil	Amazon
2	Russia	Volga

Notice that Pandas displays an index column in the output, providing a unique identifier for each row, whereas PySpark does not explicitly show an index column in its DataFrame output.

As we have seen in this section, it is easy to toggle between Pandas and PySpark, thanks to the close integration between the two libraries.

Similarity in Syntax

Another key factor that contributes to the smooth transition from small data tools (Pandas and Scikit-Learn) to big data with PySpark is the familiarity of syntax. PySpark shares a similar syntax with Pandas and Scikit-Learn in many cases. For example, square brackets ([]) can be used on Databricks or Google Colab to select columns directly from a PySpark DataFrame, just like in Pandas. Similarly, PySpark provides a DataFrame API that resembles Pandas, and Spark MLlib (Machine Learning Library) includes implementations of various machine learning algorithms found in Scikit-Learn.

In this section, we present examples of how PySpark code is similar to Pandas and Scikit-Learn, facilitating an easy transition for data scientists between these tools.

Loading Data

Let's start with loading and reading data. We can use the read_csv() function in Pandas to load a CSV file called data.csv:

```
[In]: import pandas as pd
[In]: pandas_df = pd.read_csv('data.csv')
```

The first line imports the Pandas library and renames it as pd for convenience. The second line uses the read_csv() function to read the CSV file named data.csv. The contents of the file are stored in a DataFrame called pandas_df.

In PySpark, we can use the spark.read.csv() method to do the same:

```
[In]: from pyspark.sql import SparkSession
[In]: spark = SparkSession.builder.appName("ReadCSV").getOrCreate()
[In]: spark_df = spark.read.csv('data.csv', header=True, inferSchema=True)
```

This code first imports the SparkSession class and creates a new Spark Session named spark with the application name ReadCSV. If an existing session with the same name already exists, it will be reused; otherwise, a new one will be created. The code then uses the spark.read.csv() method to read the CSV file named data.csv. The header parameter is set to True to indicate that the first row of the CSV file contains column names. The inferSchema parameter is set to True to enable PySpark to automatically infer the data types of each column. The contents of the file are stored in a DataFrame called spark_df.

We can see from the preceding examples that despite some code differences (e.g., PySpark requires us to specify the header and inferSchema options explicitly, while in Pandas, these options have default values that are used if we do not specify them), Pandas and PySpark read data in a fairly similar fashion.

Selecting Columns

Both Pandas and PySpark provide diverse approaches for selecting columns, and to some extent, their column selection syntax shares similarities. Let's start by exploring the square bracket method, beginning with Pandas:

```
[In]: import pandas as pd
[In]: pandas_df = pd.DataFrame({'President': ['George Washington', 'Abraham
      Lincoln', 'Franklin D. Roosevelt', 'John F. Kennedy', 'Barack
      Obama'], 'Year of Office': [1789, 1861, 1933, 1961, 2009]})
[In]: pandas_df[["President", "Year of Office"]]
[Out]:
```

	President	Year of Office
0	George Washington	1789
1	Abraham Lincoln	1861
2	Franklin D. Roosevelt	1933
3	John F. Kennedy	1961
4	Barack Obama	2009

The first line of code imports the Pandas library and renames it as pd for convenience. The second line creates a new Pandas DataFrame called pandas_df. The DataFrame is initialized with a dictionary where the keys represent the column names and the values represent the data for each column.

In this example, there are two columns named President and Year of Office, and each column has five rows of data. The third line of code selects specific columns from the DataFrame. The double square brackets notation `pandas_df[["President", "Year of Office"]]` is used to select both columns. The resulting DataFrame contains these selected columns.

Running PySpark code in a Databricks or Google Colab notebook, we can select columns using the double square brackets notation, just like in Pandas:

```
[In]: from pyspark.sql import SparkSession
[In]: spark = SparkSession.builder.appName("Presidents").getOrCreate()
[In]: spark_df = spark.createDataFrame([
```

```
                        ("George Washington", 1789),
                        ("Abraham Lincoln", 1861),
                        ("Franklin D. Roosevelt", 1933),
                        ("John F. Kennedy", 1961),
                        ("Barack Obama", 2009)],
                        ["President", "Year of Office"])
[In]: spark_df[["President", "Year of Office"]].show()
[Out]:
```

President	Year of Office
George Washington	1789
Abraham Lincoln	1861
Franklin D. Roosevelt	1933
John F. Kennedy	1961
Barack Obama	2009

The code first imports the SparkSession class and then creates a new Spark Session named spark with the application name Presidents. Next, it creates a new PySpark DataFrame called spark_df. This is initialized with a list of tuples where each tuple represents a row of data, and the column names are specified as a separate list. In this example, there are two columns named President and Year of Office, and each column has five rows of data. The code then selects the columns from the DataFrame using double square brackets notation. The resulting DataFrame contains the selected columns. The last line of code displays the resulting DataFrame using the `show()` method.

It is apparent from the double square brackets notation, `pandas_df[["President", "Year of Office"]]` and `spark_df[["President", "Year of Office"]]`, that Pandas and PySpark select columns in the same way.

Note The `show()` method in PySpark is used to display the results of a DataFrame, while Pandas does not have a similar method as the results are displayed immediately. This is because PySpark uses lazy evaluation to optimize data processing by delaying computation until necessary, while Pandas uses immediate evaluation. However, the `head()` method in Pandas can be used to display the first few rows of a DataFrame and can be thought of as the equivalent of the `show()` method in PySpark for sub-setting. On Databricks, PySpark provides the `display()` method, for example, `display(spark_df)`, which allows visualizing data within the Databricks notebook environment.

In PySpark, we have the option to select specific columns using the `select()` method. On the other hand, in Pandas, we can achieve the same result using the `filter()` function, as demonstrated here:

In PySpark:

```
[In]: spark_df.select("President", "Year of Office").show()
[Out]:
```

President	Year of Office
George Washington	1789
Abraham Lincoln	1861
Franklin D. Roosevelt	1933
John F. Kennedy	1961
Barack Obama	2009

In Pandas:

```
[In]: pandas_df.filter(items = ["President", "Year of Office"])
[Out]:
```

	President	Year of Office
0	George Washington	1789
1	Abraham Lincoln	1861
2	Franklin D. Roosevelt	1933
3	John F. Kennedy	1961
4	Barack Obama	2009

While the specific operations performed by select() and filter() methods are different, the resulting objects share the common trait of containing a subset of the original DataFrame's data.

Aggregating Data

Not only that Pandas and PySpark can read data and select columns in the same way, they can also aggregate data in a similar fashion.

Let's illustrate this with examples. First, we create a Pandas DataFrame with movie and revenue columns and aggregate by movie:

Step 1: Import pandas as pd

```
[In]: import pandas as pd
```

Step 2: Create example data in the form of a Python dictionary called pandas_data that contains two keys: movie and revenue

```
[In]: pandas_data = {
    'movie': ['Avengers', 'Frozen', 'Star Wars',
    'The Lion King', 'Harry Potter'],
    'revenue': [90000000, 70000000, 110000000, 80000000, 100000000]}
```

Step 3: Create a DataFrame from the dictionary

```
[In]: pandas_df = pd.DataFrame(pandas_data)
```

Step 4: Aggregate by movie using the groupby() method, and then calculate the mean revenue for each movie using the agg() method with a dictionary specifying that we want to calculate the mean for the revenue column

```
[In]: agg_pandas_df = pandas_df.groupby('movie').agg({'revenue':'mean'})
```

Step 5: Display data

```
[In]: print(agg_pandas_df)
[Out]:
```

	movie	revenue
0	Avengers	90000000
1	Frozen	70000000
2	Harry Potter	100000000
3	Star Wars	110000000
4	The Lion King	80000000

Here's the equivalent code to aggregate revenue by movie using PySpark:

Step 1: Import the SparkSession class and create a Spark Session

```
[In]: from pyspark.sql import SparkSession
[In]: spark = SparkSession.builder.appName("MovieRevenueAggregation").
    getOrCreate()
```

Step 2: Create example data

```
[In]: spark_data = [('Avengers', 90000000), ('Frozen', 70000000), ('Harry
    Potter', 100000000), ('Star Wars', 110000000), ('The Lion King',
    80000000)]
spark_df = spark.createDataFrame(spark_data, ['movie', 'revenue'])
```

Step 3: Aggregate by movie

```
[In]: agg_spark_df = spark_df.groupBy('movie').agg({'revenue': 'mean'})
```

Step 4: Display data.

```
[In]: agg_spark_df.show()
[Out]:
```

movie	revenue
Avengers	90000000
Frozen	70000000
Harry Potter	100000000
Star Wars	110000000
The Lion King	80000000

As can be seen from the code examples shown previously, both Pandas and PySpark implement the `groupby()` and `agg()` methods to aggregate data. This is no coincidence as PySpark DataFrame structure has been inspired by that of Pandas.

Filtering Data

Having demonstrated that Pandas and PySpark have similarities in syntax for loading and reading data, selecting columns, and aggregating data, we can now look at examples of how they both use similar syntax to filter data.

Let's start with Pandas. We create a DataFrame called pandas_df with four rows of data and two columns, animal and muscle_power, and then filter it to show only the rows where the muscle_power is greater than 350:

Step 1: Import the Pandas library and rename it as pd

```
[In]: import pandas as pd
```

Step 2: Create an example DataFrame

```
[In]: pandas_df = pd.DataFrame({
          'animal': ['Lion', 'Elephant', 'Tiger', 'Gorilla'],
          'muscle_power': [400, 6000, 350, 800]})
```

Step 3: Filter the DataFrame to show only rows where the muscle_power is greater than 350:

```
[In]: filtered_pandas_df = pandas_df[pandas_df['muscle_power'] > 350]
```

Step 4: Print the results

```
[In]: print(filtered_pandas_df)
[Out]:
```

	animal	muscle_power
0	Lion	400
1	Elephant	6000
2	Gorilla	800

In PySpark, we follow the following steps:

Step 1: Import the SparkSession class and create a Spark Session

```
[In]: from pyspark.sql import SparkSession
[In]: spark = SparkSession.builder.appName("FilterByMusclePower").
      getOrCreate()
```

Step 2: Create a list of tuples called spark_data, with each tuple containing an animal and muscle_power

```
[In]: spark_data = [("Lion", 400), ("Elephant", 6000), ("Tiger", 350),
      ("Gorilla", 800)]
```

Step 3: Create a PySpark DataFrame called spark_df from the spark_data list, with column names animal and muscle_power

```
[In]: spark_df = spark.createDataFrame(spark_data, ["animal", "muscle_
      power"])
```

Step 4: Filter the DataFrame to show only rows where the muscle_power is greater than 350:

```
[In]: filtered_spark_df = spark_df[spark_df['muscle_power'] > 350]
```

Step 5: Display the filtered DataFrame using the show method

```
[In]: filtered_spark_df.show()
[Out]:
```

animal	muscle_power
Lion	400
Elephant	6000
Gorilla	800

We can also use the `query()` and `filter()` methods in Pandas and PySpark, respectively, to show only rows where muscle_power is greater than 350. These are equivalent to square brackets notation, `pandas_df[pandas_df['muscle_power'] > 350]` and `spark_df[spark_df['muscle_power'] > 350]`.

In Pandas:

```
[In]: pandas_df.query("muscle_power > 350")
[Out]:
```

	animal	muscle_power
0	Lion	400
1	Elephant	6000
2	Gorilla	800

In PySpark:

```
[In]: spark_df.filter("muscle_power > 350").show()
[Out]:
```

animal	muscle_power
Lion	400
Elephant	6000
Gorilla	800

There are at least two more ways to achieve the same filtering result in Spark DataFrame:

1. Using SQL-like syntax with where clause:

```
[In]: spark_df.where("muscle_power > 350").show()
[Out]:
```

animal	muscle_power
Lion	400
Elephant	6000
Gorilla	800

2. Using column expressions:

```
[In]: from pyspark.sql.functions import col
[In]: spark_df.filter(col("muscle_power") > 350).show()
[Out]:
```

animal	muscle_power
Lion	400
Elephant	6000
Gorilla	800

The preceding examples demonstrate that both Pandas and PySpark provide various methods for data filtering and, to a significant extent, there are similarities in their respective approaches.

Joining Data

In both Pandas and PySpark, joining two DataFrames requires specifying a common column between the DataFrames and the type of join to perform (inner, outer, left, right). The code syntax for joining DataFrames in Pandas and PySpark can be similar, as demonstrated in the following examples. Here, we will demonstrate the inner join, but the syntax remains the same for other join types; only the join type needs to be modified.

In Pandas, we can use the `merge()` method to merge or join two DataFrames, pandas_df1 and pandas_df2, on a common column, and create a new DataFrame called merged_pandas_df. For example, we want to merge two DataFrames based on the job title, so we can compare and analyze the salary differences for the same job title across different US cities. The merged DataFrame allows us to see the salary information from both DataFrame 1 and DataFrame 2 side by side for the common job titles.

Step 1: Create the first DataFrame

```
[In]: import pandas as pd
[In]: pandas_data1 = {'Job Title': ['Software Engineer', 'Data Analyst',
      'Project Manager'],
         'Salary': [100000, 75000, 90000],
         'Location': ['San Francisco', 'New York', 'Seattle']}
```

```
[In]: pandas_df1 = pd.DataFrame(pandas_data1)
[In]: print(pandas_df1)
[Out]:
```

	Job Title	Salary	Location
0	Software Engineer	100000	San Francisco
1	Data Analyst	75000	New York
2	Project Manager	90000	Seattle

Step 2: Create the second DataFrame

```
[In]: pandas_data2 = {'Job Title': ['Software Engineer', 'Data Scientist',
      'Project Manager'],
          'Salary': [120000, 95000, 90000],
          'Location': ['Los Angeles', 'Chicago', 'Boston']}
[In]: pandas_df2 = pd.DataFrame(pandas_data2)
[In]: print(pandas_df2)
[Out]:
```

	Job Title	Salary	Location
0	Software Engineer	120000	Los Angeles
1	Data Scientist	95000	Chicago
2	Project Manager	90000	Boston

Step 3: Perform an inner join on the Job Title column

```
[In]: merged_pandas_df = pd.merge(pandas_df1, pandas_df2, on='Job Title',
      how='inner')
[In]: print(merged_pandas_df)
[Out]:
```

	Job Title	Salary_x	Location_x	Salary_y	Location_y
0	Software Engineer	100000	San Francisco	120000	Los Angeles
1	Project Manager	90000	Seattle	90000	Boston

In PySpark, we can use the join() method to do the same:

Step 1: Create the first DataFrame

```
[In]: from pyspark.sql import SparkSession
[In]: spark = SparkSession.builder.getOrCreate()
[In]: spark_data1 = {'Job Title': ['Software Engineer', 'Data Analyst',
      'Project Manager'],
         'Salary': [100000, 75000, 90000],
         'Location': ['San Francisco', 'New York', 'Seattle']}
[In]: Spark_df1 = spark.createDataFrame(pd.DataFrame(spark_data1))
[In]: spark_df1.show()
[Out]:
```

Job Title	Salary	Location
Software Engineer	100000	San Francisco
Data Analyst	75000	New York
Project Manager	90000	Seattle

Step 2: Create the second DataFrame

```
[In]: spark_data2 = {'Job Title': ['Software Engineer', 'Data Scientist',
      'Project Manager'],
         'Salary': [120000, 95000, 90000],
         'Location': ['Los Angeles', 'Chicago', 'Boston']}
[In]: spark_df2 = spark.createDataFrame(pd.DataFrame(spark_data2))
[In]: spark_df2.show()
[Out]:
```

Job Title	Salary	Location
Software Engineer	120000	Los Angeles
Data Scientist	95000	Chicago
Project Manager	90000	Boston

Step 3: Perform an inner join on the Job Title column

```
[In]: joined_spark_df = spark_df1.join(spark_df2, on='Job Title',
      how='inner')
[In]: joined_spark_df.show()
[Out]:
```

Job Title	Salary	Location	Salary	Location
Software Engineer	100000	San Francisco	120000	Los Angeles
Project Manager	90000	Seattle	90000	Boston

Notice that in Pandas, the suffixes _x and _y are automatically added to the column labels when there are overlapping column names in the two DataFrames being merged. This helps to differentiate between the columns with the same name in the merged DataFrame. In the preceding example, both pandas_df1 and pandas_df2 have a column named Salary and a column named Location. When performing the merge, Pandas automatically appends the suffixes _x and _y to the column labels to distinguish them.

This, however, is not the case in PySpark. It does not automatically append suffixes like _x and _y to the column labels when there are overlapping column names in the DataFrames being joined.

Saving Data

Both Pandas and PySpark provide methods to save/write the content of a DataFrame to a CSV file. In the following example, we use the Pandas `to_csv()` method to write a DataFrame named pandas_df to a CSV file called pandas_output.csv in the current working directory. The parameter index=False specifies that the row index should not be written to the output file. If the index parameter is not specified or set to True, the row index will be included in the output file as a separate column:

```
[In]: pandas_df.to_csv('pandas_output.csv', index=False)
```

In PySpark, we can use the `write()` method to write a DataFrame named spark_df to a CSV file called spark_output.csv:

```
[In]: spark_df.write.csv('spark_output.csv', header=True, mode='overwrite')
```

The header=True parameter specifies that we want to include a header row in the output file containing the column names. If this parameter is not specified or set to False, the output file will not have a header row. The mode=overwrite means that if the output file already exists, it will be overwritten. Other possible values for the mode parameter include append, which will append the data to the existing file, and ignore, which will do nothing if the file already exists.

Note When writing to a CSV file in PySpark, the data is split into multiple CSV files and written to a directory. The output file name specified is the name of the directory, not the file itself.

Modeling Steps

Both Scikit-Learn and PySpark adhere to the same fundamental modeling steps: data preparation, model training, model evaluation, and prediction. As we will see from the comparison shown here, some class or function names associated with specific steps can be identical between the two frameworks. Here, we present common machine learning functions in Scikit-Learn and their corresponding equivalents in PySpark.

1. Data preparation:

   Scikit-Learn: Functions such as `train_test_split` for splitting data into training and testing sets, `StandardScaler` for feature scaling, `OneHotEncoder` for one-hot encoding categorical variables, and `SimpleImputer` for handling missing values.

   PySpark: Operations on Spark DataFrames, such as `randomSplit` for splitting data, `StandardScaler` for feature scaling, `OneHotEncoder` for one-hot encoding categorical variables, and `Imputer` for handling missing values. Additionally, the `VectorAssembler` is used to combine multiple feature columns into a single vector column.

2. Model training:

   Scikit-Learn: Models have a `fit` method for training, where you provide the features (X_train) and target variable (y_train) as input.

   PySpark: Models have a `fit` method as well, where you provide a DataFrame (train_data) with features and a column for the target variable. The `VectorAssembler` step is required before training the model to combine the feature columns into a single vector column.

3. Model evaluation:

   Scikit-Learn: Various evaluation metrics are available, such as `mean_squared_error` and `r2_score` for regression and `accuracy_score` for classification. These metrics are applied directly to the predicted and actual target variable values (y_pred, y_test).

   PySpark: The `RegressionEvaluator` or `BinaryClassificationEvaluator` can be used to compute evaluation metrics, such as mean squared error for regression tasks or areaUnderROC for binary classification tasks. These evaluators require the predicted and actual values to be specified (predictions, y_test), along with the respective columns.

4. Prediction:

   Scikit-Learn: Models have a `predict` method to generate predictions for new data, where you provide the features (X_test) and obtain the predicted target variable values (y_pred).

   PySpark: Models have a `transform` method to generate predictions for new data, where you provide a DataFrame (test_data) with features and obtain the predicted values in a new column (prediction). The `VectorAssembler` step is required before making predictions to ensure the new data has the same feature vector structure.

Pipelines

The concept of pipeline (a way to streamline and automate the sequence of steps involved in data preprocessing, feature engineering, model training, and evaluation) is quite similar in Scikit-Learn and PySpark, even though there are differences in the specific syntax and classes used. Here is how the two libraries approach the pipeline concept:

1. Importing the `Pipeline` class:

   In Scikit-Learn, we import the `Pipeline` class from sklearn.pipeline.

   In PySpark, we import the `Pipeline` class from the pyspark.ml package.

2. Defining pipeline steps:

   In Scikit-Learn, we define the steps of the pipeline using a list of tuples, where each tuple contains a name and an instance of a transformer or estimator.

   In PySpark, we define the pipeline steps using a list of Transformer and Estimator objects. Each object represents a different stage of the pipeline, such as preprocessing and model training. However, PySpark does not require explicit naming of the steps like in Scikit-Learn.

3. Creating the pipeline:

   In Scikit-Learn, we create the pipeline by passing the steps list to the `Pipeline` class.

   In PySpark, we create the pipeline by passing the steps list to the `Pipeline` class as well.

4. Fitting the pipeline:

   In Scikit-Learn, we fit the pipeline to the training data by calling the `fit` method on the pipeline object and providing the training data (X_train and y_train).

In PySpark, we fit the pipeline to the training data by calling the `fit` method on the pipeline object and providing the training data as well.

5. Transforming and training:

   In Scikit-Learn, the pipeline applies the preprocessing step first, followed by the model training step. It sequentially transforms the data and trains the model.

   In PySpark, the pipeline also applies the transformations and model training in a sequential manner. Each stage of the pipeline processes the data and passes it to the next stage until the final model training.

6. Making predictions:

   In Scikit-Learn, we make predictions on new data (X_test) using the pipeline by calling the `predict` method.

   In PySpark, we make predictions on new data using the fitted pipeline by calling the `transform` method on the pipeline object and providing the new data.

Let's now translate these steps into code. In Scikit-Learn, we start by importing the required libraries:

Step 1: Import necessary libraries

```
[In]: from sklearn.pipeline import Pipeline
[In]: from sklearn.preprocessing import ...
[In]: from sklearn.feature_selection import ...
[In]: from sklearn.linear_model import ...
```

Step 2: Define the pipeline steps

```
[In]: steps = [
    ('preprocessing', PreprocessingStep()),
    ('feature_selection', FeatureSelectionStep()),
    ('regressor', RegressionModel())
]
```

Step 3: Create the pipeline

```
[In]: pipeline = Pipeline(steps)
```

Step 4: Fit the pipeline to the data

```
[In]: pipeline.fit(X_train, y_train)
```

Step 5: Make predictions

```
[In]: y_pred = pipeline.predict(X_test)
```

In PySpark, we also start by importing the required libraries:

Step 1: Import necessary libraries

```
[In]: from pyspark.ml import Pipeline
[In]: from pyspark.ml.feature import ...
[In]: from pyspark.ml.regression import ...
```

Step 2: Define the pipeline stages

```
[In]: stages = [
    PreprocessingStep(),
    FeatureSelectionStep(),
    ModelStep()
]
```

Step 3: Create the pipeline

```
[In]: pipeline = Pipeline(stages=stages)
```

Step 4: Fit the pipeline to the data

```
[In]: pipelineModel = pipeline.fit(trainData)
```

Step 5: Make predictions

```
[In]: predictions = pipelineModel.transform(testData)
```

Summary

In this chapter, we examined the factors that contribute to the smooth transition from small data tools (Pandas and Scikit-Learn) to big data processing with PySpark. Specifically, we focused on the integration between PySpark and Pandas, as well as the similarities in syntax among PySpark, Pandas, and Scikit-Learn. We have also shown how both libraries use similar modeling and pipeline steps, even though there are differences in the specific syntax and classes used.

In the next chapter, we embark on our journey into machine learning, which forms the core of this book. We will introduce k-fold cross-validation, a technique that helps us select the best-performing model from a range of different algorithms. Using this method, we can short-list the top-performing models for further evaluation and testing.

CHAPTER 2

Selecting Algorithms

Machine learning offers a variety of algorithms for both supervised and unsupervised learning tasks, each with numerous parameters to fine-tune. However, testing and optimizing all of these models in each category would be incredibly cumbersome and require significant computational power. To address this challenge, this chapter introduces k-fold cross-validation, a technique that helps select the best-performing model from a range of different algorithms. With this method, the top-performing models can be short-listed for further evaluation and testing.

Another reason for using cross-validation is that it can guard against overfitting (when an algorithm does well on the training set but performs poorly on new, unseen data). Instead of a single train/test split, k-fold cross-validation can be used. We divide data into k subsets (folds) and train the model k times, each time using a different fold as the test set and the remaining folds as the training set. We calculate the average performance across all folds. If the model consistently performs better on the training folds compared to the test folds, it could indicate overfitting.

To demonstrate the similarities between Scikit-Learn and PySpark machine learning libraries, we will apply them to the same dataset. We will employ four algorithms for illustration: logistic regression, decision tree classifier, random forest classifier, and support vector classifier. The metric of choice will be the Area Under the Curve (AUC), which measures a classification model's ability to distinguish between positive and negative classes.

Before building our model, let's first explore the dataset using Pandas and PySpark. This will allow us to see how these two libraries utilize similar approaches to reading and processing data.

A. Testas, *Distributed Machine Learning with PySpark*, https://doi.org/10.1007/978-1-4842-9751-3_2

The Dataset

The dataset used for this project is known as the Pima Indians Diabetes Database. Pima refers to the Pima people, who are a group of Native American tribes indigenous to the Southwestern United States, primarily living in what is now Arizona. The Pima people have a unique cultural heritage and history.

The dataset typically contains 768 records or instances. Each record represents a Pima Indian woman and includes various health-related attributes, along with a target variable indicating whether the woman developed diabetes or not. This dataset is commonly used in machine learning and data analysis for classification tasks aimed at predicting diabetes outcomes based on the provided attributes.

The dataset was originally created by the National Institute of Diabetes and Digestive and Kidney Diseases. For the purpose of this analysis, we sourced it from Kaggle—a website designed for data scientists. The following are the contributor's name, approximate upload date, dataset name, site name, and the URL from which we downloaded a copy of the CSV file:

Title: Pima Indians Diabetes Database
Source: Kaggle
URL: `www.kaggle.com/uciml/pima-indians-diabetes-database`
Contributor: UCI Machine Learning
Date: 2016

We start by defining a Python function named `load_diabetes_data()` to load the diabetes data from a CSV file named diabetes.csv located on the GitHub repository. The data is loaded using the Pandas `read_csv()` function, which reads the CSV file into a Pandas DataFrame.

```
[In]: import pandas as pd
[In]: def load_diabetes_data():
          url = ('https://raw.githubusercontent.com/abdelaziztestas/'
                 'spark_book/main/diabetes.csv')
          return pd.read_csv(url)
```

We then create a Pandas DataFrame by calling the `load_diabetes_data()` function:

```
[In]: pandas_df = load_diabetes_data()
```

Next, we convert the Pandas DataFrame we have just created (pandas_df) to a PySpark DataFrame named spark_df using the PySpark createDataFrame() method:

```
[In]: from pyspark.sql.session import SparkSession
[In]: spark = SparkSession.builder.appName("diabetes_data").getOrCreate()
[In]: spark_df = spark.createDataFrame(pandas_df)
```

In the next step, we take a look at the top five rows using the Pandas head() method:

```
[In]: pandas_df.head()
[Out]:
```

	Pregnancies	Glucose	Blood Pressure	Skin Thickness	Insulin	BMI	Diabetes Pedigree Function	Age	Outcome
0	6	148	72	35	0	33.6	0.627	50	1
1	1	85	66	29	0	26.6	0.351	31	0
2	8	183	64	0	0	23.3	0.672	32	1
3	1	89	66	23	94	28.1	0.167	21	0
4	0	137	40	35	168	43.1	2.288	33	1

In PySpark, we use the show() method to display the contents of the spark_df:

```
[In]: spark_df.show(5)
[Out]:
```

Pregnancies	Glucose	BloodPressure	Skin Thickness	Insulin	BMI	Diabetes Pedigree Function	Age	Outcome
6	148	72	35	0	33.6	0.627	50	1
1	85	66	29	0	26.6	0.351	31	0
8	183	64	0	0	23.3	0.672	32	1
1	89	66	23	94	28.1	0.167	21	0
0	137	40	35	168	43.1	2.288	33	1

We can see from the output of both Pandas and PySpark that there are nine columns (eight attributes and one outcome variable). Pregnancies refer to the count of pregnancies a woman has experienced. Glucose pertains to the concentration of glucose in the bloodstream two hours after an oral glucose tolerance test. BloodPressure signifies the diastolic blood pressure, measured in millimeters of mercury (mm Hg). SkinThickness is the thickness of the triceps skinfold, measured in millimeters. Insulin represents the serum insulin level after two hours, measured in micro international units per milliliter (mu U/ml). BMI, or body mass index, is calculated as an individual's weight in kilograms divided by the square of their height in meters, offering insights into their body composition. DiabetesPedigreeFunction assesses the likelihood of diabetes based on familial history and is assigned a score. Age stands for the person's age in years. Lastly, Outcome serves as a class variable, taking values of either 0 or 1, indicating the presence or absence of a particular condition.

We can use the Pandas `info()` method to display a summary of pandas_df, including the number of non-null values and data types of each column:

```
[In]: pandas_df.info()
[Out]:

<class 'pandas.core.frame.DataFrame'>
RangeIndex: 768 entries, 0 to 767
Data columns (total 9 columns):
 #   Column                    Non-Null Count  Dtype
---  ------                    --------------  -----
 0   Pregnancies               768 non-null    int64
 1   Glucose                   768 non-null    int64
 2   BloodPressure             768 non-null    int64
 3   SkinThickness             768 non-null    int64
 4   Insulin                   768 non-null    int64
 5   BMI                       768 non-null    float64
 6   DiabetesPedigreeFunction  768 non-null    float64
 7   Age                       768 non-null    int64
 8   Outcome                   768 non-null    int64
dtypes: float64(2), int64(7)
memory usage: 54.1 KB
```

The Pandas output shows that there are 9 columns (as we already know) and 768 instances, all of which are numerical and none of which are null. The data types are either int64 (integer) or float64 (decimal).

In PySpark, we can use the `printSchema()` method to achieve a similar functionality, where the *long* and *double* data types are the equivalent of Pandas *int64* and *float64*, respectively:

```
[In]: spark_df.printSchema()
[Out]:

root
 |-- Pregnancies: long (nullable = true)
 |-- Glucose: long (nullable = true)
 |-- BloodPressure: long (nullable = true)
 |-- SkinThickness: long (nullable = true)
 |-- Insulin: long (nullable = true)
 |-- BMI: double (nullable = true)
 |-- DiabetesPedigreeFunction: double (nullable = true)
 |-- Age: long (nullable = true)
 |-- Outcome: long (nullable = true)
```

While the `printSchema()` method doesn't show the number of observations, we can use the `count()` method to accomplish this task:

```
[In]: spark_df.count()
[Out]: 768
```

We can use the Pandas `describe()` method to compute summary statistics (count, mean, standard deviation, minimum, and maximum values) for each of the eight attributes. The `drop()` method allows us to exclude the Outcome column since it's a categorical variable that is either 1 (woman has diabetes) or 0 (woman does not have diabetes). The axis=1 parameter specifies that we want to drop a column, not a row:

```
[In]: pandas_df.drop('Outcome', axis=1).describe()
[Out]:
```

	Pregnancies	Glucose	Blood Pressure	Skin Thickness	Insulin	BMI	Diabetes Pedigree Function	Age
count	768	768	768	768	768	768	768	768
mean	3.85	120.89	69.11	20.54	79.80	31.99	0.47	33.24
std	3.37	31.97	19.36	15.95	115.24	7.88	0.33	11.76
min	0.00	0.00	0.00	0.00	0.00	0.00	0.08	21.00
25%	1.00	99.00	62.00	0.00	0.00	27.30	0.24	24.00
50%	3.00	117.00	72.00	23.00	30.50	32.00	0.37	29.00
75%	6.00	140.25	80.00	32.00	127.25	36.60	0.63	41.00
max	17.00	199.00	122.00	99.00	846.00	67.10	2.42	81.00

In PySpark, we can obtain the same results using the `summary()` method, which would give us the same summary statistics, including the 25%, 50%, and 75% percentiles. While we could use the `describe()` method as in Pandas, it wouldn't show these percentiles:

```
[In]: spark_df.drop('Outcome').summary().show()
[Out]:
```

summary	Pregnancies	Glucose	Blood Pressure	BMI	Diabetes Pedigree Function	Age
count	724	724	724	724	724	724
mean	3.87	121.88	72.40	32.47	0.47	33.35
stddev	3.36	30.75	12.38	6.89	0.33	11.77
min	0	44	24	18.2	0.078	21
25%	1	99	64	27.5	0.245	24
50%	3	117	72	32.4	0.378	29
75%	6	142	80	36.6	0.627	41
max	17	199	122	67.1	2.42	81

Turning now to the Outcome variable, we can find out how many women have diabetes and how many do not have diabetes using the `value_counts()` method in Pandas:

```
[In]: pandas_df["Outcome"].value_counts()
[Out]:

0 500
1 268
Name: Outcome, dtype: int64
```

In PySpark, we can use the `groupBy()` and `count()` methods to get the same results:

```
[In]: spark_df.groupBy("Outcome").count().show()
[Out]:

+-------+-----+
|Outcome|count|
+-------+-----+
|      0|  500|
|      1|  268|
+-------+-----+
```

Both Pandas and PySpark return 268 women for category 1 (i.e., woman has diabetes) and 500 women for category 0 (i.e., woman does not have diabetes). This, as expected, adds up to the total of 768 observations.

In the next step, we will check if there are missing values. This is important because machine learning algorithms generally require complete data sets to be trained effectively. This is because the algorithms use statistical methods to find patterns in the data, and missing data can disrupt the accuracy of those patterns.

We saw earlier, using the `info()` and `printSchema()` methods, that there are no null values. We can replicate this by counting the number of null values in each column of the pandas_df using the Pandas `isnull()` and `sum()` methods:

```
[In]: pandas_df.isnull().sum()
[Out]:

Pregnancies                0
Glucose                    0
BloodPressure              0
SkinThickness              0
```

```
Insulin                     0
BMI                         0
DiabetesPedigreeFunction    0
Age                         0
Outcome                     0
dtype: int64
```

In PySpark, we can also use the `isNull()` and `sum()` methods but in a slightly different way. More precisely, we use `sum(col(c).isNull().cast("int")).alias(c)` to generate a PySpark expression that will count the number of null values in a single column c of the dataframe (c is a variable that is used in a list comprehension to generate a list of expressions, where each expression calculates the number of null values in the corresponding column and gives it an alias with the same column name):

```
[In]: from pyspark.sql.functions import sum, col
[In]: spark_df.select([sum(col(c).isNull().cast("int")).alias(c) for c in
      spark_df.columns]).show()
[Out]:
```

Pregnancies	Glucose	Blood Pressure	BMI	Diabetes Pedigree Function	Age	Outcome
0	0	0	0	0	0	0

Both Pandas and PySpark outputs suggest that there are no null values in the dataset. However, having 0 nulls does not always indicate that there are no missing values in a dataset. This is because in some cases, missing values are replaced with 0s.

Let's use the following Pandas code to calculate the total number of 0 values in each of the nine columns, including the Outcome variable:

```
[In]: print((pandas_df == 0).sum())
[Out]:

Pregnancies                 111
Glucose                       5
BloodPressure                35
SkinThickness               227
```

```
Insulin                     374
BMI                          11
DiabetesPedigreeFunction      0
Age                           0
Outcome                     500
dtype: int64
```

In PySpark, we first import the sum() and col() functions then generate expressions that count the number of zeros in each column:

```
[In]: from pyspark.sql.functions import sum, col
[In]: exprs = [sum((col(c) == 0).cast("int")).alias(c) for c in spark_
      df.columns]
```

Next, we apply the expressions to the dataframe and show the result:

```
[In]: spark_df.select(exprs).show()
[Out]:
```

Pregnancies	Glucose	Blood Pressure	Skin Thickness	Insulin	BMI	Diabetes Pedigree Function	Age	Outcome
111	5	35	227	374	11	0	0	500

Pandas and PySpark both indicate that there are no 0 cases for Diabetes Pedigree Function and Age, but there are many cases with 0 values for the other variables. For example, 111 cases have 0 values for the Pregnancy variable, which makes sense since some females in the sample have no kids. Additionally, 500 cases have 0 values for the Outcome variable, indicating that these women do not have diabetes. However, it is illogical for Glucose, Blood Pressure, Skin Thickness, Insulin, or BMI to have 0 values.

There are three common ways to deal with invalid readings: exclude columns or features with 0 values, exclude rows with 0 values, or impute 0 values with mean or average values. For this chapter, we have chosen options 1 and 2:

- Option 1: Exclude Skin Thickness and Insulin as the number of cases with 0 values is too large (227 and 374, respectively). Excluding rows instead of columns would make the sample too small.

- Option 2: Exclude rows with 0 values for Glucose, Blood Pressure, and BMI. The number of invalid cases is not too large, so excluding rows won't significantly impact the sample size.

The following is the code to achieve this in Pandas:

```
[In]: pandas_df = pandas_df.loc[(pandas_df['Glucose'] != 0)
& (pandas_df['BloodPressure'] != 0)
& (pandas_df['BMI'] != 0), ['Pregnancies', 'Glucose', 'BloodPressure',
'BMI', 'DiabetesPedigreeFunction', 'Age', 'Outcome']]
```

In PySpark:

```
[In]: from pyspark.sql.functions import col
[In]: spark_df = spark_df.filter((col('Glucose') != 0)
& (col('BloodPressure') != 0)
& (col('BMI') != 0)).select(['Pregnancies', 'Glucose', 'BloodPressure',
'BMI', 'DiabetesPedigreeFunction', 'Age', 'Outcome'])
```

We end up with 7 columns (6 features and 1 class variable) and 724 rows. We can confirm these counts by using the shape attribute in Pandas:

```
[In]: print(pandas_df.shape)
[Out]: (724, 7)
```

PySpark doesn't have the shape attribute, but we can accomplish the same task by combining the count() and len() functions:

```
[In]: print((spark_df.count(), len(spark_df.columns)))
[Out]: (724, 7)
```

Selecting Algorithms with Cross-Validation

Cross-validation is a method of evaluating the performance of a machine learning model by splitting the dataset into k-folds. The model is then trained on k-1 folds and tested on the remaining fold. This process is repeated k times, with each fold serving as the test set once.

This method is superior to training and testing the model on just one dataset for a number of reasons:

1. K-fold cross-validation allows for a better estimation of the model's performance by reducing the risk of overfitting to a particular training set. By using multiple folds of the data to train and evaluate the model, we can get a better sense of how well the model will perform on unseen data.

2. With k-fold cross-validation, we can obtain multiple estimates of the model performance, and therefore, we can make a more informed decision about which model to choose. This approach reduces the likelihood of selecting a model that performs well on a specific training set but poorly on unseen data.

3. By splitting the data into multiple folds, we can use more data to train the model, which can lead to better performance. This approach is especially useful when the size of the dataset is limited.

To illustrate how k-fold cross-validation works, we will compare the performance of four classifiers (even though we can use any number of models): logistic regression, decision tree classifier, random forest classifier, and support vector classifier. The metric of choice will be the area under the ROC curve (receiver operating characteristic curve), also known as AUC (Area Under the Curve). We will explain this important metric in more detail once we evaluate the models.

Before computing the ROC AUC for each classifier, we will first standardize the features using the StandardScaler function. Standard scaling, also known as z-score normalization or standardization, is a data preprocessing technique used to transform numerical features to have a mean of 0 and a standard deviation of 1. It involves subtracting the mean of each feature from the data and then dividing by the standard deviation.

The standard scaling formula for a feature x is

x_scaled = (x - mean(x)) / std(x)

where mean(x) is the mean of feature x and std(x) is its standard deviation.

Standard scaling is an essential step in data preprocessing to make features comparable and improve the performance and stability of machine learning algorithms. It ensures that features have a consistent scale, which is especially important when dealing with algorithms sensitive to the scale of input data or when interpreting the model's coefficients. We will use Scikit-Learn and PySpark to perform classification on the same diabetes dataset described in the previous section. We will use the same list of classifiers and evaluate each of them using ten-fold cross-validation.

For the purposes of this chapter, we will use default hyperparameters. While both Scikit-Learn and PySpark provide default hyperparameters for their models, these default values are not always identical. We will demonstrate how to fine-tune these hyperparameters in Chapter 16 using grid search for improved model performance.

There are other differences between Scikit-Learn and PySpark. The first is that Scikit-Learn separates cross-validation from hyperparameter tuning while PySpark combines them by default. The second is that PySpark combines the features in a single vector while Scikit-Learn does not.

Scikit-Learn

In this subsection, we demonstrate how to preprocess data, define classification models, and perform k-fold cross-validation to evaluate the models' performance using the Scikit-Learn platform. We will be using the same pandas_df DataFrame that was generated earlier with the following code:

```
[In]: url = 'https://raw.githubusercontent.com/abdelaziztestas/
      spark_book/main/diabetes.csv'
[In]: pandas_df = pd.read_csv(url)
[In]: pandas_df = pandas_df.loc[(pandas_df['Glucose'] != 0)
                            & (pandas_df['BloodPressure'] != 0)
                            & (pandas_df['BMI'] != 0),
                            ['Pregnancies',
                             'Glucose',
                             'BloodPressure',
                             'BMI',
                             'DiabetesPedigreeFunction',
                             'Age',
                             'Outcome']]
```

After generating this DataFrame that filters out invalid diabetes readings, the first step in the k-fold cross-validation process is to import the required libraries:

Step 1: Import necessary packages

- StandardScaler for preprocessing
- cross_val_score for cross-validation
- LogisticRegression, DecisionTreeClassifier, RandomForestClassifier, and LinearSVC (Linear Support Vector Classification) for classification

```
[In]: from sklearn.preprocessing import StandardScaler
[In]: from sklearn.model_selection import cross_val_score
[In]: from sklearn.linear_model import LogisticRegression
[In]: from sklearn.tree import DecisionTreeClassifier
[In]: from sklearn.ensemble import RandomForestClassifier
[In]: from sklearn.svm import LinearSVC
```

Step 2: Define the feature and target columns. Assign the feature columns to the variable x and the target column to the variable y:

```
[In]: x_cols = ['Pregnancies', 'Glucose', 'BloodPressure', 'BMI',
      'DiabetesPedigreeFunction', 'Age']
[In]: x = pandas_df[x_cols]
[In]: y = pandas_df["Outcome"]
```

Step 3: Standardize the features using the StandardScaler() function, which scales the features to have a mean of 0 and a standard deviation of 1

```
[In]: scaler = StandardScaler()
[In]: x_scaled = scaler.fit_transform(x)
```

Step 4: Define a list of classifiers:

```
[In]: classifiers = [LogisticRegression(), DecisionTreeClassifier(),
      RandomForestClassifier(), LinearSVC(max_iter=1500)]
```

The *max_iter* argument in LinearSVC is a hyperparameter that controls the maximum number of iterations the algorithm should run for convergence. We set it to 1500 because the algorithm did not converge within the default maximum number of iterations, which is 1000. As we will see in Chapter 16, experimenting with different iterations is part of the hyperparameter tuning process.

The word "Linear" in LinearSVC means we are using an SVM (support vector machine)-based classifier with a linear kernel, which uses linear decision boundaries to separate classes. There is another classifier in Scikit-Learn, SVC(), which can implement SVMs with a polynomial kernel. This is a better choice when nonlinear decision boundaries are needed. We didn't experiment with this algorithm as it is currently unavailable in PySpark due to the fact that Kernelized SVM training is hard to distribute. Note, however, that SVC() can also implement a linear kernel, so an SVC(kernel='linear') would give us fairly similar results as LinearSVC().

Step 5: Use a *for loop* to perform ten-fold cross-validation for each classifier. This splits the data into ten folds and computes the mean of the model's accuracy (as measured by roc_auc scoring) for each fold. The name of the model and the mean ROC AUC are printed for each classifier:

```
[In]: for classifier in classifiers:
          result = cross_val_score(
              classifier,
              x_scaled,
              y,
              cv=10,
              scoring= "roc_auc" )
          print(f"{classifier.__class__.__name__}
            mean: {result.mean():.2f}")
[Out]:
LogisticRegression mean: 0.84
DecisionTreeClassifier mean: 0.67
RandomForestClassifier mean: 0.83
LinearSVC mean: 0.84
```

Let's explain in detail what these numbers mean. In the output, the mean ROC AUC indicates the average ROC AUC score across the ten cross-validation folds. The ROC AUC ranges from 0 to 1, with higher values indicating better performance.

The ROC curve is a graphical representation of the performance of a binary classifier across different discrimination thresholds, that is, the different values that a classifier uses to determine whether a sample belongs to the positive class or the negative class. The curve is created by plotting the true positive rate (TPR) against the false positive rate (FPR) at various threshold values.

The following is a breakdown of the key components of the ROC curve:

1) True positive rate (TPR) or sensitivity:

   TPR is the proportion of positive samples (actual positives) that are correctly identified as positive by the classifier. It is calculated as follows:

   TPR = True Positives / (True Positives + False Negatives)

   TPR is also known as sensitivity or recall.

2) False positive rate (FPR) or fall-out:

   FPR is the proportion of negative samples (actual negatives) that are incorrectly classified as positive by the classifier. It is calculated as follows:

   FPR = False Positives / (False Positives + True Negatives)

   FPR is also known as the fall-out.

The ROC curve is generated through the plotting of TPR (sensitivity) along the vertical y axis and FPR (1-specificity) along the horizontal x axis. This curve visually portrays how alterations in the discrimination threshold impact the classifier's efficacy. In an ideal scenario, the ROC curve for a classifier would ascend directly from the origin (FPR = 0) and continue horizontally to the upper-right corner (TPR = 1), signifying a perfect differentiation between the two classes.

In real-world scenarios, most classifiers yield curves that curve toward the upper-left corner. The proximity of the ROC curve to this corner is directly proportional to the classifier's performance quality—closer alignment indicating superior performance. The area beneath the ROC curve (AUC) serves as a single metric encapsulating the classifier's performance across all conceivable threshold values. An AUC value of 1 signifies a perfect classifier, while an AUC value of 0.5 implies a classifier performing no better than random chance.

Let's go ahead and plot the ROC curves of the decision tree and random forest classifiers to better understand what they are. We can achieve this with the following code, using the same pandas_df DataFrame we generated earlier to train and evaluate the models:

Step 1: Import necessary libraries

```
[In]: import matplotlib.pyplot as plt
[In]: from sklearn.model_selection import cross_val_predict
[In]: from sklearn.tree import DecisionTreeClassifier
[In]: from sklearn.ensemble import RandomForestClassifier
[In]: from sklearn.metrics import RocCurveDisplay, roc_curve, auc
```

Step 2: Define the decision tree and random forest models

```
[In]: decision_tree_model = DecisionTreeClassifier()
[In]: random_forest_model = RandomForestClassifier()
```

Step 3: Perform cross-validation predictions for the models

```
[In]: y_scores_decision_tree = cross_val_predict(
      decision_tree_model,
      x,
      y,
      cv=10,
      method="predict_proba"
      )
[In]: y_scores_random_forest = cross_val_predict(
      random_forest_model,
      x,
      y,
      cv=10,
      method="predict_proba"
      )
```

Step 4: Calculate ROC curves and AUCs for the models

```
[In]: fpr_decision_tree, tpr_decision_tree, _ = roc_curve(y, y_scores_
      decision_tree[:, 1])
[In]: roc_auc_decision_tree = auc(fpr_decision_tree, tpr_decision_tree)
```

```
[In]: fpr_random_forest, tpr_random_forest, _ = roc_curve(y, y_scores_
      random_forest[:, 1])
[In]: roc_auc_random_forest= auc(fpr_random_forest, tpr_random_forest)
```

Step 5: Plot the ROC curves

```
[In]: plt.figure()
[In]: plt.plot(fpr_decision_tree, tpr_decision_tree, color='darkorange',
      lw=2, label='Decision Tree (area = %0.2f)' % roc_auc_decision_tree)
[In]: plt.plot(fpr_random_forest, tpr_random_forest, color='blue', lw=2,
      label='Random Forest (area = %0.2f)' % roc_auc_random_forest)
[In]: plt.plot([0, 1], [0, 1], color='navy', lw=2, linestyle='--')
[In]: plt.xlim([0.0, 1.0])
[In]: plt.ylim([0.0, 1.05])
[In]: plt.xlabel('False Positive Rate')
[In]: plt.ylabel('True Positive Rate')
[In]: plt.title('Receiver Operating Characteristic')
[In]: plt.legend(loc="lower right")
[In]: plt.show()
[Out]:
```

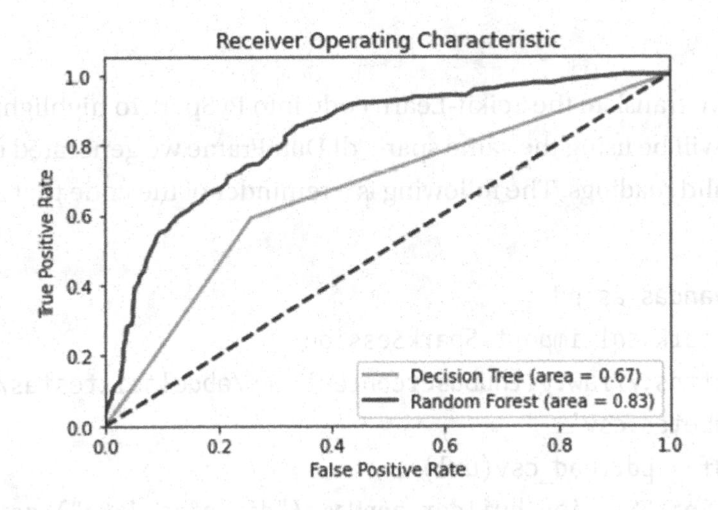

The ROC curve for the random forest classifier with an ROC AUC of 0.83 is positioned closer to the top-left corner of the plot than the decision tree classifier. This suggests that the random forest classifier has a better ability to correctly classify positive and negative instances, with a more moderate false positive rate.

This is not unexpected as random forest classifiers are known to generally perform better than single decision tree classifiers. This is due to the ensemble nature of the random forests and their reduction of overfitting. A random forest combines the predictions of multiple decision trees, each of which is trained on a subset of the data and makes an independent prediction. The final prediction is then determined by majority voting the predictions of individual trees. This ensemble approach helps in reducing the impact of overfitting.

Additionally, the majority vote or averaging in random forest helps in reducing the influence of noisy data points or outliers on the final prediction. Decision trees can be sensitive to individual data points, which may lead to poor generalization.

More importantly, random forest can model complex relationships in the data more effectively than a single decision tree. By combining predictions from multiple trees, it can approximate nonlinear decision boundaries better. Looking back at the output of the k-fold cross-validation for the four classifiers, the results indicate that both logistic regression and linear SVC had the best performance with an AUC of 0.84 each, followed by random forest with an AUC of 0.83. The decision tree classifier did not perform as well, with an AUC of only 0.67. These findings suggest that we may consider logistic regression or linear SVC as the model of choice for further training and evaluation.

PySpark

In this section, we translate the Scikit-Learn code into PySpark to highlight their similarities. We will be using the same spark_df DataFrame we generated earlier, after filtering out invalid readings. The following is a reminder of the code that achieves this task:

```
[In]: import pandas as pd
[In]: from pyspark.sql import SparkSession
[In]: url = 'https://raw.githubusercontent.com/abdelaziztestas/spark_book/
      main/diabetes.csv'
[In]: pandas_df = pd.read_csv(url)
[In]: spark = SparkSession.builder.appName("diabetes_data").getOrCreate()
[In]: spark_df = spark.createDataFrame(pandas_df)
[In]: spark_df = spark_df.filter((col('Glucose') != 0)
                             & (col('BloodPressure') != 0)
                             & (col('BMI') != 0)) \
```

```
                    .select(['Pregnancies',
                       'Glucose',
                       'BloodPressure',
                       'BMI',
                       'DiabetesPedigreeFunction',
                       'Age',
                       'Outcome'])
```

In a nutshell, our first step involves reading the CSV file into a Pandas DataFrame. We then convert this DataFrame into a PySpark DataFrame and proceed to filter out the invalid records. Once we have generated the clean DataFrame, the initial stage of the k-fold cross-validation process with PySpark requires us to import the necessary libraries:

Step 1: Import the required libraries

- `VectorAssembler` and `StandardScaler` for feature engineering
- `LogisticRegression`, `DecisionTreeClassifier`, `RandomForestClassifier`, and `LinearSVC` for classification
- `BinaryClassificationEvaluator` for evaluating the models' performance
- `CrossValidator` for cross-validation
- `col` function for renaming the label column
- `ParamGridBuilder` for the hyperparameters

```
[In]: from pyspark.ml.feature import VectorAssembler, StandardScaler
[In]: from pyspark.ml.classification import LogisticRegression,
      DecisionTreeClassifier, RandomForestClassifier, LinearSVC
[In]: from pyspark.ml.evaluation import BinaryClassificationEvaluator
[In]: from pyspark.ml.tuning import CrossValidator, ParamGridBuilder
[In]: from pyspark.sql.functions import col
```

Step 2: Combine the feature columns into a single vector column named features using `VectorAssembler`:

```
[In]: assembler = VectorAssembler(inputCols=[
            'Pregnancies',
            'Glucose',
```

```
        'BloodPressure',
        'BMI',
        'DiabetesPedigreeFunction',
        'Age', ], outputCol="features")
[In]: data = assembler.transform(spark_df)
```

Step 3: Use VectorAssembler to convert the set of input features into a single vector feature, which is then used as input for the next step

This step is necessary in distributed computing, where data is stored in a distributed fashion and needs to be transformed to a vector format before it can be used for machine learning. Scikit-Learn, on the other hand, is designed for single-machine use and doesn't require the use of VectorAssembler. Input features in Scikit-Learn are typically stored in a NumPy array or pandas dataframe and can be directly used for training machine learning models.

To see what the VectorAssembler does, let's print the top five rows of the "data" DataFrame (the argument truncate=False means that we want to display the full content of each column, even if it exceeds the default display length, which is normally 20 characters):

```
[In]: data.show(5, truncate=False)
[Out]:
```

Pregnancies	Glucose	Blood Pressure	BMI	Diabetes Pedigree Function	Age	Outcome	features
6	148	72	33.6	0.627	50	1	[6,148,72,33.6,0.627,50]
1	85	66	26.6	0.351	31	0	[1,85,66,26.6,0.351,31]
8	183	64	23.3	0.672	32	1	[8,183,64,23.3,0.672,32]
1	89	66	28.1	0.167	21	0	[1,89,66,28.1,0.167,21]
0	137	40	43.1	2.288	33	1	[0,137,40,43.1,2.288,33]

We can see from the preceding post-vector assembling output that the features have been combined into a single vector named features. The first row, for example, combines the individual values of pregnancies, glucose, blood pressure, BMI, diabetes pedigree function, and age into the numerical vector [6,148,72,33.6,0.627,50].

Notice that the label column (Outcome) is not included in this assembly process, as vector assembling is exclusively meant for the features, excluding the target variable.

Step 4: Rename the label column Outcome to label using `withColumnRenamed`

This is for convenience as in machine learning, the column containing the target variable is typically called "label". Furthermore, the `BinaryClassificationEvaluator` used later in the code expects the target variable to be in a column named label, so renaming the Outcome column to label ensures that the evaluator can correctly evaluate the performance of the trained models:

```
[In]: data = data.withColumnRenamed("Outcome", "label")
```

Step 5: Standardize the feature columns using `StandardScaler`:

```
[In]: scaler = StandardScaler(inputCol="features", outputCol="scaled_
      features")
[In]: data = scaler.fit(data).transform(data)
```

Scaling features is an important engineering/preprocessing step in classification models, including logistic regression, LinearSVC (support vector classifier), and tree-based classifiers, such as decision trees and random forests. This is to ensure that the algorithms perform optimally and consistently.

Logistic regression and LinearSVC models use gradient-based optimization algorithms to find the best coefficients for the features. Scaling helps the optimization process by ensuring that all features have a similar scale, which can prevent one feature from dominating the optimization process due to its larger magnitude. It can also improve convergence speed.

While decision trees are not directly affected by feature scaling, ensemble methods like random forests involve aggregating multiple decision trees. Scaling can help improve the stability and performance of these ensemble methods by ensuring that the tree-building process is not influenced by varying scales of features. This can lead to more accurate predictions and better generalization.

Step 6: Define a list of classification models to evaluate

```
[In]: classifiers = [
    LogisticRegression(),
    DecisionTreeClassifier(),
    RandomForestClassifier(),
    LinearSVC(maxIter=1500)])]
```

Step 7: Build an empty parameter grid using `ParamGridBuilder`:

```
[In]: paramGrid = ParamGridBuilder().build()
```

Here, an empty parameter grid is built using `ParamGridBuilder().build()`. The `ParamGridBuilder` class constructs a grid of hyperparameters to be searched over during cross-validation to train a model. The `build()` method returns an empty parameter grid, which can then be used to define the search space for the hyperparameters of each model.

This is necessary in PySpark as the MLlib incorporates the `ParamGridBuilder` by default. In Scikit-Learn, this is not required as it separates cross-validation from the process of hyperparameter tuning. We will examine hyperparameter tuning in more depth in Chapter 16.

Step 8: Define a `BinaryClassificationEvaluator` to evaluate the models based on the area under the ROC curve:

```
[In]: evaluator = BinaryClassificationEvaluator(labelCol="label",
          rawPredictionCol="rawPrediction", metricName="areaUnderROC")
```

Step 9: For each classifier in the list of classifiers:

- Define a `CrossValidator` with the current classifier, the parameter grid, the evaluator, and the number of folds (10)
- Fit the `CrossValidator` to the data and obtain the model
- Transform the data with the model and evaluate the performance using the `BinaryClassificationEvaluator`
- Print the name of the classifier and its mean accuracy as measured by the ROC AUC.

```
[In]: for classifier in classifiers:
          cv = CrossValidator(
              estimator=classifier, estimatorParamMaps=paramGrid,
              evaluator=evaluator, numFolds=10)
              cvModel = cv.fit(data)
              results = cvModel.transform(data)
              accuracy = evaluator.evaluate(results)
```

```
        print(f"{classifier.__class__.__name__} mean:
            {accuracy:.2f}")
[Out]:
LogisticRegression mean: 0.84
DecisionTreeClassifier mean: 0.78
RandomForestClassifier mean: 0.91
LinearSVC mean: 0.84
```

The results indicate that random forest had the best performance with an AUC of 0.91, followed by logistic regression and support vector machine with an AUC of 0.84 each. The decision tree classifier did not perform as well, with an AUC of 0.78.

These findings are not exactly the same as those from Scikit-Learn as the two frameworks use different default hyperparameters, which can cause differences in the accuracy metric (we will explore hyperparameter tuning in great detail in Chapter 16). Additionally, the k-fold sampling is random, so each time the k-fold cross-validation is performed, a different set of data is used for training and validation, resulting in a different model and evaluation metric.

Bringing It All Together

After describing each piece of code separately, we combine them all together in a single block for both Scikit-Learn and PySpark so that the data scientist can see how the code works as a whole.

Scikit-Learn

```
# Import necessary packages
[In]: from sklearn.preprocessing import StandardScaler
[In]: from sklearn.model_selection import cross_val_score
[In]: from sklearn.linear_model import LogisticRegression
[In]: from sklearn.tree import DecisionTreeClassifier
[In]: from sklearn.ensemble import RandomForestClassifier
[In]: from sklearn.svm import LinearSVC
[In]: import pandas as pd
```

```
# Read the csv file and create a Pandas df
[In]: url = 'https://raw.githubusercontent.com/abdelaziztestas/spark_book/
      main/diabetes.csv'
[In]: pandas_df = pd.read_csv(url)
# Filter out invalid records
[In]: pandas_df = pandas_df.loc[(pandas_df['Glucose'] != 0)
                        & (pandas_df['BloodPressure'] != 0)
                        & (pandas_df['BMI'] != 0),
                        ['Pregnancies',
                         'Glucose',
                         'BloodPressure',
                         'BMI',
                         'DiabetesPedigreeFunction',
                         'Age',
                         'Outcome']]
# Define the feature columns and target column
[In]: x_cols = ['Pregnancies',
          'Glucose',
          'BloodPressure',
          'BMI',
          'DiabetesPedigreeFunction',
          'Age']
[In]: x = pandas_df[x_cols]
[In]: y = pandas_df["Outcome"]

# Standardize the features using a StandardScaler object
[In]: scaler = StandardScaler()
[In]: x_scaled = scaler.fit_transform(x)

# Define a list of models to be used for cross-validation
[In]: classifiers = [
      LogisticRegression(),
      DecisionTreeClassifier(),
      RandomForestClassifier(),
      LinearSVC(max_iter=1500)])
       ]
```

```
# Perform 10-fold cross-validation for each model and print the mean of
accuracy
[In]: for classifier in classifiers:
          result = cross_val_score(
            classifier,
            x_scaled,
            y,
            cv=10,
            scoring= "roc_auc"  #"accuracy"
              )
         # Print the name of the model, mean accuracy, and standard
         deviation of accuracy
         print(f"{classifier.__class__.__name__} mean:
                 {result.mean():.2f}")
```

PySpark

```
[In]: from pyspark.sql.session import SparkSession
[In]: from pyspark.ml.feature import VectorAssembler, StandardScaler
[In]: from pyspark.ml.classification import LogisticRegression,
      DecisionTreeClassifier, RandomForestClassifier, LinearSVC
[In]: from pyspark.ml.evaluation import BinaryClassificationEvaluator
[In]: from pyspark.ml.tuning import CrossValidator, ParamGridBuilder
[In]: from pyspark.sql.functions import col
[In]: import pandas as pd

# read data from csv file and create Pandas df
[In]: url = 'https://raw.githubusercontent.com/abdelaziztestas/spark_book/
      main/diabetes.csv'
[In]: pandas_df = pd.read_csv(url)
[In]: spark = SparkSession.builder.appName("diabetes_data").getOrCreate()
[In]: spark_df = spark.createDataFrame(pandas_df)
# filter out invalid records
[In]: spark_df = spark_df.filter((col('Glucose') != 0)
                            & (col('BloodPressure') != 0)
```

```
                                        & (col('BMI') != 0)) \
                                        .select(['Pregnancies',
                                            'Glucose',
                                            'BloodPressure',
                                            'BMI',
                                            'DiabetesPedigreeFunction',
                                            'Age',
                                            'Outcome'])

# combine the feature columns into a single vector column
[In]: assembler = VectorAssembler(inputCols=[
          'Pregnancies',
          'Glucose',
          'BloodPressure',
          'BMI',
          'DiabetesPedigreeFunction',
          'Age'], outputCol="features")
[In]: data = assembler.transform(spark_df)

# specify the name of the label column
[In]: data = data.withColumnRenamed("Outcome", "label")

# standardize the feature columns
[In]: scaler = StandardScaler(inputCol="features", outputCol="scaled_
      features")
[In]: data = scaler.fit(data).transform(data)

# define the classification models to evaluate
[In]: classifiers = [
        LogisticRegression(),
        DecisionTreeClassifier(),
        RandomForestClassifier(),
        LinearSVC(maxIter=1500)]

# define the parameter grid for the models
[In]: paramGrid = ParamGridBuilder().build()
```

```
# define the evaluator
[In]: evaluator = BinaryClassificationEvaluator(labelCol="label",
      rawPredictionCol="rawPrediction", metricName="areaUnderROC")

# perform cross-validation for each model
[In]: for classifier in classifiers:
          # define the cross-validator
          cv = CrossValidator(estimator=classifier,
               estimatorParamMaps=paramGrid,
               evaluator=evaluator, numFolds=10)
          # evaluate the model using cross-validation
          cvModel = cv.fit(data)
          results = cvModel.transform(data)
          accuracy = evaluator.evaluate(results)
          # print the results
          print(f"{classifier.__class__.__name__} mean:
          {accuracy:.2f}")
```

Summary

This chapter introduced the machine learning modeling technique of classification. It demonstrated the utilization of Scikit-Learn and PySpark for performing classification on diabetes data and selecting the best-performing algorithm from a list of four supervised learning classifiers (logistic regression, decision tree, random forest, and support vector machines). Additionally, this chapter illustrated the processes of cross-validation and hyperparameter tuning—both of which are vital steps in a machine learning pipeline.

In the next chapter, we demonstrate how to build, train, evaluate, and use a different type of supervised machine learning: multiple linear regression. We will show that the steps involved in machine learning, including splitting data, model training, model evaluation, and prediction, are the same in both Scikit-Learn and PySpark. Furthermore, we will demonstrate how Pandas and PySpark have similar approaches to data manipulation, which simplifies tasks like exploring data.

CHAPTER 3

Multiple Linear Regression with Pandas, Scikit-Learn, and PySpark

This chapter demonstrates how to build, train, evaluate, and use a multiple linear regression model in both Scikit-Learn and PySpark. It shows that the steps involved in machine learning, including splitting data, model training, model evaluation, and prediction, are the same in both frameworks. Furthermore, Pandas and PySpark have similar approaches to data manipulation, which simplifies tasks like exploring data.

These similarities aid the data scientist in switching from in-memory data processing tools such as Pandas and Scikit-Learn to PySpark, which is designed for distributed processing of large datasets across multiple machines. However, it is important to note that these general similarities do not override the unique nuances of each tool. Another goal of this chapter is to explain these differences.

When it comes to executing PySpark, there are multiple approaches available. The code can be executed in cluster environments, such as Apache Hadoop YARN, Apache Mesos, or Kubernetes. Additionally, it can run within cloud environments like Amazon Web Services (AWS), Google Cloud Platform (GCP), or Microsoft Azure. PySpark code can also be containerized using technologies like Docker and executed on container orchestration platforms, such as Kubernetes. Moreover, users have the option to utilize Jupyter notebooks for interactive writing and running of PySpark code. Integrated Development Environments (IDEs) like PyCharm, IntelliJ IDEA, and Visual Studio Code also support PySpark development, enabling users to write and run code within these environments. Alternatively, Databricks provides a unified analytics platform that streamlines the deployment of Spark and PySpark applications. This platform offers an integrated workspace catering to data engineering, data science, and machine learning needs.

A. Testas, *Distributed Machine Learning with PySpark*, https://doi.org/10.1007/978-1-4842-9751-3_3

For the purpose of this book, as suggested in the previous chapter, two of the easiest ways to test the code are either by signing up for an online Databricks Community Edition account and creating a workspace or through the Google Colab. PySpark is available on both platforms by default. On Google Colab, it can be installed through the !pip install pyspark command. PySpark code can also be run locally by installing PySpark on a local machine using the same command.

The Dataset

The first step in our project is to generate some data to demonstrate how data exploration can be done using Pandas and PySpark, followed by machine learning using Scikit-Learn and PySpark for multiple linear regression.

The following code generates this data for Scikit-Learn:

```
[In]: import pandas as pd
[In]: from sklearn.datasets import make_regression
[In]: X, y = make_regression(n_samples=100, n_features=2, noise=10, random_
      state=42)
[In]: pandas_df = pd.DataFrame({'Feature 1': X[:, 0], 'Feature 2': X[:, 1],
      'Target': y})
```

This code generates regression data with two features and random noise using the `make_regression` function from Scikit-Learn. The n_samples parameter specifies the number of samples, n_features specifies the number of features, noise controls the amount of random noise added to the data, and random_state ensures reproducibility.

The generated features are stored in the variable X, and the corresponding target variable is stored in the variable y. The code converts the NumPy arrays X and y into a Pandas DataFrame named pandas_df. This is constructed using a dictionary where the keys are column names (Feature 1, Feature 2, Target) and the values are extracted from the X and y arrays. X[:, 0] represents the first column of X, X[:, 1] represents the second column, and y represents the target values. The values from X are assigned to the respective columns Feature 1 and Feature 2, while the values from y are assigned to the Target column.

Let's do the same with PySpark:

```
[In]: from pyspark.sql import SparkSession
[In]: spark = SparkSession.builder.appName("PySparkRegressionData").
      getOrCreate()
[In]: spark_df = [(float(X[i, 0]), float(X[i, 1]), float(y[i])) for i in
      range(len(X))]
[In]: spark_df = spark.createDataFrame(spark_df, ["Feature 1", "Feature 2",
      "Target"])
```

To work with the data in a PySpark environment, the code converts the NumPy X and y arrays, which were previously generated, to a PySpark DataFrame. Each row of the DataFrame is created as a tuple, where the feature values and target values are casted to float and included in the tuple. The loop iterates over the range of the length of X, ensuring that each row is created for the corresponding feature and target values. The PySpark DataFrame, named spark_df, is created with the same three columns: Feature 1, Feature 2, and Target.

The following is a breakdown of the code line by line.

The first line imports the `SparkSession` class, which is the entry point for working with structured data in Spark and is used to create a DataFrame. The second line initializes a SparkSession with the name "PySparkRegressionData". `SparkSession.builder` creates a builder object, and `getOrCreate()` either retrieves an existing SparkSession or creates a new one if it doesn't exist. The third line converts the NumPy arrays X and y into a list of tuples. Each tuple represents a row of data in the form (Feature 1, Feature 2, Target). The values are converted to floats using the `float()` function. The last line converts the list of tuples into a PySpark DataFrame named spark_df using the `createDataFrame()` method of the spark object (SparkSession). The first argument is the spark_df, and the second argument is a list of column names [Feature 1, Feature 2, Target].

Now, let's begin by inspecting the top five rows of each DataFrame as an initial step in our exploratory data analysis. The `head(n)` method in Pandas and the `show(n)` method in PySpark are used to display the first n rows of a DataFrame. By default, n=5 in Pandas and n=20 in PySpark. We have specified the first five rows to override the default values.

In Pandas, using the head() method:

```
[In]: pandas_df.head()
[Out]:
```

	Feature 1	Feature 2	Target
0	-1.19	0.66	-49.27
1	0.06	-1.14	-85.16
2	0.59	2.19	211.23
3	0.47	-0.07	29.20
4	0.74	0.17	84.35

In PySpark, using the show() method to display the same five rows:

```
[In]: spark_df.show(5)
[Out]:
```

Feature 1	Feature 2	Target
-1.19	0.66	-49.27
0.06	-1.14	-85.16
0.59	2.19	211.23
0.47	-0.07	29.20
0.74	0.17	84.35

We can see that each DataFrame has three columns: Feature 1, Feature 2, and Target. We can investigate the data a bit further using Pandas and PySpark to understand how they compare. We can check the shape of the datasets as follows:

In Pandas, using the shape attribute:

```
[In]: print(pandas_df.shape)
[Out]: (100, 3)
```

PySpark doesn't have the shape attribute, but we can accomplish the same task by combining the `count()` and `len()` methods:

```
[In]: print((spark_df.count(), len(spark_df.columns)))
[Out]: (100, 3)
```

In both Pandas and PySpark, the output is a tuple containing the number of rows and columns of the DataFrame. We can see that each DataFrame has 100 rows and 3 columns.

We can call the `dtypes` attribute in both Pandas and PySpark to check the data type of each column in the dataset:

In Pandas:

```
[In]: print(pandas_df.dtypes)
[Out]:
Feature 1     float64
Feature 2     float64
Target        float64
dtype: object
```

We can see that all the columns are of the decimal type (float64). In PySpark, the equivalent data type is *double*, as shown in the output of the following code:

```
[In]: print(spark_df.dtypes)
[Out]: [('Feature 1', 'double'), ('Feature 2', 'double'), ('Target',
       'double')]
```

We can call the `columns` attribute to return the name of each column in the DataFrame.

In Pandas:

```
[In]: print(pandas_df.columns)
[Out]: Index(['Feature 1', 'Feature 2', 'Target'], dtype='object')
```

The output is an Index object that contains three labels or column names in the following order: Feature 1, Feature 2, and Target. The Index object is an important component of a Pandas DataFrame, as it allows to select specific rows or columns of data using the labels as a reference.

For example, we can select the Feature 1 column by calling `pandas_df['Feature 1']`. We can also select multiple columns by passing a list of labels to the DataFrame: `pandas_df[['Feature 1', 'Feature 2']]`.

In PySpark:

```
[In]: print(spark_df.columns)
[Out]: ['Feature 1', 'Feature 2', 'Target']
```

There is no index in PySpark output, just an array of the three labels or columns. PySpark does not use the concept of indexing like Pandas because PySpark is designed to handle distributed computing across multiple machines, while Pandas is designed for in-memory data processing on a single machine. In PySpark, the data is typically stored in a distributed manner across multiple machines, and the computations are executed in parallel on those machines. This makes the indexing concept less efficient in PySpark, as indexing can result in a bottleneck when working with large datasets that are distributed across multiple machines.

Instead of indexing, PySpark uses a partitioning concept to divide data into partitions, which can be processed in parallel on multiple machines. Partitioning can be done based on various criteria, such as range partitioning, hash partitioning, or custom partitioning. Partitioning allows PySpark to achieve high performance and scalability when working with large datasets, while minimizing the impact of indexing overhead.

The `info()` method in Pandas can be used to provide a summary of pandas_df, including its shape, column data types, and the number of non-null values in each column:

```
[In]: pandas_df.info()
[Out]:

<class 'pandas.core.frame.DataFrame'>
RangeIndex: 100 entries, 0 to 99
Data columns (total 3 columns):
 #   Column     Non-Null Count  Dtype
---  ------     --------------  -----
 0   Feature 1  100 non-null    float64
 1   Feature 2  100 non-null    float64
 2   Target     100 non-null    float64
dtypes: float64(3)
```

There are 3 columns and 100 instances, all of which are of the decimal data type, and none of them contain null values.

In PySpark, we can use the printSchema() method to achieve a similar functionality, where the double data type is the equivalent of Pandas float64:

```
[In]: spark_df.printSchema()
[Out]:
root
 |-- Feature 1: double (nullable = true)
 |-- Feature 2: double (nullable = true)
 |-- Target: double (nullable = true)
```

We can use the describe() function in Pandas to produce some summary statistics:

```
[In]: pandas_df.describe()
[Out]:
```

	Feature 1	Feature 2	Target
Count	100.00	100.00	100.00
Mean	-0.12	0.03	-7.20
Std	0.86	1.00	106.64
Min	-2.62	-1.99	-201.72
25%	-0.78	-0.71	-84.39
50%	-0.04	0.11	-17.45
75%	0.34	0.66	70.90
Max	1.89	2.72	211.23

The output includes the count, mean, standard deviation (a measure of how dispersed the values are), min, and max values of each of the numerical columns. Included are also the 25%, 50%, and 75% percentiles (often referred to as the 1st quartile, the median, and the 3rd quartile, respectively). Each percentile indicates the value below which a given percentage of observations falls. For example, 75% of Feature 1 have a value lower than 0.34.

In PySpark, we can use the `summary()` method to provide similar functionality:

```
[In]: spark_df.summary().show()
[Out]:
```

Summary	Feature 1	Feature 2	Target
Count	100.00	100.00	100.00
Mean	-0.12	0.03	-7.20
Stddev	0.86	1.00	106.64
Min	-2.62	-1.99	-201.72
25%	-0.78	-0.71	-84.39
50%	-0.04	0.11	-17.45
75%	0.34	0.66	70.90
Max	1.89	2.72	211.23

PySpark also offers the `describe()` method as in Pandas, but its output only includes count, mean, std, min, and max values for each numerical column. The PySpark `summary()` function, on the other hand, shows all the summary statistics including the percentiles.

Multiple Linear Regression

Now that we have generated some data, we can proceed to demonstrate how to perform multiple linear regression using Scikit-Learn and PySpark. We will build, train, and evaluate a multiple linear regression model using the same dataset we generated in the previous section. The goal is to model the relationship between the predictors (Feature 1, Feature 2) and the response variable (Target) by fitting a linear equation to the data.

However, let's begin by briefly reviewing some theory before highlighting the main similarities between Scikit-Learn and PySpark.

Theory

Linear regression fits within the category of supervised learning. In this type of learning, the algorithm is trained on a labeled dataset. The labeled dataset consists of input features and their corresponding output labels. The algorithm learns to map the input features to the output labels and then uses this mapping to make predictions on new, unseen data.

In multiple linear regression, the goal is to model the relationship between two or more independent variables (predictors) and a dependent variable (response) by fitting a linear equation to the observed data. It is an extension of simple linear regression, which involves modeling the relationship between a single independent variable and a dependent variable.

The equation of the multiple linear regression line has the following form:

$$y = w_0 + w_1x_1 + w_2x_2 + \ldots + w_nx_n$$

where y is the dependent variable, $x_1, x_2, \ldots, x_n$ are the independent variables, w_0 is the intercept, and $w_1, w_2, \ldots, w_n$ are the regression coefficients (or weights) that represent the change in the dependent variable for a one-unit change in the corresponding independent variable while holding all other independent variables constant.

The goal of multiple linear regression is to estimate the values of the regression coefficients that minimize the sum of the squared differences between the observed values of the dependent variable and the predicted values of the dependent variable based on the independent variables.

Scikit-Learn/PySpark Similarities

Both Scikit-Learn and PySpark follow the same modeling steps: data preparation, training and evaluating the model, and predicting the target variable using new, unseen data.

The following are the key machine learning functions in Scikit-Learn and their equivalents in PySpark.

1. Data preparation:

   Scikit-Learn: Functions such as `train_test_split` for splitting data into training and testing sets, `StandardScaler` for feature scaling, `OneHotEncoder` for one-hot encoding categorical variables, and `SimpleImputer` for handling missing values.

PySpark: Operations on Spark DataFrames, such as `randomSplit` for splitting data, `StandardScaler` for feature scaling, `OneHotEncoder` for one-hot encoding categorical variables, and `Imputer` for handling missing values. Additionally, the `VectorAssembler` is used to combine multiple feature columns into a single vector column.

2. Model training:

   Scikit-Learn: Models have a `fit()` method for training, where you provide the features (X_train) and target variable (y_train) as input.

   PySpark: Models have a `fit()` method as well, where you provide a DataFrame (train_data) with features and a column for the target variable. The `VectorAssembler` step is required before training the model to combine the feature columns into a single vector column.

3. Model evaluation:

   Scikit-Learn: Various evaluation metrics are available, such as `mean_squared_error` and `r2_score` for regression. These metrics are applied directly to the predicted and actual target variable values (y_pred, y_test).

   PySpark: The `RegressionEvaluator` can be used to compute evaluation metrics, such as mean squared error for regression tasks. These evaluators require the predicted and actual values to be specified (predictions, y_test), along with the respective columns.

4. Prediction:

   Scikit-Learn: Models have a `predict()` method to generate predictions for new data, where you provide the features (X_test) and obtain the predicted target variable values (y_pred).

PySpark: Models have a `transform()` method to generate predictions for new data, where you provide a DataFrame (test_data) with features and obtain the predicted values in a new column (prediction). The `VectorAssembler` step is required before making predictions to ensure the new data has the same feature vector structure.

Let's now write some code to convert these descriptive steps into practice beginning with Scikit-Learn.

Multiple Linear Regression with Scikit-Learn

Our first step in building, training, and evaluating a multiple linear regression model in Scikit-Learn is to import the necessary libraries.

Step 1: Import necessary libraries

```
[In]: from sklearn.datasets import make_regression
[In]: from sklearn.model_selection import train_test_split
[In]: from sklearn.linear_model import LinearRegression
[In]: from sklearn.metrics import mean_squared_error, r2_score
```

Step 2: Generate data

```
[In]: X, y = make_regression(n_samples=100, n_features=2, noise=10,
      random_state=42)
```

Step 3: Split data into training and testing sets

```
[In]: X_train, X_test, y_train, y_test = train_test_split(X, y,
      test_size=0.2, random_state=42)
```

Step 4: Train the model

```
[In]: model = LinearRegression()
[In]: model.fit(X_train, y_train)
```

Step 5: Make predictions on test data

```
[In]: y_pred = model.predict(X_test)
```

Step 6: Evaluate the model by calculating MSE and r2

```
[In]: mse = mean_squared_error(y_test, y_pred)
[In]: r2 = r2_score(y_test, y_pred)
```

Let's break down the code line by line:

Step 1: Importing necessary libraries

The code imports specific modules from the sklearn library as they are needed for implementing the machine learning algorithm:

- `make_regression`: This function generates regression data for testing models.
- `train_test_split`: This function splits the data into training and testing sets.
- `LinearRegression`: This class is a linear regression model.
- `mean_squared_error`, `r2_score`: These functions are used for evaluating the performance of the regression model.

Step 2: Generating data

This code uses the `make_regression` function to generate a simulated dataset for the multiple linear regression model. It creates a dataset with 100 samples and 2 features and adds some random noise to the data to mimic real-world scenarios. The random state is set to 42 for reproducibility.

Step 3: Data preparation

This step involves splitting the data into training and testing sets. The code splits the dataset into training and testing sets, with 80% of the data used for training (X_train and y_train) and 20% for testing (X_test and y_test).

Step 4: Model training

The code creates and trains a linear regression model using the training data. It first creates an instance of the LinearRegression model and then uses the `fit()` method to train the model using the training data.

We can print the model's coefficients and intercept as follows:

```
[In]: print("Coefficients:", model.coef_)
[In]: print("Intercept:", model.intercept_)
[Out]: Coefficients: [86.31 73.67], Intercept: 0.066
```

The Coefficients are the weights assigned to each feature in the model. In this example, there are two features [Feature 1, Feature 2], and the corresponding coefficients are [86.31, 73.67], respectively. These coefficients represent the estimated impact or importance of each feature on the target variable. A positive coefficient indicates that an increase in the feature value leads to an increase in the predicted output, while a negative coefficient indicates an inverse relationship.

The Intercept represents the bias term of the linear regression model. The intercept is an additional constant term that is added to the weighted sum of the features. It represents the expected or average value of the target variable when all the features are zero or have no impact. In this case, the intercept is 0.066, meaning that when all the features have zero values, the model predicts an output value of approximately 0.066.

Step 5: Prediction

The trained model is used to make predictions on the test data. It predicts the output (y) based on the input (X_test) using the `predict()` method.

We can compare the actual and predicted values side by side by printing the top five rows as follows:

```
[In]: print("Actual\tPredicted")
[In]: for i in range(5):
          print(y_test[i], "\t", y_pred[i])
[Out]:
```

Actual	Predicted
-49.72	-56.82
4.15	-0.48
82.05	84.59
-34.16	-20.40
69.78	66.99

We can see that there are differences between the actual and predicted values. For example, the first observation indicates that the model's predictions are lower than the actual values since -56.82 is less than -49.72. A sample of five data points, however, does not provide a comprehensive evaluation of the model's overall performance. To assess the model's quality, we need to calculate evaluation metrics such as mean squared error (MSE) and R-squared.

Step 6: Evaluation

The code evaluates the performance of the model using mean squared error (MSE) and R-squared. It calculates the MSE between the predicted values (y_pred) and the true values (y_test) as well as the R-squared value, which measures how well the linear regression model fits the data.

We can print the results using the built-in `print()` function as follows:

```
[In]: print("Mean Squared Error (MSE):", mse)
[In]: print("R-squared:", r2)
[Out]: Mean Squared Error (MSE): 154.63, R-squared: 0.98
```

The MSE measures the average squared difference between the predicted values and the actual values of the target variable. It quantifies the overall accuracy of the model's predictions, with lower values indicating better performance. In our case, an MSE of 154.63 means that, on average, the squared difference between the predicted and actual values is 154.63.

R-squared, also known as the coefficient of determination, is a statistical measure that represents the proportion of the variance in the dependent variable (target variable) that can be explained by the independent variables (features) used in the model. It ranges from 0 to 1, with higher values indicating a better fit of the model to the data. An R-squared of 0.98 means that approximately 98% of the variance in the dependent variable is explained by the independent variables in the model.

While the obtained MSE of 154.63 and R-squared of 0.98 provide initial insights into the model's performance, it's important to note that these values are based on randomly generated data. Real-world scenarios often involve more complex and diverse datasets. Therefore, it becomes crucial to fine-tune the model and validate its performance on real and representative data. The process of fine-tuning may involve adjusting model hyperparameters (explained in detail in Chapter 16), optimizing feature selection, and applying regularization techniques.

For the sake of illustration, let's assume that we are satisfied with the performance of this model. What should we do next? The next step would be to utilize the trained model for making predictions on new, unseen data.

We can demonstrate this using the following code:

```
[In]: new_data = [[1.0, 2.0], [3.0, 4.0], [5.0, 6.0]]
[In]: new_predictions = model.predict(new_data)
```

```
[In]: print(new_predictions)
[Out]: [233.71 553.67 873.63]
```

The new_data variable represents a list of input feature values for which we want to obtain predictions. Each inner list within new_data corresponds to a different set of input features. In our example, there are three sets of input features: [1.0, 2.0], [3.0, 4.0], and [5.0, 6.0].

By calling the `predict()` method, we pass the new_data to the trained model to obtain predictions for each set of input features. The resulting predictions, [233.71, 553.67, 873.63], are stored in the new_predictions variable and then printed to the screen using the `print()` function.

Multiple Linear Regression with PySpark

Now, let's demonstrate how to convert the data preparation, model training, model evaluation, and prediction steps into PySpark code to leverage distributed computing. Similar to Scikit-Learn, our starting point in PySpark remains the same–importing the necessary libraries.

Step 1: Import necessary libraries

```
[In]: from sklearn.datasets import make_regression
[In]: from pyspark.ml.regression import LinearRegression
[In]: from pyspark.ml.evaluation import RegressionEvaluator
[In]: from pyspark.ml.feature import VectorAssembler
[In]: from pyspark.sql import SparkSession
```

Step 2: Initialize SparkSession

```
[In]: spark = SparkSession.builder.appName("RegressionExample").
getOrCreate()
```

Step 3: Generate data

```
[In]: X, y = make_regression(n_samples=100, n_features=2, noise=10, random_
      state=42)
[In]: data = [(float(X[i, 0]), float(X[i, 1]), float(y[i])) for i in
      range(len(X))]
[In]: data = spark.createDataFrame(data, ["Feature 1", "Feature 2",
      "Target"])
```

Step 4: Split data into training and testing sets

```
[In]: train_data, test_data = data.randomSplit([0.8, 0.2], seed=42)
```

Step 5: Combine features into a single vector column

```
[In]: assembler = VectorAssembler(inputCols=["Feature 1", "Feature 2"],
      outputCol="features")
[In]: train_data = assembler.transform(train_data)
[In]: test_data = assembler.transform(test_data)
```

Step 6: Train the model

```
[In]: lr = LinearRegression(featuresCol="features", labelCol="Target")
[In]: model = lr.fit(train_data)
```

Step 7: Make predictions on test data

```
[In]: predictions = model.transform(test_data)
```

Step 8: Evaluate the model using MSE and r2

```
[In]: evaluator = RegressionEvaluator(labelCol="Target",
      predictionCol="prediction", metricName="mse")
[In]: mse = evaluator.evaluate(predictions)
[In]: r2 = model.summary.r2
```

Let's dive into the code line by line:

Step 1: Import necessary libraries

The following are the imported classes and functions and their role in the modeling process:

- `make_regression`: This function generates simulated regression data for testing.
- `LinearRegression`: This class represents a linear regression model in PySpark.
- `RegressionEvaluator`: This class is used to evaluate the model's performance.

- `VectorAssembler`: This class is used to assemble the input features into a single vector column.
- `SparkSession`: This class is used to create a SparkSession, which is the entry point for working with DataFrame in PySpark.

Step 2: Initialize SparkSession

The code in this step establishes a SparkSession named "RegressionExample". A Spark Session is essential for interacting with Spark, although this requirement varies based on the platform being used. For example, when utilizing Databricks notebooks, the Spark Session is automatically created and initialized. Consequently, we can directly employ the "spark" object to interact with Spark features, execute queries, generate DataFrames, and utilize a variety of Spark libraries

Step 3: Generate data

This step has been explained in detail in the "The Dataset" section. The code uses the `make_regression` function to generate simulated regression data. It creates a two-dimensional feature matrix X and a target variable y. The code converts the NumPy arrays X and y into a Spark DataFrame data by creating a list of tuples and specifying the column names as [Feature 1, Feature 2, Target].

Step 4: Split data into training and testing sets

The code randomly splits the data DataFrame into training and testing DataFrames with an 80:20 ratio. The `randomSplit` method is used, and the seed is set to 42 for reproducibility.

Step 5: Combine features into a single vector column

The first line of the code in this step creates a `VectorAssembler` object, specifying the input columns [Feature 1, Feature 2] and the output column name as features. The next two lines transform the training and testing DataFrames by applying the assembler to combine the input features into a single vector column named features.

In PySpark, the `VectorAssembler` step is required to transform the individual input features into a single vector column. This is necessary because PySpark's machine learning algorithms expect input data to be in a specific format where all the features are combined into a single vector column. This `VectorAssembler` step helps in streamlining the workflow and maintaining consistency in data representation across different machine learning tasks in PySpark.

In contrast, Scikit-Learn does not require a separate step for combining features because it accepts input data in a different format. Scikit-Learn's machine learning algorithms typically work with NumPy arrays or Pandas DataFrames, where each feature is represented by a separate column.

Let's print the top five rows of the train_data to see how the data looks:

```
[In]: train_data.show(5)
[Out]:
```

Feature 1	Feature 2	Target	Features
-1.92	-0.03	-177.39	[-1.92,-0.03]
-1.19	0.66	-49.27	[-1.19,0.66]
-0.89	-0.82	-132.59	[-0.89,-0.82]
-0.45	0.86	27.34	[-0.45,0.86]
0.06	-1.14	-85.16	[0.06,-1.14]

We can see from the Features column that the values of Feature 1 and Feature 2 (e.g., -1.92 and -0.03, respectively) have been combined into the numerical vector [-1.92, -0.03] using the `VectorAssembler`. This transformation allows the machine learning model to interpret and process the input features together as a single entity.

Step 6: Train the model

The code in this step creates a `LinearRegression` object, specifying the input feature column as features and the label column as Target. It fits the linear regression model on the training data (train_data) using the `fit()` method.

We can print the estimated coefficients and intercept of the regression model with the following code:

```
[In]: print("Coefficients: ", model.coefficients)
[In]: print("Intercept: ", model.intercept)
[Out]: Coefficients: [86.03,73.24], Intercept: -0.33
```

The output represents the coefficients and intercept of the trained linear regression model. The coefficients indicate the weights assigned to each feature in the regression model, which, in turn, indicate the impact or importance of each feature on the target variable. The first coefficient of 86.03 corresponds to Feature 1, and the second coefficient of 73.24 corresponds to Feature 2.

The intercept, also known as the bias term, is an additional constant term added to the regression model. It represents the expected target variable value when all the features have a value of zero. In this example, the intercept is -0.33.

These coefficients and the intercept are the learned parameters of the linear regression model. They represent the relationship between the features and the target variable, allowing the model to make predictions based on new input data. This prediction step is explained next.

Step 7: Make predictions on test data

The code in this step applies the trained model to the test data (test_data) to make predictions. The `transform` method adds a prediction column to the DataFrame predictions with the predicted values.

We can print the top five rows of the actual and predicted values as follows:

```
[In]: actual_vs_pred = predictions.select("Target", "prediction")
[In]: actual_vs_pred.show(5)
[Out]:
```

Target	prediction
-192.19	-183.11
29.20	35.05
-14.49	-9.28
-62.52	-57.03
-51.66	-37.62

We can observe from the preceding output that there are differences between the actual target and predicted values, most of which are quite significant. To get a better understanding of the model's performance, however, we need to look at metrics such as the MSE and r2, which utilize the entire sample.

Step 8: Evaluate the model

This step creates a `RegressionEvaluator` object, specifying the label column as Target, the prediction column as prediction, and the metric as mse (mean squared error). The mean squared error (MSE) between the predicted values and the actual values in the predictions DataFrame is calculated. The code also retrieves the R-squared value from the model summary. The r2 variable will contain the R-squared value, which represents the goodness of fit of the model.

We can print the MSE and R-squared values as follows:

```
[In]: print("Mean Squared Error (MSE):", mse)
[In]: print("R-squared:", r2)
[Out]: Mean Squared Error (MSE): 154.63, R-squared: 0.99
```

The output consists of two evaluation metrics for the linear regression model:

1. The MSE is a measure of the average squared difference between the predicted values and the actual values. In this example, the MSE is 154.63, which indicates, on average, the squared difference between the predicted values and the actual values is 154.63. A lower MSE indicates better model performance, as it represents a smaller average squared difference between the predicted and actual values.

2. The R-squared value, also known as the coefficient of determination, measures the proportion of the variance in the target variable that can be explained by the linear regression model. In this example, an R-squared value of 0.99 indicates that approximately 99% of the variance in the target variable is accounted for by the linear regression model. A higher R-squared value suggests that the model provides a good fit to the data.

Assuming that we are satisfied with the model, we can now use it to predict the target variable using new data:

```
[In]: new_data = [(1.0, 2.0), (3.0, 4.0), (5.0, 6.0)]
[In]: new_data = spark.createDataFrame(new_data, ["Feature 1",
      "Feature 2"])
[In]: new_data = assembler.transform(new_data)
[In]: new_predictions = model.transform(new_data)
[In]: new_predictions.show()
[Out]:
```

Feature 1	Feature 2	Features	prediction
1	2	[1,2]	232.18
3	4	[3,4]	550.72
5	6	[5,6]	869.26

The steps of applying the model to new data are explained as follows:

1. New data points, represented as a Python list of tuples, are converted into a PySpark DataFrame called new_data. The DataFrame has two columns named Feature 1 and Feature 2.

2. The `VectorAssembler` is applied to the new_data DataFrame to transform the individual feature columns into a single vector column called features. This step aligns the new data with the same structure as the training data.

3. The trained model is applied to the new_data DataFrame using the `transform()` method. This generates predictions for the target variable based on the features in the features column.

4. The `show()` method is used to display the new_predictions DataFrame, which contains the predicted values for the new data points, along with the individual as well as the assembled features.

Summary

In this chapter, we have demonstrated how to build, train, and evaluate a multiple linear regression model and use it to make predictions on new data in both Scikit-Learn and PySpark. We have illustrated that the various stages of machine learning, encompassing data preparation, model training, evaluation, and prediction, are consistent in both frameworks. Moreover, Pandas and PySpark exhibit comparable methodologies for data manipulation, facilitating tasks such as data exploration.

These similarities provide assistance to data scientists who wish to transition from in-memory data processing tools like Pandas and Scikit-Learn to PySpark, which is specifically designed for distributed processing of extensive datasets across multiple machines. However, it is important to acknowledge that the general similarities discussed in this chapter do not undermine the unique nuances inherent in each tool.

CHAPTER 4

Decision Tree Regression with Pandas, Scikit-Learn, and PySpark

In this chapter, we continue with our exploration of supervised learning with a focus on regression tasks. Specifically, we will be building a regression model using the decision tree algorithm—an alternative to the multiple linear regression model we used in the previous chapter. We will use both Scikit-Learn and PySpark to train and evaluate the model and then use it to predict the sale price of houses based on several features such as the size of property and the number of bedrooms, bathrooms, and stories, among others. Additionally, we will compare the performance of Pandas and PySpark in data loading and exploration tasks to better understand their similarities and differences.

By the end of this chapter, our aim is to provide a deeper understanding of the nuances inherent in these tools and how they approach data processing, model training, and evaluation. While these tools share similarities, each has its own unique subtleties. This chapter will assist data scientists in better understanding and navigating these distinctions. The ultimate goal is to facilitate a smoother transition for those accustomed to small data tools such as Pandas and Scikit-Learn, enabling them to shift toward big data processing and machine learning with PySpark, thereby reaping the benefits of distributed computing.

We will be using an open source dataset that we have harvested from the Kaggle website. As we have done in the previous chapter, we will start by reading the dataset and then delve into it to understand more about the features and the target variable.

A. Testas, *Distributed Machine Learning with PySpark*, https://doi.org/10.1007/978-1-4842-9751-3_4

The Dataset

The dataset we are using to train and evaluate the decision tree regression model in Scikit-Learn and PySpark is widely known as the housing dataset. This contains 545 records and 12 features, with each record representing a house and the target variable indicating its price. The features represent the attributes of a house, including its total area, the number of bedrooms, bathrooms, stories, and parking spaces, as well as whether the house is connected to the main road, has a guest room, a basement, a hot water heating system, and an air-conditioning system, or is located in a preferred area, and whether it is fully furnished, semi-furnished, or unfurnished.

The dataset is publicly accessible on Kaggle. Provided here are the contributor's name, approximate upload date, dataset name, site name, and the URL from which we downloaded a copy of the CSV file:

Title: Housing Prices Dataset
Site: Kaggle
URL: `www.kaggle.com/datasets/yasserh/housing-prices-dataset`
Contributor: Yasser H.
Date: 2021

Similar to what we have done in the previous chapter, we first define a Python function to load the data from a CSV file using the Pandas `read_csv()` function:

```
[In]: import pandas as pd
[In]: def load_housing_data():
          url = ('https://raw.githubusercontent.com/abdelaziztestas/'
                 'spark_book/main/housing.csv')
          return pd.read_csv(url)
```

We then create a Pandas DataFrame by calling the `load_housing_data()` function:

```
[In]: pandas_df = load_housing_data()
```

In the next step, we convert the pandas_df DataFrame to a PySpark DataFrame named spark_df using the PySpark `createDataFrame()` method:

```
[In]: from pyspark.sql import SparkSession
[In]: spark = SparkSession.builder.appName("housing_data").getOrCreate()
[In]: spark_df = spark.createDataFrame(pandas_df)
```

To take a look at the top five rows of the pandas_df DataFrame, we use the head() method. To enhance readability due to the size of the table, we have split the output into two separate tables. This allows us to present the data more clearly without overwhelming a single page.

```
[In]: pandas_df.head()
[Out]:
```

	price	area	bedrooms	bathrooms	stories	mainroad	guestroom
0	13300000	7420	4	2	3	yes	no
1	12250000	8960	4	4	4	yes	no
2	12250000	9960	3	2	2	yes	no
3	12215000	7500	4	2	2	yes	no
4	11410000	7420	4	1	2	yes	yes

	basement	hotwaterheating	airconditioning	parking	prefarea	furnishingstatus
0	no	no	yes	2	yes	furnished
1	no	no	yes	3	no	furnished
2	yes	no	no	2	yes	semi-furnished
3	yes	no	yes	3	yes	furnished
4	yes	no	yes	2	no	furnished

We can do the same in PySpark, using the show() method:

```
[In]: spark_df.show(5)
[Out]:
```

price	area	bedrooms	bathrooms	stories	mainroad	guestroom
13300000	7420	4	2	3	yes	no
12250000	8960	4	4	4	yes	no
12250000	9960	3	2	2	yes	no
12215000	7500	4	2	2	yes	no
11410000	7420	4	1	2	yes	yes

basement	hotwaterheating	airconditioning	parking	prefarea	furnishingstatus
no	no	yes	2	yes	furnished
no	no	yes	3	no	furnished
yes	no	no	2	yes	semi-furnished
yes	no	yes	3	yes	furnished
yes	no	yes	2	no	furnished

The output from both Pandas and PySpark is identical except that PySpark doesn't show an index. We can also see from the output that each of the datasets contains 13 columns or attributes, which represent different aspects of houses. These attributes include the sale price and the total area in square feet that the house covers. Additionally, the dataset specifies the number of bedrooms and bathrooms in each house, as well as the total number of stories or levels the house comprises. Information about the house's connectivity to the main road, the presence of a guest room, and the availability of a basement are denoted by "Yes" or "No" entries in the respective columns. Moreover, the dataset indicates whether the house is equipped with a hot water heating system and air conditioning, along with the count of available parking spaces. The preferred location of the house is indicated by a "Yes" or "No" response under the "Prefarea" category. Lastly, the furnishing status of the house is categorized as "Fully Furnished," "Semi-Furnished," or "Unfurnished," offering insights into the house's interior setup.

The price in the dataset is the target variable, while the other features are the predictors. The goal is to make predictions by following the decision tree structure that will be learned during the training phase.

We can learn more about the data using the Pandas shape attribute:

```
[In]: print(pandas_df.shape)
[Out]: (545, 13)
```

In PySpark, we can use the count() and len() functions:

```
[In]: print((spark_df.count(), len(spark_df.columns)))
[Out]: (545, 13)
```

The output from both Pandas and PySpark confirms that we have 545 rows and 13 columns. In Pandas, we can count the number of null values in each column using the isnull() and sum() methods:

```
[In]: pandas_null_counts = pandas_df.isnull().sum()
[In]: print(pandas_null_counts)
[Out]:
price               0
area                0
bedrooms            0
bathrooms           0
stories             0
mainroad            0
guestroom           0
basement            0
hotwaterheating     0
airconditioning     0
parking             0
prefarea            0
furnishingstatus    0
dtype: int64
```

In PySpark, we can use the isNull() and sum() methods but in a slightly different way:

```
[In]: from pyspark.sql.functions import sum, col
[In]: spark_df.select([sum(col(c).isNull().cast("int")).alias(c) for c in
      spark_df.columns]).show()
[Out]:
```

price	area	bedrooms	bathrooms	stories	mainroad	guestroom
0	0	0	0	0	0	0

basement	hotwaterheating	airconditioning	parking	prefarea	furnishingstatus
0	0	0	0	0	0

Since the PySpark output is horizontal, unlike that of Pandas that fits vertically within the page, we have split it into two tables to enhance readability.

We can see from both Pandas and PySpark output that there are 0 null values. This is good news as machine learning algorithms don't usually work well with missing data.

Next, we can use the `describe()` method in Pandas and the `summary()` method in PySpark to produce key summary statistics for the numerical variables.

We start with Pandas:

```
[In]: pandas_df.describe()
[Out]:
```

	price	area	bedrooms	bathrooms	stories	parking
count	545	545	545	545	545	545
mean	4766729	5150.54	2.97	1.29	1.81	0.69
std	1870440	2170.14	0.74	0.50	0.87	0.86
min	1750000	1650	1	1	1	0
25%	3430000	3600	2	1	1	0
50%	4340000	4600	3	1	2	0
75%	5740000	6360	3	2	2	1
max	13300000	16200	6	4	4	3

PySpark doesn't automatically exclude the categorical variables from the calculation, and it shows them as null. We can specifically select the numerical features as follows:

```
[In]: spark_df.select("price", "area", "bedrooms", "bathrooms", "stories",
      "parking").summary().show()
[Out]:
```

summary	price	area	bedrooms	bathrooms	stories	parking
count	545	545	545	545	545	545
mean	4766729	5150.54	2.97	1.29	1.81	0.69
stddev	1870440	2170.14	0.74	0.50	0.87	0.86
min	1750000	1650	1	1	1	0
25%	3430000	3600	2	1	1	0
50%	4340000	4600	3	1	2	0
75%	5740000	6360	3	2	2	1
max	13300000	16200	6	4	4	3

The difference between Pandas and PySpark outputs is purely cosmetic. PySpark's presentation includes a summary header to indicate the statistics column, while the standard deviation metric is denoted as stddev rather than std as seen in Pandas. Both show the count, mean, standard deviation, min, and max values of each of the numerical columns, as well as the 25%, 50%, and 75% percentiles (also known as the 1st quartile, the median, and the 3rd quartile, respectively).

Finally, we can check the categorical variables in the housing dataset using a for loop. To get the Pandas output, we loop over each column of the categorical variables using a list of column names. For each column, we perform the following operations:

- We use the `unique()` method of the Pandas Series to get an array of unique values. The method returns an array of all unique values in a Pandas Series.
- We then print the column name along with its unique values.

```
[In]: for col in ['mainroad', 'guestroom', 'basement', 'hotwaterheating',
     'airconditioning', 'prefarea', 'furnishingstatus']:
          unique_values = pandas_df[col].unique()
          print(f"Unique values in {col}: {unique_values}")
[Out]:
Unique values in mainroad: ['yes' 'no']
Unique values in guestroom: ['no' 'yes']
Unique values in basement: ['no' 'yes']
Unique values in hotwaterheating: ['no' 'yes']
Unique values in airconditioning: ['yes' 'no']
Unique values in prefarea: ['yes' 'no']
Unique values in furnishingstatus: ['furnished' 'semi-furnished'
'unfurnished']
```

To extract the same categorical column values from the PySpark DataFrame, we employ the `col()` function available in the pyspark.sql.functions module. Within our loop, each categorical column is processed individually, following these steps:

- We initiate a new DataFrame by employing the `select()` method, focusing solely on the desired column.
- The `distinct()` method is then applied to the selected column to gather unique values.
- By utilizing the `collect()` method, we retrieve all the records as a list of Row objects.
- To isolate the unique values from the list, we leverage `flatMap()` with a lambda function.
- The resulting column name is associated with its unique values in the printed output.

```
[In]: from pyspark.sql.functions import col
[In]: for col_name in ['mainroad', 'guestroom', 'basement',
     'hotwaterheating', 'airconditioning', 'prefarea', 'furnishingstatus']:
          unique_values = spark_df.select(col(col_name)).distinct()\
                           .rdd.flatMap(lambda x: x).collect()
          print(f"Unique values in {col_name}: {unique_values}")
[Out]:
```

```
Unique values in mainroad: ['yes', 'no']
Unique values in guestroom: ['no', 'yes']
Unique values in basement: ['no', 'yes']
Unique values in hotwaterheating: ['no', 'yes']
Unique values in airconditioning: ['no', 'yes']
Unique values in prefarea: ['no', 'yes']
Unique values in furnishingstatus: ['semi-furnished', 'unfurnished',
'furnished']
```

The output of both Pandas and PySpark shows that 7 out of the 12 features are categorical variables. This suggests that we need to perform one-hot encoding on these variables to convert them into numerical values before feeding them to the algorithm.

Decision Tree Regression

In this section, we build, train, and evaluate a decision tree regressor with default hyperparameters and use it to predict house sale prices. Similar to multiple linear regression in the previous chapter, this model is an example of supervised learning. This is because we will be providing the target variable along the features.

Decision tree regressors are widely regarded as powerful and effective algorithms in machine learning and are utilized in popular libraries such as Scikit-Learn and PySpark. One of the key advantages of decision tree regressors is their interpretability as it allows for easier understanding and communication of the model's workings. Furthermore, decision tree regressors are capable of capturing nonlinear relationships within the data, making them especially useful when dealing with complex datasets. They are also robust to missing values and outliers and do not require feature standardization, enabling them to handle real-world datasets effectively.

However, decision tree regressors come with their own set of limitations. For example, their vulnerability to overfitting is a known issue, especially if their hyperparameters are not set up properly. For example, if the number of trees is too large to the extent that they are allowed to extend deeply into the data and capture noise, this can lead to inadequate performance on new, unseen data. Another aspect to consider is the inherent susceptibility to minor data fluctuations, which can trigger significant alterations in the tree's structure, thereby diminishing its resilience and dependability. As a result, decision trees can be inclined to generate considerably distinct trees when trained on different subsets of the training data, making the model susceptible to noise in the data. Since overfitting is an important topic in data science, it will be covered in detail in this section.

The prediction process in decision tree regressors is accomplished by recursively splitting the data into smaller subsets based on the values of the predictor variables. The splitting is done in a way that maximizes the homogeneity of the subsets with respect to the target variable. This is achieved by selecting the feature that offers the greatest information gain or the greatest reduction in impurity to split on. Once the best feature is selected, the data is partitioned into two or more subsets based on the values of that feature, and the process is repeated recursively for each subset until a stopping criterion is met. Finally, the algorithm produces a tree-like model where the leaves represent the predicted values of the target variable for the corresponding subsets of the input data.

Scikit-Learn/PySpark Similarities

Both Scikit-Learn and PySpark are similar in terms of their approach to machine learning as they both follow similar steps for building, training, and evaluating the decision tree regressor. Let's go through a list of these similarities by examining the key machine learning functions in both platforms:

1. Data preparation:

   Scikit-Learn: The Scikit-Learn library uses functions such as train_test_split for splitting decision tree regression data into training and testing sets, StandardScaler for feature scaling, OneHotEncoder for one-hot encoding categorical variables, and SimpleImputer for handling missing values.

   PySpark: In PySpark, similar operations can be performed on DataFrames. Use randomSplit for data splitting, StandardScaler for feature scaling, OneHotEncoder for categorical encoding, and Imputer for managing missing values. Additionally, the VectorAssembler is used to combine feature columns into a single vector column (a step that is not required in Scikit-Learn).

2. Model training:

   Scikit-Learn: For decision tree regression, you can create a DecisionTreeRegressor model and use the fit method by providing the feature matrix (X_train) and target variable (y_train).

   PySpark: Use the DecisionTreeRegressor model and its fit method as well. Provide a DataFrame (train_data) with features and a column for the target variable. Remember to apply the VectorAssembler step before training the model.

3. Model evaluation:

   Scikit-Learn: Evaluate the performance of the decision tree regression model using metrics such as mean_squared_error and r2_score. Calculate these metrics by comparing the predicted (y_pred) and actual (y_test) target variable values.

   PySpark: Utilize the RegressionEvaluator class to compute metrics like mean squared error and r2. Specify the predicted values column (predictions) and the actual values column (y_test) for evaluation.

4. Prediction:

   Scikit-Learn: To make predictions with the trained decision tree regressor, utilize the predict method. Provide the feature matrix (X_test) to obtain the predicted target variable values (y_pred).

   PySpark: Employ the transform method of the trained model to generate predictions for new data. Provide a DataFrame (test_data) containing the features and retrieve the predicted values in a new column (prediction). Ensure the VectorAssembler step is applied to maintain the necessary feature vector structure.

As can be seen from this comparison, the modeling steps in both Scikit-Learn and PySpark are similar. The implementation details and syntax may differ, however, as Scikit-Learn is a Python-based library while PySpark is a distributed computing framework based on Apache Spark.

Let's now proceed with implementing these steps using decision tree regression in both Scikit-Learn and PySpark, beginning with Scikit-Learn.

Decision Tree Regression with Scikit-Learn

Before we start the modeling process, a reminder of the housing dataframe we will be using. We first define a function for loading the data:

```
[In]: def load_housing_data():
          url = ('https://raw.githubusercontent.com/abdelaziztestas/'
                 'spark_book/main/housing.csv')
          return pd.read_csv(url)
```

then create a Pandas DataFrame by calling the function:

```
[In]: pandas_df = load_housing_data()
```

Next, we split the data into features and target variable:

```
[In]: X = pandas_df.drop('price', axis=1)
[In]: y = pandas_df['price']
```

The Modeling Steps

We can now proceed with the modeling steps: importing necessary libraries, preparing data, training the model, making predictions on test data, and evaluating the model.

Step 1: Importing necessary libraries

The first step in the modeling process is importing the necessary libraries. This step allows us to access a wide range of pre-built functions, classes, and tools that can significantly simplify and speed up our development process compared to building, training, and evaluating the model by writing code from scratch.

```
[In]: import pandas as pd
[In]: from sklearn.tree import DecisionTreeRegressor
[In]: from sklearn.model_selection import train_test_split
[In]: from sklearn.metrics import r2_score, mean_squared_error
[In]: from sklearn.preprocessing import OneHotEncoder
[In]: import numpy as np
```

Step 2: Data preparation

In this step, we prepare data in a format that is suitable for the machine learning algorithm. It includes one-hot encoding categorical variables and splitting data into training and testing sets.

2.1: Performing one-hot encoding on categorical variables

```
[In]: cat_cols = ['mainroad', 'guestroom', 'basement', 'hotwaterheating',
      'airconditioning', 'prefarea', 'furnishingstatus']
[In]: onehot_encoder = OneHotEncoder(sparse=False)
[In]: X_encoded = onehot_encoder.fit_transform(X[cat_cols])
```

```
[In]: X_encoded_df = pd.DataFrame(X_encoded, columns=onehot_encoder.get_
      feature_names_out(cat_cols))
[In]: X.drop(cat_cols, axis=1, inplace=True)
[In]: X = pd.concat([X, X_encoded_df], axis=1)
```

2.2: Splitting the data into training and testing sets

```
[In]: X_train, X_test, y_train, y_test = train_test_split(X, y,
      test_size=0.2, random_state=42)
```

Step 3: Model training

This is the step where we build and train the model.

```
[In]: scikit_model = DecisionTreeRegressor()
[In]: scikit_model.fit(X_train, y_train)
```

Step 4: Prediction

In this step, we make predictions on the test set using the model we have built and trained in the previous step.

```
[In]: y_pred = scikit_model.predict(X_test)
```

Step 5: Model evaluation

This is our final step in the modeling process where we evaluate the model's performance using previously unseen (test) data. We calculate two key performance statistics: the root mean squared error (RMSE) and the coefficient of determination, R-squared.

```
[In]: r2 = r2_score(y_test, y_pred)
[In]: rmse = np.sqrt(mean_squared_error(y_test, y_pred))
```

Let's go through the code step by step:

Step 1: Importing necessary libraries

In this step, the following libraries are imported:

- `Pandas`: This library will help us create and concatenate DataFrames.
- `DecisionTreeRegressor`: This class is the decision tree model.
- `train_test_split`: This function splits data into training and testing sets.

- `r2_score, mean_squared_error`: These functions are used for evaluation metrics.
- `OneHotEncoder`: This class encodes categorical variables.
- `numpy`: This library will help us calculate the RMSE based on the MSE.

Step 2: Data preparation

For the purpose of this project, data preparation in Scikit-Learn has two substeps. In the first substep, the categorical columns specified in cat_cols (mainroad, guestroom, basement, hot water heating, air conditioning, preferred area, and furnishing status) are one-hot encoded using `OneHotEncoder`. We start by initializing this class with the sparse parameter set to False. This means that the resulting encoded matrix will have all the 1s and 0s for the encoded categories instead of just the 1s if we were to set the parameter to True.

The subsequent step involves applying the `fit_transform` method of the encoder to the selected categorical columns within the feature data X. The encoded categorical features are stored in the X_encoded DataFrame. Each unique category within a column is transformed into an individual binary column, where a value of 1 signifies the presence of that category in a specific row. This DataFrame is then converted into X_encoded_df, where the encoded columns are labeled using the `get_feature_names_out` method. This method of the Scikit-Learn's `OneHotEncoder` class retrieves the names of the newly created binary columns generated during one-hot encoding, each of which corresponds to a unique category within the original categorical feature.

After one-hot encoding, the original categorical columns in X are no longer needed. Thus, we utilize the `drop` method to remove them (the specified axis = 1 means we are dropping columns, not rows, while the parameter inplace=True is used to indicate that the operation should modify the original DataFrame X directly, without creating a new DataFrame).

Subsequently, using the Pandas `concat` method, we merge X and X_encoded_df, combining the original noncategorical features with the one-hot encoded binary columns. This produces a transformed X that is now ready for use as an input into the decision tree algorithm.

To get a clearer understanding of the transformation carried out by the `OneHotEncoder` on categorical features, we can examine the following case. This example focuses on a single record prior to the application of the one-hot encoding substep:

```
[In]: (X[(X['area'] == 6000)
      & (X["bedrooms"] == 3)
      & (X["bathrooms"] == 2)
      & (X["stories"] == 4)
      & (X["parking"] == 1)
      & (X["guestroom"] == 'no')]
    .filter([
        "mainroad",
        "basement",
        "hotwaterheating",
        "airconditioning",
        "prefarea",
        "furnishingstatus"]))
[Out]:
```

mainroad	basement	hotwaterheating	airconditioning	prefarea	furnishingstatus
yes	no	no	yes	no	furnished

The initial section of the code uses the AND (&) operator to set the following conditions:

- The area column has a value of 6000.
- The bedrooms column has a value of 3.
- The bathrooms column has a value of 2.
- The stories column has a value of 4.
- The parking column has a value of 1.
- The guestroom column has a value of no.

Afterward, the following categorical columns are selected using the `filter` method in the latter part of the code:

- mainroad
- basement
- hotwaterheating

- airconditioning
- prefarea
- furnishingstatus

Now, if we query the same features DataFrame X after the `OneHotEncoder` has been applied, we will find that one-hot encoding has transformed the categorical variables into binary columns, effectively expanding each original categorical column into two separate columns as shown here:

mainroad_no	mainroad_yes	basement_no	basement_yes
0	1	1	0

hotwaterheating_no	hotwaterheating_yes	airconditioning_no	airconditioning_yes
1	0	0	1

prefarea_no	prefarea_yes	furnishingstatus_furnished	furnishingstatus_semi-furnished	furnishingstatus_unfurnished
1	0	1	0	0

Let's take the mainroad categorical feature as an example. In the original dataset, each row has a value for this variable, which can be either yes or no, depending on whether the house is on a mainroad or not. One-hot encoding converts each category of this feature into a separate binary column (0 or 1) in the dataset. For the mainroad feature, two new columns are created: mainroad_no and mainroad_yes. If the original mainroad value for a row was no, the mainroad_no column for that row is set to 1, and mainroad_yes is set to 0. If the original mainroad value for a row was yes, the mainroad_yes column for that row is set to 1, and mainroad_no is set to 0.

Since the original mainroad value was yes for the specific row we have selected, the mainroad_yes column for that row is set to 1, and the mainroad_no column is set to 0 after one-hot encoding is applied. This is the same scenario with air conditioning. Since the original value for the selected record was yes, the new columns are set to 1 for airconditioning_yes and 0 for airconditioning_no.

For basement, hot water heating, and preferred area, the opposite is true: all these variables had values of no in the original X DataFrame, so after one-hot encoding is applied, the _no columns are set to 1 while the _yes columns are set to 0.

The furnishing status might look a bit more complicated, since it has three values instead of a binary yes or no, but the one-hot encoding process works in exactly the same way. Since the original value was furnished for the furnished status, the encoded furnishingstatus_furnished column is set to 1, while the furnishingstatus_semi-furnished and furnishingstatus_unfurnished columns are both set to 0.

This one-hot encoding process allows machine learning algorithms to work better with categorical data because it transforms categorical features into a numerical representation. This helps algorithms understand the relationships between different categories without assuming any ordinal relationship between them.

A common mistake in data science is to use `LabelEncoder` in Scikit-Learn to encode categorical features, but this class is reserved for encoding the label or target variable, as the name suggests. The Scikit-Learn website also recommends against it. The reason for this is primarily rooted in the way `LabelEncoder` works. This assigns a unique numerical label to each unique category in a categorical feature. However, these numerical labels have an inherent ordinal relationship, which implies a certain order or ranking among categories.

When applied to categorical features that don't have a natural ordinal relationship, this encoding can inadvertently introduce unintended patterns and relationships into the data that might mislead machine learning algorithms. This can lead to incorrect model assumptions and poorer predictive performance. For example, if we used the `LabelEncoder` to encode the furnishing status feature, it would result in an encoded column with the values 0, 1, and 2, implying an order that doesn't actually exist in the original data. So if the label encoded furnished as 0, semi-furnished as 1, and unfurnished as 2, the model might incorrectly interpret this as unfurnished being greater than semi-furnished, and semi-furnished being greater than furnished. Such unintended ordering could lead to erroneous conclusions and potentially degrade the model's predictive accuracy.

In contrast, one-hot encoding, which creates separate binary columns for each category, avoids introducing such ordinal relationships and ensures that categorical variables are treated as nominal variables without any implied ordering.

In the next substep of the data preparation process (post one-hot encoding), the dataset is divided into training and testing sets employing the `train_test_split` function. This separation is typically performed in an 80-20 ratio, where 80% of the data is allocated for training and the remaining 20% is reserved for testing.

Additionally, the parameter random_state is set to 42 within the `train_test_split` function. This ensures that the same data split will be consistently obtained whenever the code is executed with the same random state. The resulting training and testing subsets are labeled as X_train and X_test for the feature matrices and y_train and y_test for the target variables.

Step 3: Model training

In this step, a decision tree regressor model is created using `DecisionTreeRegressor()`. The model is trained on the training data using the `fit()` method.

After the model is trained, we can print the importance of each of its features as follows:

```
[In]: importances = scikit_model.feature_importances_
[In]: feature_names = X.columns
[In]: indices = np.argsort(importances)[::-1]
[In]: print('Feature importances:')
[In]: for i in indices:
          print(feature_names[i], ':', importances[i])
[Out]:
area : 0.4976
bathrooms : 0.1759
stories : 0.0512
parking : 0.0412
bedrooms : 0.0393
furnishingstatus_unfurnished : 0.0318
airconditioning_no : 0.0242
prefarea_no : 0.0233
airconditioning_yes : 0.0206
hotwaterheating_no : 0.0191
basement_no : 0.0158
guestroom_yes : 0.0134
guestroom_no : 0.0123
prefarea_yes : 0.0099
```

```
basement_yes : 0.0085
mainroad_yes : 0.0063
furnishingstatus_semi-furnished : 0.0046
mainroad_no : 0.0023
furnishingstatus_furnished : 0.0017
hotwaterheating_yes : 0.0007
```

These results offer a comprehensive overview of the feature importances in the model. Among these features, area holds the highest importance with a value of 0.4976, suggesting that it significantly influences the model's predictions. Following closely is the bathrooms feature with an importance of 0.1759. These two features seem to be the primary drivers of the model's outcomes.

On the other hand, features like furnishingstatus_furnished and hotwaterheating_yes exhibit lower importances (0.0017 and 0.0007, respectively), indicating that they have relatively minor impact on the predictions. Understanding the hierarchy of feature importances helps in focusing on the most influential factors and potentially simplifying the model by considering only the most crucial features for accurate predictions.

Step 4: Prediction

In this step of the modeling process in Scikit-Learn, the trained model (scikit_model) is used to make predictions on the testing data (X_test). The predictions are stored in the variable y_pred.

We can now compare the top five rows of the actual and predicted target values to have some idea about the model's predictions. We can achieve this with the following code:

```
[In]: results_df = pd.DataFrame({'Price': y_test, 'Prediction': y_pred})
[In]: print(results_df.head())
[Out]:
```

Price	Prediction
4060000	5600000
6650000	7840000
3710000	3850000
6440000	4935000
2800000	2660000

We can see that there are differences between the actual and predicted values. For example, the first observation indicates that the model predicts a higher price than the actual price since 5,600,000 is larger than 4,060,000. This means that the model overestimates the target variable for this particular sample by $1,540,000. On the contrary, the last observation shows that the model underestimates the target (price) variable since 2,660,000 < 2,800,000.

A sample of five data points, however, does not provide a comprehensive evaluation of the model's overall performance. To assess the model's quality, we need to calculate evaluation metrics such as the root mean squared error (RMSE) and R-squared score.

Step 5: Model evaluation

In this step, evaluation metrics are calculated using the predicted values y_pred and the actual values y_test. The R-squared score and root mean squared error (RMSE) are calculated. They can be printed with the following code:

```
[In]: print('R-squared score:', r2)
[In]: print('Root mean squared error:', rmse)
[Out]: R-squared score: 0.42 Root mean squared error: 1715690.65
```

The decision tree regression model's RMSE of approximately 1715691 means that the root mean squared error value for this model's predictions, when compared to the actual values, is approximately 1715691. RMSE is a metric used to measure the average magnitude of the differences between predicted values and actual values. In this example, a higher RMSE value suggests that, on average, the predictions made by the model are off by around 1715691 units from the actual values. Lower RMSE values are desirable, as they indicate that the model's predictions are closer to the true values and that the model has better accuracy and precision.

An acceptable percentage of the root mean squared error is one that is less than 10% of the mean of the actual target variable. In the specific example, the mean price across all houses in the sample is approximately 4,766,729:

```
[In]: y.mean()
[Out]: 4766729.25
```

This indicates that the RMSE exceeds 35% of that value, implying that the model's performance is not as high as one would have hoped.

The R-squared (also known as the coefficient of determination) is a statistical metric that measures the proportion of variability in the target variable that the model's predictions can explain. It serves as a measure of how well the model fits the data. R2 values range from 0 to 1, with 0 indicating that the model explains none of the variability and 1 signifying that the model perfectly predicts the target variable. An R2 value closer to 1 signifies a high degree of variance explanation and a stronger model fit.

Here, the R2 value is approximately 0.42, indicating that the model accounts for roughly 42% of the variability in the target variable. This suggests a moderate fit, indicating that the model can be significantly improved.

So what explains the model's modest performance? Well, decision trees are known to overfit because a standard model is typically trained on a single tree, and we cannot directly increase the number of trees as we would in ensemble methods like random forest or gradient boosting. There is also the curse of dimensionality: the current regression model originally had 12 features, but after applying one-hot encoding, it added 8 new features, resulting in a total of 20 columns. This increase in dimensionality could lead to unnecessary complexity, a concern that is amplified by the fact that our sample size is relatively small, comprising only 545 records.

Let's delve deeper into the concept of overfitting, as it holds significant importance in the field of machine learning. In the following subsection, we explore methods to detect this issue and strategies for effectively addressing it.

Overfitting

One way to detect overfitting is the train/test split. We divide the data into two separate sets: a training set and a test set. We train the model on the training set and evaluate its performance on the test set. If the model performs significantly better on the training set compared to the test set, it may be overfitting.

We have already done most of this. We only need to calculate the train r2 and compare it with that of the test subsample. We can print the r2 of the training sample as follows:

```
[In]: y_train_pred = scikit_model.predict(X_train)
[In]: r2_train = r2_score(y_train, y_train_pred)
[In]: print('R-squared score on training data:', r2_train)
[Out]: R-squared score on training data: 0.9985
```

The r2 on the train set is over 0.99, meaning that the model's predictions can explain almost 100% of the variability in the target variable. This is massively higher than the 42% on the test set, a strong indicator of overfitting.

Another way to detect overfitting is learning curves. The curve plots the model's performance (in this case, r2) on the training and testing sets as a function of the number of training examples. If the model's performance on the training set continues to improve while the performance on the test set plateaus or declines, it suggests overfitting. Conversely, if the performance on both sets converges and reaches a plateau, the model may have achieved an optimal level of generalization.

We can plot the learning curve for the current DecisionTreeRegressor with the following code:

Step 1: Import necessary libraries

```
[In]: import numpy as np
[In]: import matplotlib.pyplot as plt
[In]: from sklearn.model_selection import learning_curve
```

Step 2: Define a function to plot the learning curve

```
[In]: def plot_learning_curve(estimator, title, X, y, ylim=None,
        cv=None,
        n_jobs=None, train_sizes=np.linspace(0.1, 1.0, 5)):
        plt.figure()
        plt.title(title)
        if ylim is not None:
        plt.ylim(*ylim)
        plt.xlabel("Training examples")
        plt.ylabel("Score")
        train_sizes, train_scores, test_scores = learning_curve(
        estimator, X, y, cv=cv, n_jobs=n_jobs,
        train_sizes=train_sizes)
        train_scores_mean = np.mean(train_scores, axis=1)
        train_scores_std = np.std(train_scores, axis=1)
        test_scores_mean = np.mean(test_scores, axis=1)
        test_scores_std = np.std(test_scores, axis=1)
        plt.grid()
        plt.fill_between(train_sizes, train_scores_mean -
        train_scores_std,
```

```
    train_scores_mean + train_scores_std, alpha=0.1,
            color="r")
    plt.fill_between(train_sizes, test_scores_mean -
    test_scores_std,
    test_scores_mean + test_scores_std, alpha=0.1, color="g")
    plt.plot(train_sizes, train_scores_mean, 'o-', color="r",
       label="Training score")
    plt.plot(train_sizes, test_scores_mean, 'o-', color="g",
    label="Cross-validation score")
    plt.legend(loc="best")
    return plt
```

Step 3: Plot the learning curve

```
plot_learning_curve(scikit_model, "Learning Curve", X_train, y_train, cv=5)
plt.show()
[Out]:
```

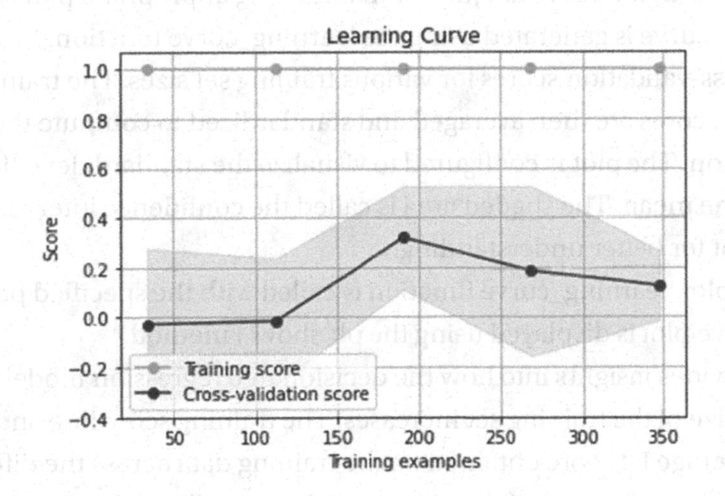

The code demonstrates the creation of a learning curve plot to assess the performance of the decision tree regressor on the housing dataset. The goal is to observe how the model's performance changes as the size of the training dataset increases.

The code begins by importing the necessary libraries, including NumPy for numerical operations, Matplotlib for visualization, and the learning_curve module for generating learning curves. A custom function named plot_learning_curve is then defined. This takes several parameters, including the estimator (the decision tree regression model being evaluated), a title for the plot, the feature matrix X, the target variable y, y-axis limits (ylim), cross-validation configuration (cv), and an array of training sizes (train_sizes) that specifies the proportions of the dataset used for training. For this, we have set the default (0.1, 1.0, 5), meaning the starting point of the range is 0.1, or 10% of the dataset size, the ending point of the range is 1.0 representing 100% of the dataset size, and 5 representing the number of evenly spaced values within the specified range.

Inside the function, a Matplotlib figure is created, and the title is set. If y-axis limits are provided, they are applied to the plot. The x axis represents the number of training examples used for model training. The x-axis values are generated based on the train_sizes parameter, which is an array of values indicating the proportions of the dataset used for training. In this example, we have used a maximum of approximately 65% of the sample (350 out of 545).

The y axis represents the r2 score, which measures the proportion of variance in the target variable that can be explained by the model. This normally ranges from 0 to 1, so the negative values in the curve are just for maintaining appropriate plot scaling.

The learning curve is generated using the learning_curve function. It calculates the training and cross-validation scores for various training set sizes. The training scores and cross-validation scores are then averaged and standardized to compute the mean and standard deviation. The plot is configured to visualize the standard deviation as a shaded region around the mean. The shaded area is called the confidence interval. A legend is added to the plot for better understanding.

Finally, the plot_learning_curve function is called with the specified parameters, and the learning curve plot is displayed using the plt.show() method.

The plot provides insights into how the decision tree regression model's performance changes as the size of the training set increases. The training score line and points represent the average R2 score obtained on the training data across the different training set sizes. The line's trajectory indicates how well the model's predictions fit the training data as the training set size increases. The points show the R2 score for each individual training set size.

The cross-validation score line and points represent the average R2 score obtained through cross-validation across the different training set sizes. Cross-validation involves training the model on different subsets of the data and testing it on the remaining data. The line's trajectory shows how well the model generalizes to new, unseen data as the training set size increases.

The shaded area around the cross-validation score line represents the standard deviation of the R2 score. The area provides a sense of the variability in the scores obtained during the cross-validation process. Wider shaded areas indicate higher variability in performance.

In terms of interpretation, the learning curve indicates that the model's performance on the training set is consistently high with an r2 close to 1, which is likely indicative of overfitting. However, as we increase the training set size, the cross-validation score starts to rise initially, suggesting that the model's generalization improves with more diverse data. However, at a certain point (close to 200 training examples), the cross-validation score begins to decrease. This indicates that the model is starting to overfit even on the cross-validation data, causing its performance to degrade as more training examples are used.

So now that we know that the model overfits, what are the options? There are several strategies to address overfitting:

- More training data: Our dataset is relatively small with only 545 records. A larger dataset can be beneficial in addressing overfitting. By incorporating a larger volume of data, the model gains exposure to a wide array of examples, thereby diminishing the risk of memorizing noise or outliers that might be present in a smaller dataset.
- Hyperparameter tuning: Fine-tuning hyperparameters of the model can significantly affect its performance. We can experiment with different parameter values, such as max_depth, min_samples_split, and max_features, along with cross-validation and grid/random search to find the optimal combination. We explore this option in Chapter 16.
- Different models: We can also consider trying different algorithms beyond decision trees, like random forests and gradient-boosted trees. Different algorithms have different strengths and weaknesses, and one might perform better for our specific data. We will explore this option in the next two chapters as we delve into regression with random forest and gradient-boosted trees.
- Feature selection: One potential reason for overfitting is that the model is too complex. This is likely the case here as our model has a relatively small sample size (545 observations) with 20 features,

most of which contribute little to the model's predictions. Carefully choosing relevant features can help reduce overfitting. An effective method for feature selection in this example is utilizing the feature_importances_ attribute. As we have seen earlier in this section, decision tree implementation allows us to gauge the importance of each feature using this attribute. As we have seen, only a small number of features, such as the area of the house and the number of bathrooms, drive the decision tree predictions. Eliminating some weak features can boost the model's performance.

- Outliers: Identifying and addressing outliers in the data can also impact the model's performance. There are three common methods to identify and remove outliers:
    - Z-score: This involves calculating the z-score for each data point and identifying outliers based on a specified threshold. Data points with z-scores greater than a certain value (typically 2 or 3) are considered outliers.
    - Percentiles/quartiles: This involves calculating the lower and upper quartiles (or percentiles) of the data and defining a range within which data points are considered normal. Data points outside this range are classified as outliers.
    - Interquartile range (IQR): This method defines a range based on the difference between the upper quartile (Q3) and the lower quartile (Q1) of the data. Data points outside the range Q1 - 1.5 * IQR to Q3 + 1.5 * IQR are considered outliers.

Decision Tree Regression with PySpark

Just as we did with Scikit-Learn, we need to remind ourselves of the housing dataset we will be using in PySpark before beginning the modeling process.

We start by importing the SparkSession class and then use it to create a Spark Session to be able to access the spark functionality:

```
[In]: from pyspark.sql import SparkSession
[In]: spark = SparkSession.builder.appName("DecisionTreeRegression").
      getOrCreate()
```

Next, we define a function to read the CSV file and then create a Pandas DataFrame by calling the function:

```
[In]: def load_housing_data():
          url = ('https://raw.githubusercontent.com/abdelaziztestas/'
                 'spark_book/main/housing.csv')
          return pd.read_csv(url)
[In]: pandas_df = load_housing_data()
```

Finally, we convert the Pandas DataFrame to a PySpark DataFrame using the `createDataFrame` method:

```
[In]: spark_df = spark.createDataFrame(pandas_df)
```

The Modeling Steps

The modeling steps in PySpark are largely similar to the steps in Scikit-Learn: importing necessary libraries, data preparation, model training, prediction, and model evaluation. Two additional substeps in data preparation, however, are needed: using `StringIndexer` before applying `OneHotEncoder` and combining features into a vector using `VectorAssembler`.

Step 1: Importing necessary libraries

```
[In]: from pyspark.ml.feature import StringIndexer, OneHotEncoder,
      VectorAssembler
[In]: from pyspark.ml.regression import DecisionTreeRegressor
[In]: from pyspark.ml.evaluation import RegressionEvaluator
```

Step 2: Data preparation

There are four substeps here compared to only two in Scikit-Learn:

2.1. Label encoding categorical columns with `StringIndexer`

```
[In]: cat_columns = ['mainroad', 'guestroom', 'basement',
      'hotwaterheating','airconditioning', 'prefarea', 'furnishingstatus']
[In]: indexers = [StringIndexer(inputCol=col, outputCol=col+'_label',
      handleInvalid='keep') for col in cat_columns]
[In]: for indexer in indexers:
          spark_df = indexer.fit(spark_df).transform(spark_df)
```

2.2. One-hot encoding for all categorical columns

```
[In]: encoder = OneHotEncoder(inputCols=[col+'_label' for col in cat_
      columns], outputCols=[col+'_encoded' for col in cat_columns])
[In]: spark_df = encoder.fit(spark_df).transform(spark_df)
```

2.3. Combining all features into a single vector

```
[In]: feature_cols = ['area', 'bedrooms', 'bathrooms', 'stories',
      'parking'] + [col+'_label' for col in cat_columns]
[In]: assembler = VectorAssembler(inputCols=feature_cols,
      outputCol='features')
[In]: spark_df = assembler.transform(spark_df)
```

2.4. Splitting data into training and testing sets

```
[In]: (training_data, test_data) = spark_df.randomSplit([0.8, 0.2], seed=42)
```

3. Model training

```
[In]: dt = DecisionTreeRegressor(featuresCol='features', labelCol='price')
[In]: spark_model = dt.fit(training_data)
```

4. Prediction

```
[In]: predictions = spark_model.transform(test_data)
```

5. Model evaluation

```
[In]: evaluator_rmse = RegressionEvaluator(labelCol='price',
      predictionCol='prediction', metricName='rmse')
[In]: rmse = evaluator_rmse.evaluate(predictions)
[In]: evaluator_r2 = RegressionEvaluator(labelCol='price',
      predictionCol='prediction', metricName='r2')
[In]: r2 = evaluator_r2.evaluate(predictions)
```

Let's go through the code step by step to mimic what we did with Scikit-Learn:

Step 1: Importing necessary libraries

In this step, the required libraries for working with PySpark's machine learning components are imported. These include modules for

- Features transformation (`StringIndexer`, `OneHotEncoder`, `VectorAssembler`)
- Regression (`DecisionTreeRegressor`)
- Evaluation (`RegressionEvaluator`)

Step 2: Data preparation

This step involves preparing the data for training and testing the model. It has four substeps:

2.1. Label encoding categorical columns

The categorical columns in the dataset are specified in the cat_columns list. For each categorical column, a `StringIndexer` is created, which converts categorical values into numerical indices. The handleInvalid parameter is set to keep, meaning that if a category unseen during training appears in the test set, it won't throw an error but will be handled appropriately.

The need for `StringIndexer` before applying `OneHotEncoder` in PySpark but not in Scikit-Learn arises from the differences in how these libraries handle categorical data and encoding. In PySpark, we need to convert categorical string values into numerical indices before feeding the data into `OneHotEncoder`. The Indexer assigns a unique index to each distinct category in a column.

The following is how the indexed data looks like using the mainroad categorical column as an example:

```
[In]: spark_df.select("mainroad", "mainroad_label").distinct().show()
[Out]:
+--------+--------------+
|mainroad|mainroad_label|
+--------+--------------+
|     yes|           0.0|
|      no|           1.0|
+--------+--------------+
```

In the output, we can see that the `StringIndexer` has successfully transformed the original mainroad values into numeric labels, with yes being represented by 0.0 and no by 1.0.

2.2. One-hot encoding for all categorical columns

In this substep of the data preparation step, an instance of OneHotEncoder is created. It takes the labeled columns from the previous step and transforms them into one-hot encoded vectors. Each original categorical column is now represented by a set of binary columns indicating the presence of a specific category.

Here is a glimpse of the mainroad categorical variable, both indexed and one-hot encoded:

```
[In]: spark_df.select("mainroad", "mainroad_label", "mainroad_encoded").
      distinct().show()
[Out]:

+--------+--------------+----------------+
|mainroad|mainroad_label|mainroad_encoded|
+--------+--------------+----------------+
|     yes|           0.0|   (2,[0],[1.0])|
|      no|           1.0|   (2,[1],[1.0])|
+--------+--------------+----------------+
```

The first observation to make is that PySpark's OneHotEncoder doesn't create two columns for each of the one-hot encoded features. In Scikit-Learn, OneHotEncoder created two columns for the mainroad feature, mainroad_yes and mainroad_no, while in PySpark, there is only one column: mainroad_encoded.

In terms of output interpretation, in the one-hot encoded mainroad notation of the first row, (2,[0],[1.0]), the number 2 represents the count of unique categories (yes and no) within the categorical feature. The notation [0] is a reference to the first category (indexed as 0 from the indexed mainroad_label column). The value [1.0] signifies that the first category (index 0) is assigned the value 1.0, indicating the presence of yes.

In the encoded mainroad notation of the second row, (2,[1],[1.0]), the number 2 still represents the count of unique categories (yes and no) within the categorical feature. The notation [1] is a reference to the second category (indexed as 1 from the indexed mainroad_label column). The value [1.0] signifies that the second category (index 1) is assigned the value 1.0, indicating the presence of no.

2.3. Combining all features into a single feature vector

In this substep of the data preparation step, a list of feature column names (feature_cols) is created, consisting of both numerical features (e.g., area, bedrooms) and the one-hot encoded categorical features. A VectorAssembler is then used to combine all these features into a single column named features, which the model will use as input. The assembler uses the `transformation` method to make the transformation.

Let's pull one single row of the features column, which contains the assembled features, to see how the assembled vector looks:

```
[In]: spark_df.select("features").show(1, truncate=False)
[Out]:
(20,[0,1,2,3,4,5,7,9,11,14,16,19],
[7420.0,4.0,2.0,3.0,2.0,1.0,1.0,1.0,1.0,1.0,1.0,1.0])
```

The number 20, before the first bracket, refers to the total number of features, including both original numerical features (12) and those that have been one-hot encoded (8):

- Five original numerical features (area, bedrooms, bathrooms, stories, parking).
- Seven one-hot encoded features (for the seven categorical columns with two unique values each). These are mainroad, guestroom, basement, hot water heating, air conditioning, preferred area, and furnishing status.
- One additional slot for the furnishing status column with three unique values: furnished, semi-furnished, and unfurnished.

There are 12 indices and 12 corresponding values in the first bracket and second bracket, respectively. These indices and values correspond to the 12 features in the dataset, including both original numerical features and one-hot encoded features.

2.4. Splitting data into training and testing sets

In this substep of the data preparation step, the data is split into training and testing sets using the `randomSplit` method. Here, 80% of the data is allocated for training and 20% for testing. A seed value of 42 is provided to ensure reproducibility.

Step 3: Model training

In this step, a `DecisionTreeRegressor` model is instantiated with the specified input (featuresCol) and target (labelCol) columns. The model is then fitted to the training data using the `fit` method.

We can print the feature importances of the model as follows:

a) Get feature importances.

```
[In]: feature_importances = spark_model.featureImportances
```

b) Get column names from feature_cols.

```
[In]: feature_names = feature_cols
```

c) Create a dictionary to store feature importances with names.

```
[In]: feature_importances_dict = {}
[In]: for i in range(len(feature_names)):
          feature_name = feature_names[i]
          importance = feature_importances[i]
          feature_importances_dict[feature_name] = importance
```

d) Sort the feature importances dictionary by importance values in descending order.

```
[In]: sorted_feature_importances = sorted(feature_importances_
dict.items(), key=lambda x: x[1], reverse=True)
```

e) Print sorted feature importances along with names.

```
[In]: print("Sorted Feature Importances:")
[In]: for feature_name, importance in sorted_feature_importances:
          print(f"Feature '{feature_name}': {importance}")
[Out]:

area: 0.5012
bathrooms: 0.1887
stories: 0.0595
parking: 0.0495
basement_encoded: 0.0241
furnishingstatus_encoded: 0.0084
airconditioning_encoded: 0.0079
bedrooms: 0.0044
mainroad_encoded: 0.0
guestroom_encoded: 0.0
```

```
hotwaterheating_encoded: 0.0
prefarea_encoded: 0.0
```

We observe from the output that the OneHotEncoder has not created multiple binary columns like in Scikit-Learn where each one-hot encoded variable was expanded into two columns (_yes and _no). Instead, PySpark has created sparse vectors for the encoded categorical columns. For example, for the furnishing status, there is only one-hot encoded column named furnishingstatus_encoded.

Step 4: Prediction

In this step of the PySpark modeling process, the trained model is used to make predictions on the test data using the transform method. The resulting predictions are added as a new column in the predictions DataFrame.

We can print the top five rows of the actual vs. predicted target values as follows:

```
[In]: predictions.select('price', 'prediction').show(5)
[Out]:
+-------+-----------------+
|  price|       prediction|
+-------+-----------------+
|6930000|          4566692|
|7070000|          5197333|
|7210000|          6940863|
|7350000|          5197333|
|7455000|        2791250.0|
+-------+-----------------+
only showing top 5 rows
```

We can see significant differences between the actual target variable (price) and the predicted target variable (prediction). For example, the first row indicates that the model underestimates the true price by close to 35% (4,566,692 vs. 6,930,000). This suggests that the model's performance is likely to be modest, just as it was in Scikit-Learn. We can confirm this by looking at the RMSE and r2 statistics.

Step 5: Model evaluation

Two different evaluation metrics are computed for the model's performance on the test data. The RegressionEvaluator is used to calculate the root mean squared error (RMSE) and the R-squared (r2) values, both of which provide insights into how well the model's predictions align with the actual target values.

The following code prints those metrics:

```
[In]: print(f"RMSE: {rmse}")
[In]: print(f"R2: {r2}")
[Out]: RMSE: 1463511.73 R2: 0.31
```

We can observe from the value of r2 that only about 30% of the variations in the target variable (price) are explained by the variations in the predicting variables. The RMSE (1,463,512) is also more than 10% of the mean value of the actual price variable (4,766,729), indicating that the model's performance is quite modest.

Based on our earlier investigation (comparing r2 between test and train sets and the learning curve in Scikit-Learn), this is likely due to the model overfitting because of the relatively large number of features and the relatively small sample size.

Let's calculate the r2 for the train set:

```
[In]: train_predictions = spark_model.transform(training_data)
[In]: evaluator_r2 = RegressionEvaluator(labelCol='price',
      predictionCol='prediction', metricName='r2')
[In]: train_r2 = evaluator_r2.evaluate(train_predictions)

[In]: print(f"R2 score for the training set: {train_r2}")
[Out]: R2 score for the training set: 0.72
```

The code to calculate and print the r2 for the train set begins by using the trained spark_model to make predictions on the training data, resulting in the train_predictions DataFrame. Then, a `RegressionEvaluator` object is created, specifying that it should evaluate the R2 metric between the "price" column (actual target) and the "prediction" column (model's prediction) within the train_predictions DataFrame. The `evaluate()` method of the evaluator is applied to the train_predictions, computing the R2 score for the training set. Finally, the calculated R2 score is printed, revealing that the model's predictive performance on the training data achieves an R2 score of 0.72.

Clearly, there is a substantial difference between the R2 score of the test set (0.30) and the R2 score of the training set (0.72), which is indicative of overfitting. This confirms what we have already known from the Scikit-Learn results.

Bringing It All Together

In this section, we have combined all the modeling steps into one code block to make it easier for the data scientist to replicate the results in this chapter and to understand better how all the modeling steps fit together.

Scikit-Learn

The following code demonstrates how to perform decision tree regression in Scikit-Learn using the open source housing dataset:

```
# Import necessary libraries
[In]: import pandas as pd
[In]: from sklearn.tree import DecisionTreeRegressor
[In]: from sklearn.model_selection import train_test_split
[In]: from sklearn.metrics import r2_score, mean_squared_error
[In]: from sklearn.preprocessing import OneHotEncoder
[In]: import numpy as np

# Define function for loading data
[In]: def load_housing_data():
          url = ('https://raw.githubusercontent.com/abdelaziztestas/'
                 'spark_book/main/housing.csv')
          return pd.read_csv(url)

# Create a Pandas DataFrame by calling the function:
[In]: pandas_df = load_housing_data()

# Split the data into features and target variable
[In]: X = pandas_df.drop('price', axis=1)
[In]: y = pandas_df['price']

# Perform one-hot encoding on categorical variables
[In]: cat_cols = ['mainroad', 'guestroom', 'basement', 'hotwaterheating',
      'airconditioning', 'prefarea', 'furnishingstatus']
[In]: onehot_encoder = OneHotEncoder(sparse=False)
[In]: X_encoded = onehot_encoder.fit_transform(X[cat_cols])
```

```
[In]: X_encoded_df = pd.DataFrame(X_encoded, columns=onehot_encoder.get_
      feature_names_out(cat_cols))
[In]: X.drop(cat_cols, axis=1, inplace=True)
[In]: X = pd.concat([X, X_encoded_df], axis=1)

# Split the data into training and testing sets
[In]: X_train, X_test, y_train, y_test = train_test_split(X, y, test_
      size=0.2, random_state=42)

# Train the model
[In]: scikit_model = DecisionTreeRegressor()
[In]: scikit_model.fit(X_train, y_train)

# Print feature importances
[In]: importances = scikit_model.feature_importances_
[In]: feature_names = X.columns
[In]: indices = np.argsort(importances)[::-1]
[In]: print('Feature importances:')
[In]: for i in indices:
    print(feature_names[i], ':', importances[i])

# Make predictions on the testing set
[In]: y_pred = scikit_model.predict(X_test)

# Print actual and predicted values
[In]: results_df = pd.DataFrame({'Price': y_test, 'Prediction': y_pred})
[In]: print(results_df.head())

# Calculate evaluation metrics
[In]: r2 = r2_score(y_test, y_pred)
[In]: rmse = np.sqrt(mean_squared_error(y_test, y_pred))

# Print evaluation metrics
[In]: print('R-squared score:', r2)
[In]: print('Root mean squared error:', rmse)
```

PySpark

The following code shows how decision tree regression can be done in PySpark using the same open source housing dataset used to train the Scikit-Learn decision tree regressor:

```
# Import necessary libraries
[In]: from pyspark.sql import SparkSession
[In]: from pyspark.ml.feature import StringIndexer, OneHotEncoder,
      VectorAssembler
[In]: from pyspark.ml.regression import DecisionTreeRegressor
[In]: from pyspark.ml.evaluation import RegressionEvaluator
[In]: import pandas as pd
[In]: import numpy as np

# Create a Spark session
[In]: spark = SparkSession.builder.appName("DecisionTreeRegressorExample").
      getOrCreate()

[In]: def load_housing_data():
          url = ('https://raw.githubusercontent.com/abdelaziztestas/'
                 'spark_book/main/housing.csv')
          return pd.read_csv(url)

[In]: pandas_df = load_housing_data()
[In]: spark_df = spark.createDataFrame(pandas_df)

# Label encoding categorical columns
[In]: cat_columns = ['mainroad', 'guestroom', 'basement',
      'hotwaterheating', 'airconditioning', 'prefarea', 'furnishingstatus']
[In]: indexers = [StringIndexer(inputCol=col, outputCol=col+'_label',
      handleInvalid='keep') for col in cat_columns]
[In]: for indexer in indexers:
          spark_df = indexer.fit(spark_df).transform(spark_df)

# One-hot encoding for all categorical columns
[In]: encoder = OneHotEncoder(inputCols=[col+'_label' for col in cat_
      columns], outputCols=[col+'_encoded' for col in cat_columns])
[In]: spark_df = encoder.fit(spark_df).transform(spark_df)
```

```
# Combining all features into a single feature vector
[In]: feature_cols = ['area', 'bedrooms', 'bathrooms', 'stories',
      'parking'] + [col+'_encoded' for col in cat_columns]
[In]: assembler = VectorAssembler(inputCols=feature_cols,
      outputCol='features')
[In]: spark_df = assembler.transform(spark_df)

# Splitting data into training and test sets
[In]: (training_data, test_data) = spark_df.randomSplit([0.8, 0.2],
      seed=42)

# Training the model
[In]: dt = DecisionTreeRegressor(featuresCol='features', labelCol='price')
[In]: spark_model = dt.fit(training_data)

# printing feature importances
# i. Get feature importances
[In]: feature_importances = spark_model.featureImportances

# ii. Get column names from feature_cols
[In]: feature_names = feature_cols

# iii. Create a dictionary to store feature importances with names
[In]: feature_importances_dict = {}
[In]: for i in range(len(feature_names)):
          feature_name = feature_names[i]
          importance = feature_importances[i]
          feature_importances_dict[feature_name] = importance

# iv. Sort the feature importances dictionary by importance values in
      descending order
[In]: sorted_feature_importances = sorted(feature_importances_dict.items(),
      key=lambda x: x[1], reverse=True)

# v. Printing sorted feature importances along with names
[In]: print("Sorted Feature Importances:")
[In]: for feature_name, importance in sorted_feature_importances:
          print(f"Feature '{feature_name}': {importance}")
```

```
# Making predictions on test data
[In]: predictions = spark_model.transform(test_data)

# Printing actual and predicted values
[In]: predictions.select('price', 'prediction').show(5)

# Evaluating the model using RMSE and R2
[In]: evaluator_rmse = RegressionEvaluator(labelCol='price',
      predictionCol='prediction', metricName='rmse')
[In]: rmse = evaluator_rmse.evaluate(predictions)

[In]: evaluator_r2 = RegressionEvaluator(labelCol='price',
      predictionCol='prediction', metricName='r2')
[In]: r2 = evaluator_r2.evaluate(predictions)

# Printing RMSE and R2 values
[In]: print(f"RMSE: {rmse}")
[In]: print(f"R2: {r2}")
```

Summary

In this chapter, we introduced decision tree regression and demonstrated the process of constructing a regression model using the decision tree algorithm. This was done in both Scikit-Learn and PySpark. We also showed how to transform the data, encode the categorical variables, apply feature scaling, and build, train, and evaluate the model.

Furthermore, we provided insights into overfitting, which is a potential issue when applying decision tree regressors to small datasets with a relatively large number of features. We examined ways to detect overfitting (including train/split r2 comparisons and learning curves) and provided a list of strategies to mitigate the issue. One of these strategies is the use of random forest models, a topic we introduce in the next chapter.

CHAPTER 5

Random Forest Regression with Pandas, Scikit-Learn, and PySpark

In the preceding chapter, we developed a decision tree regression model to predict house prices. In this chapter, we introduce an alternative model known as random forest. Despite both regression models utilizing decision trees, they exhibit notable distinctions.

First, decision trees are simpler models characterized by a single tree structure, whereas random forests are more intricate, comprising multiple decision trees. Furthermore, decision trees are prone to overfitting, whereas random forests counteract this tendency by averaging predictions from numerous trees. Moreover, owing to ensemble learning, random forests generally offer superior accuracy compared to individual decision trees.

From an interpretability standpoint, decision trees are more straightforward to understand because they can be visualized, while random forests pose greater challenges in interpretation due to their combination of multiple trees. In terms of speed, decision trees demonstrate faster training, while random forests are slower due to their ensemble approach. More importantly, decision trees exhibit high variance (meaning they overfit or perform well on the training data but poorly on new, unseen data) and low bias (meaning they underfit or perform poorly on both training and new, unseen data), whereas random forests achieve a balanced trade-off between bias and variance, ultimately reducing both overfitting and underfitting.

The objectives of this chapter are twofold. First, we will use Scikit-Learn and PySpark to build, train, and evaluate a random forest regression model, concurrently drawing parallels between the two frameworks. Subsequently, we will assess the hypothesis that random forests outperform decision trees by applying the random forest model to the

A. Testas, *Distributed Machine Learning with PySpark*, https://doi.org/10.1007/978-1-4842-9751-3_5

same housing dataset we used in the preceding chapter for the decision tree algorithm. This means that if this hypothesis is substantiated, we should see a random forest with higher r2 and lower RMSE than the decision tree.

Before we build the random forest regression model, let's revisit the housing dataset to refresh our understanding of its fundamental attributes. Additionally, we can demonstrate that we can perform the same exploratory data analysis with PySpark as we usually do with Pandas, all while harnessing the power of PySpark's distributed computing.

The Dataset

Similar to what we did in the previous chapter, we can define a Python function to load and read the housing dataset. We then create a Pandas DataFrame and convert it to a PySpark DataFrame. The following code achieves these objectives:

```
[In]: import pandas as pd
[In]: def load_housing_data():
          url = ('https://raw.githubusercontent.com/abdelaziztestas/’
                 'spark_book/main/housing.csv')
          return pd.read_csv(url)
[In]: pandas_df = load_housing_data()
[In]: spark_df = spark.createDataFrame(pandas_df)
```

Recall from the previous chapter that this dataset had 545 rows and 13 columns. We can confirm this by using the Pandas shape attribute:

```
[In]: print(pandas_df.shape)
[Out]: (545, 13)
```

In PySpark, we use a combination of the `count()` and `len(columns)` methods:

```
[In]: print((spark_df.count(), len(spark_df.columns)))
[Out]: (545, 13)
```

We can list the name of the columns or labels using the `columns` attribute for both Pandas and PySpark:

Pandas:

```
[In]: print(pandas_df.columns)
[Out]: Index(['price', 'area', 'bedrooms', 'bathrooms', 'stories',
       'mainroad', 'guestroom', 'basement', 'hotwaterheating',
       'airconditioning', 'parking', 'prefarea', 'furnishingstatus'],
       dtype='object')
```

PySpark:

```
[In]: print(spark_df.columns)
[Out]: ['price', 'area', 'bedrooms', 'bathrooms', 'stories', 'mainroad',
       'guestroom', 'basement', 'hotwaterheating', 'airconditioning',
       'parking', 'prefarea', 'furnishingstatus']
```

Remember that the column attribute doesn't show an index in the PySpark output as the two libraries handle indexing differently.

We can choose to turn off the display of the index in Pandas to align with PySpark with the following code:

```
[In]: column_labels = '[' + ', '.join([f"'{col}'" for col in pandas_
      df.columns]) + ']'
[In]: print(column_labels)
[Out]: ['price', 'area', 'bedrooms', 'bathrooms', 'stories', 'mainroad',
       'guestroom', 'basement', 'hotwaterheating', 'airconditioning',
       'parking', 'prefarea', 'furnishingstatus']
```

The price variable is our label or target variable, while the remaining variables are the features.

As we recall from the previous chapter, the dataset is a mixture of numerical and categorical variables. To check which columns are numerical and which are categorical, we can use the dtypes attribute in both Pandas and PySpark:

Pandas:

```
[In]: print(pandas_df.dtypes)
[Out]:
price                 int64
area                  int64
bedrooms              int64
```

```
bathrooms             int64
stories               int64
mainroad             object
guestroom            object
basement             object
hotwaterheating      object
airconditioning      object
parking               int64
prefarea             object
furnishingstatus     object
dtype: object
```

PySpark:

```
[In]: for col_name, col_type in spark_df.dtypes:
          print(col_name, col_type)
[Out]:
price bigint
area bigint
bedrooms bigint
bathrooms bigint
stories bigint
mainroad string
guestroom string
basement string
hotwaterheating string
airconditioning string
parking bigint
prefarea string
furnishingstatus string
```

The numerical variables have the *int64* data type in Pandas and *bigint* in PySpark, while the categorical variables have the *object* and *string* data types, respectively. There are, therefore, seven categorical variables that will need to be one-hot encoded (mainroad, guestroom, basement, hotwaterheating, airconditioning, prefarea, furnishingstatus).

Six of these categorical variables have two categories (yes or no, or vice versa), while the furnishing status column has three categories (furnished, semi-furnished, and unfurnished). We can confirm this by printing the unique values for each categorical variable with the following code:

Pandas:

```
[In]: for col in ['mainroad', 'guestroom', 'basement', 'hotwaterheating',
      'airconditioning', 'prefarea', 'furnishingstatus']:
          unique_values = pandas_df[col].unique()
          print(f"Unique values in {col}: {unique_values}")
[Out]:
Unique values in mainroad: ['yes' 'no']
Unique values in guestroom: ['no' 'yes']
Unique values in basement: ['no' 'yes']
Unique values in hotwaterheating: ['no' 'yes']
Unique values in airconditioning: ['yes' 'no']
Unique values in prefarea: ['yes' 'no']
Unique values in furnishingstatus: ['furnished' 'semi-furnished'
'unfurnished']
```

PySpark:

```
[In]: from pyspark.sql.functions import col
[In]: for col_name in ['mainroad', 'guestroom', 'basement',
      'hotwaterheating', 'airconditioning', 'prefarea', 'furnishingstatus']:
      unique_values=spark_df.select(col(col_name)).distinct().rdd.
      flatMap(lambda x: x).collect()
          print(f"Unique values in {col_name}: {unique_values}")
[Out]:
Unique values in mainroad: ['yes', 'no']
Unique values in guestroom: ['no', 'yes']
Unique values in basement: ['no', 'yes']
Unique values in hotwaterheating: ['no', 'yes']
Unique values in airconditioning: ['no', 'yes']
Unique values in prefarea: ['no', 'yes']
Unique values in furnishingstatus: ['semi-furnished', 'unfurnished',
'furnished']
```

Now that we have reviewed the main attributes of the housing dataset and demonstrated the similar approaches of Pandas and PySpark, we can proceed to build the random forest regression model.

Random Forest Regression

Random forest models belong to the family of ensemble methods because they combine predictions from multiple decision trees, each trained on a different subset of the data and features. While random forests can be used for both classification and regression problems, in this chapter, we will be focusing on regression as our target variable (house price) is continuous, not categorical. In Chapter 9, we will demonstrate how to apply random forests to classification problems involving categorical target variables.

Before building the random forest regression model, let's first compare the modeling steps in Scikit-Learn and PySpark.

Scikit-Learn/PySpark Similarities

Scikit-Learn and PySpark share a comparable approach to machine learning, as they both adhere to similar procedures for building, training, and evaluating the decision tree regressor. We can explore the commonalities between these platforms by exploring the primary machine learning functions they offer:

1. Data preparation:

   Scikit-Learn: In Scikit-Learn, functions like train_test_split are employed to divide decision tree regression data into training and testing subsets. Additional functions include StandardScaler for feature scaling, OneHotEncoder for encoding categorical variables, and SimpleImputer for addressing missing data.

   PySpark: Within PySpark, similar tasks can be performed using DataFrames. The randomSplit function serves for partitioning data into training and testing sets, StandardScaler handles feature scaling, OneHotEncoder performs categorical encoding, and Imputer addresses missing data. Additionally, the VectorAssembler is utilized to combine feature columns into a single vector column, a step that diverges from Scikit-Learn's approach.

2. Model training:

   Scikit-Learn: For random forest regression in Scikit-Learn, the RandomForestRegressor model can be instantiated, followed by the utilization of the fit method. This requires providing the feature matrix (X_train) and the target variable (y_train).

   PySpark: In PySpark, we employ the RandomForestRegressor model and its corresponding fit method as well. A DataFrame (train_data) containing the features and a column for the target variable are required. It is important to apply the VectorAssembler step before beginning the training process.

3. Model evaluation:

   Scikit-Learn: The assessment of random forest regression model performance involves two key metrics: mean_squared_error and r2_score. These metrics are computed by comparing predicted (y_pred) and actual (y_test) target variable values.

   PySpark: The RegressionEvaluator class in PySpark facilitates metric calculations of the same evaluation metrics (squared error and r2). This is accomplished by specifying the columns containing predicted values (predictions) and actual values (y_test) for evaluation.

4. Prediction:

   Scikit-Learn: Making predictions using the trained random forest regressor requires using the predict method. By providing the feature matrix (X_test), we obtain predicted target variable values (y_pred).

   PySpark: Within PySpark, the transform method of the trained model generates predictions for new data instances. A DataFrame (test_data) comprising features is supplied, and predictions are retrieved in a new column (prediction). Ensuring the application of the VectorAssembler step maintains the necessary feature vector structure.

As we can see from the preceding comparison, the modeling steps between Scikit-Learn and PySpark are strikingly similar. As we will see in the next subsection, however, the nuances of implementation and syntax may vary due to Scikit-Learn being Python based, while PySpark is a distributed computing framework built on Apache Spark.

Random Forest with Scikit-Learn

In this subsection, we translate the preceding descriptive modeling steps to Python code. First, a reminder of the data we will be using to train the random forest regressor in Scikit-Learn:

```
[In]: import pandas as pd
[In]: def load_housing_data():
          url = ('https://raw.githubusercontent.com/abdelaziztestas/'
                 'spark_book/main/housing.csv')
      return pd.read_csv(url)
[In]: pandas_df = load_housing_data()
[In]: X = pandas_df.drop('price', axis=1)
[In]: y = pandas_df['price']
```

The code defines a function called `load_housing_data()`. The URL within the function, which points to the CSV file location, is passed to the Pandas `read_csv()` method. This reads the CSV file and returns a Pandas DataFrame. The pandas_df DataFrame is created once the `load_housing_data()` function is called.

In the next step, the code creates two new DataFrames: X containing the 12 input features and y holding the target variable (price). This separation between X and y allows for supervised learning tasks where the goal is to predict the target values based on the input features in X.

Moving on to the modeling steps, as is the case in all machine learning algorithms, the first step is to import the necessary libraries.

Step 1: Import necessary libraries

In this step, we import the classes and functions required for data preparation, model training, model evaluation, and data manipulation:

```
[In]: from sklearn.ensemble import RandomForestRegressor
[In]: from sklearn.model_selection import train_test_split
[In]: from sklearn.metrics import r2_score, mean_squared_error
```

```
[In]: from sklearn.preprocessing import OneHotEncoder
[In]: import pandas as pd
[In]: import numpy as np
```

Step 2: Prepare data

There are two substeps involved here: encoding categorical variables into numerical values and splitting the dataset into training and testing sets:

2.1. Perform one-hot encoding on categorical variables

```
[In]: cat_cols = ['mainroad', 'guestroom', 'basement', 'hotwaterheating',
      'airconditioning', 'prefarea', 'furnishingstatus']
[In]: onehot_encoder = OneHotEncoder(sparse=False)
[In]: X_encoded = onehot_encoder.fit_transform(X[cat_cols])
[In]: X_encoded_df = pd.DataFrame(X_encoded, columns=onehot_encoder.get_
      feature_names_out(cat_cols))
[In]: X.drop(cat_cols, axis=1, inplace=True)
[In]: X = pd.concat([X, X_encoded_df], axis=1)
```

2.2. Split the data into training and testing sets

```
[In]: X_train, X_test, y_train, y_test = train_test_split(X, y, test_
      size=0.2, random_state=42)
```

Step 3: Train the random forest model

```
[In]: random_forest_model = RandomForestRegressor()
[In]: random_forest_model.fit(X_train, y_train)
```

Step 4: Make predictions on the testing set

```
[In]: y_pred = random_forest_model.predict(X_test)
```

Step 5: Evaluate the model

```
[In]: r2 = r2_score(y_test, y_pred)
[In]: rmse = np.sqrt(mean_squared_error(y_test, y_pred))
```

The aforementioned five modeling steps indicate that in just a few lines of code, we have prepared the data, built the random forest model, trained the model, evaluated it, and made predictions on the testing set.

Let's go through the steps in more detail:

Step 1: Importing necessary libraries

In this step, the libraries for preparing data, building, training, and evaluating the random forest regression model are imported. These modules include `RandomForestRegressor` for building a random forest model, `train_test_split` for splitting the dataset into training and testing sets, and evaluation metrics such as `r2_score` and `mean_squared_error`. Additionally, `Pandas` and `NumPy` libraries are imported for data manipulation and numerical computations.

Step 2: Data preparation

2.1. Perform one-hot encoding on categorical variables

To handle categorical variables in the housing dataset, a list of categorical column names (cat_cols) is defined. The list includes seven categorical variables: mainroad, guestroom, basement, hot water heating, air conditioning, preferred area, and furnishing status.

The `OneHotEncoder` class is instantiated with the parameter sparse=False to generate a dense array after one-hot encoding. The categorical variables in the original feature matrix X are transformed using this encoder, resulting in a new array X_encoded. The encoded features are then stored in a new DataFrame X_encoded_df with appropriate column names obtained from the encoder's `get_feature_names_out()` method. The original categorical columns are dropped from the X DataFrame using the `drop()` method, and the encoded features are concatenated back to X using the Pandas `concat()` method.

2.2. Split the data into training and testing sets

The dataset is divided into training and testing subsets using the `train_test_split()` method. The parameter test_size specifies the proportion of data to be allocated for testing (0.2 or 20%), while random_state ensures reproducibility of the split. The resulting training and testing feature sets (X_train and X_test) and corresponding target sets (y_train and y_test) are created for subsequent model training and evaluation.

Step 3: Training the random forest model

A random forest regression model is instantiated as random_forest_model. This model belongs to the ensemble learning family and combines multiple decision trees to make predictions. It's trained using the training data (X_train and y_train) using the `fit()` method.

Now that the random forest regression model has been trained, we can print the number of trees in the forest as follows:

```
[In]: print(random_forest_model.n_estimators)
[Out]: 100
```

This indicates that the model has been built using 100 trees, which is the default number of trees in Scikit-Learn. This is a hyperparameter that can be changed as we will see in Chapter 16 where we explore the topic of hyperparameter tuning in greater detail.

We can also print the feature importances of the model, just as we did in the previous chapter for the decision tree algorithm:

```
[In]: importances = random_forest_model.feature_importances_
[In]: feature_names = X.columns
[In]: indices = np.argsort(importances)[::-1]
[In]: print('Feature importances:')
[In]: for i in indices:
          print(feature_names[i], ':', importances[i])
[Out]:
area: 0.4571
bathrooms: 0.1592
stories: 0.0580
parking: 0.0542
bedrooms: 0.0463
furnishingstatus_unfurnished: 0.0350
airconditioning_yes: 0.0282
airconditioning_no: 0.0281
basement_no: 0.0188
basement_yes: 0.0167
prefarea_yes: 0.0145
prefarea_no: 0.0140
furnishingstatus_furnished: 0.0110
hotwaterheating_no: 0.0110
furnishingstatus_semi furnished: 0.0109
hotwaterheating_yes: 0.0099
guestroom_no: 0.0094
```

```
guestroom_yes: 0.0073
mainroad_no: 0.0055
mainroad_yes: 0.0049
```

Similar to our conclusion in the previous chapter regarding the decision tree regression model, we find that the area of the house and the number of bathrooms remain the primary drivers for the model's predictions. Conversely, from the preceding output, we observe that residing on the main road (yes or no) has the least impact on the model's predictions.

We can also print the depth of each tree:

```
[In]: tree_depths = [estimator.tree_.max_depth for estimator in random_
forest_model.estimators_]
[In]: print('Depth of each tree:', tree_depths)
[Out]:
Depth of each tree: [16, 18, 17, 19, 16, 18, 15, 15, 18, 15, 18, 17, 17,
18, 18, 15, 19, 17, 17, 16, 19, 19, 18, 17, 16, 15, 17, 15, 19, 18, 16, 19,
17, 16, 14, 14, 17, 17, 15, 18, 18, 16, 19, 14, 17, 16, 18, 18, 19, 17, 15,
19, 18, 18, 17, 16, 17, 18, 16, 17, 15, 17, 16, 19, 20, 17, 15, 18, 17, 19,
16, 15, 17, 17, 17, 15, 16, 17, 16, 14, 20, 17, 17, 16, 14, 17, 19, 15, 14,
17, 18, 16, 17, 18, 17, 20, 19, 18, 16, 17]
```

The output represents the maximum depth of each of the 100 individual decision trees in the random forest model. The depth of a decision tree refers to the number of levels or layers of splits it has in its structure. Each level represents a decision point based on a feature, and the tree's leaves hold the final predictions.

There are seven distinct values (14, 15, 16, 17, 18, 19, 20) in the preceding output, meaning that multiple trees share the same value. The maximum depths of the individual decision trees are typically set to None in Scikit-Learn, which means that the trees will be expanded until they contain less than the minimum samples required to split a node. This contrasts with PySpark, as we will see in the next subsection, as the default maximum depth of a tree is set to 5.

Let's take a look at a few values from the list and understand their meaning. For example, the first four values, [16, 18, 17, 19], indicate that the first decision tree in the ensemble has a maximum depth of 16. This means that the tree has been split into a series of decisions up to 16 levels deep. It can potentially create a tree structure with 2^16 (65,536) leaf nodes. Similarly, the second decision tree has a maximum depth of 18. This indicates that it's slightly deeper than the first tree, potentially capturing more intricate patterns in the data. It can have up to 2^18 (262,144) leaf nodes. The third decision tree has a maximum depth of 17, and the fourth has a maximum depth of 19. These values suggest that these trees are capturing complex relationships in the data, potentially leading to a greater number of decision paths and more nuanced predictions.

In general, a deeper decision tree can model the training data more accurately, as it can capture finer details and interactions between features. The ensemble nature of a random forest helps mitigate overfitting by aggregating the predictions of multiple decision trees. (Note, however, that excessively deep trees can also lead to overfitting, where the model fits the training data too closely and doesn't generalize well to new, unseen data.)

Step 4: Making predictions on the testing set

In this step, the trained random forest model is used to predict the target variable for the testing feature set (X_test). The predicted values are stored in the y_pred array.

We can compare a sample of the actual and predicted target values as follows:

```
[In]: results_df = pd.DataFrame({'Price': y_test, 'Prediction': y_pred})
[In]: print(results_df.head())
```

Price	Prediction
4060000	5499270
6650000	7122220
3710000	3826900
6440000	4461100
2800000	3644757

If the model were 100% accurate, the Price column (actual price of a house) and the Prediction column (predicted price) would be exactly the same. However, there are differences between the two columns. In the first row, for example, the actual price is $4,060,000, while the predicted price is $5,499,270. This indicates that the model overestimates the price by $1,439,270 or 35% of the actual price.

Step 5: Evaluating the model

In this final step, the performance of the random forest regression model is assessed using two evaluation metrics: the coefficient of determination (R-squared, r2) and the root mean squared error (RMSE, rmse). The `r2_score()` function computes the

R-squared value, which indicates the proportion of the variance in the target variable that is predictable from the features. The `mean_squared_error()` function computes the mean squared error between the predicted and actual target values, and then the square root is taken to obtain RMSE. These metrics provide insights into how well the model's predictions align with the actual target values, helping to assess the model's performance.

We can print the values of these two metrics with the following code:

```
[In]: print('R-squared score:', r2)
[In]: print('Root mean squared error:', rmse)
[Out]: R-squared score: 0.62 Root mean squared error: 1388213
```

With the evaluation metrics generated from the random forest regression model, we can now compare their values with those we obtained in the previous chapter for the decision tree regression model. This is to test the hypothesis that random forest regression models perform better than decision trees. Here's the comparison:

Model	RMSE	r2
Random forest	1,388,213	0.62
Decision tree	1,715,691	0.42

The preceding output indicates that the random forest regression model indeed outperforms the decision tree regression model. This is evident in its substantially lower RMSE (1,388,213 vs. 1,715,691) and notably higher r2 (0.62 vs. 0.42). In the context of the random forest model, this signifies that approximately 62% of the variations in house prices can be attributed to variations in the features of the house. In contrast, the decision tree regression model explains around 42% of the variations.

This outcome is unsurprising, as the random forest combines multiple decision trees, harnessing their collective predictive power to deliver more accurate and robust results.

It remains true, however, that even the random forest model's r2 of 62% isn't considered high. This suggests that even though this model has significantly mitigated the overfitting of the decision tree model, in a real-world scenario, more steps will be taken to improve the model, such as reducing its complexity through feature extraction or feature selection. As we will see in Chapter 16, fine-tuning a model's hyperparameters to find the most optimal combination of values is another way to improve its performance.

Random Forest with PySpark

We will now proceed to build the random forest regression model in PySpark and train it using the same housing dataset. First, let's have a quick reminder of the code that we used to generate the PySpark DataFrame.

In the following code, we first create a Pandas DataFrame and then convert it to PySpark using the `createDataFrame()` method.

```
[In]: import pandas as pd
[In]: from pyspark.sql import SparkSession
[In]: spark = SparkSession.builder.appName("Random Forest Regression").
      getOrCreate()
[In]: def load_housing_data():
          url = ('https://raw.githubusercontent.com/abdelaziztestas/'
                 'spark_book/main/housing.csv')
          return pd.read_csv(url)
[In]: pandas_df = load_housing_data()
[In]: spark_df = spark.createDataFrame(pandas_df)
```

In this code, the first line imports the Pandas library and names it pd. The second line imports the SparkSession class from the PySpark library, which is the entry point to utilizing Spark functionalities. The third line establishes a Spark Session named "Random Forest Regression" using the `SparkSession.builder.appName()` method. This session serves as the connection to the Spark cluster and provides the environment to perform distributed data processing.

The next step defines a function named `load_housing_data()` that retrieves housing data from a URL using the `read_csv()` method. Following this, the code reads the housing data into a Pandas DataFrame named pandas_df using the defined `load_housing_data()` function. Finally, the code converts the Pandas DataFrame pandas_df into a Spark DataFrame named spark_df using the `createDataFrame()` method. This conversion allows the data to be processed and analyzed using Spark's distributed computing capabilities.

We will now write PySpark code to prepare the housing data, build and train the random forest regression model using the training subsample, and then evaluate it using the test set. As a first step, we will need to import the necessary libraries:

Step 1: Importing necessary libraries

```
[In]: from pyspark.ml.feature import StringIndexer, OneHotEncoder,
      VectorAssembler
[In]: from pyspark.ml.regression import RandomForestRegressor
[In]: from pyspark.ml.evaluation import RegressionEvaluator
```

Step 2: Data preparation

2.1. Label encoding categorical columns

```
[In]: cat_columns = ['mainroad', 'guestroom', 'basement',
      'hotwaterheating', 'airconditioning', 'prefarea', 'furnishingstatus']
[In]: indexers = [StringIndexer(inputCol=col, outputCol=col+'_label',
      handleInvalid='keep') for col in cat_columns]
[In]: for indexer in indexers:
          spark_df = indexer.fit(spark_df).transform(spark_df)
```

2.2. One-hot encoding categorical columns

```
[In]: encoder = OneHotEncoder(inputCols=[col+'_label' for col in cat_
      columns], outputCols=[col+'_encoded' for col in cat_columns])
[In]: spark_df = encoder.fit(spark_df).transform(spark_df)
```

2.3 Combining all features into a single feature vector

```
[In]: feature_cols = ['area', 'bedrooms', 'bathrooms', 'stories',
      'parking'] + [col+'_encoded' for col in cat_columns]
[In]: assembler = VectorAssembler(inputCols=feature_cols,
      outputCol='features')
[In]: spark_df = assembler.transform(spark_df)
```

2.4. Splitting data into training and testing sets

```
[In]: (training_data, test_data) = spark_df.randomSplit([0.8, 0.2],
      seed=42)
```

Step 3: Model training

```
[In]: rf = RandomForestRegressor(featuresCol='features', labelCol='price')
[In]: spark_model = rf.fit(training_data)
```

Step 4: Prediction

```
[In]: predictions = spark_model.transform(test_data)
```

Step 5: Model evaluation

```
[In]: evaluator_rmse = RegressionEvaluator(labelCol='price',
      predictionCol='prediction', metricName='rmse')
[In]: rmse = evaluator_rmse.evaluate(predictions)
[In]: evaluator_r2 = RegressionEvaluator(labelCol='price',
      predictionCol='prediction', metricName='r2')
[In]: r2 = evaluator_r2.evaluate(predictions)
```

The preceding steps demonstrate how we can efficiently prepare data, build the model, train it, and evaluate its performance using just a few lines of code. With these steps outlined, let's now provide a detailed explanation for each.

Step 1: Importing necessary libraries

In this step, the code imports libraries that enable various stages of model construction:

- Data preparation using `StringIndexer`, `OneHotEncoder`, and `VectorAssembler`
- Model building using `RandomForestRegressor`
- Model evaluation using `RegressionEvaluator`

These classes and functions enable data preparation processes such as feature indexing, one-hot encoding, and data transformation. For example, categorical columns are labeled using the `StringIndexer` class, followed by one-hot encoding with the `OneHotEncoder` class. The `VectorAssembler` class combines features into a single vector. Furthermore, the code employs the `RandomForestRegressor` class to build the regression model and the `RegressionEvaluator` class to assess the model's performance using RMSE and R-squared.

Step 2: Data preparation

2.1. Label encoding categorical columns

To prepare the data for modeling, categorical columns in the housing dataset are encoded using string indexing (also known as label encoding). The seven categorical features (mainroad, guestroom, basement, hot water heating, air conditioning, preferred area, and furnishing status) are indexed using the `StringIndexer` class. Each categorical value is assigned a numerical label, and a new column with _label suffix is added to the dataset. This transformation makes categorical data compatible with the `OneHotEncoder` used in the next substep.

2.2. One-hot encoding categorical columns

In this substep, one-hot encoding is applied to the previously labeled categorical columns. The `OneHotEncoder` class is utilized to convert each categorical label into a binary vector representation. The original categorical columns are replaced with their corresponding one-hot encoded versions, enhancing the model's ability to capture nonlinear relationships between features.

The following is the representation of categorical features after one-hot encoding, using the furnishing status as an illustration:

```
+----------------+----------------------+------------------------+
|furnishingstatus|furnishingstatus_label|furnishingstatus_encoded|
+----------------+----------------------+------------------------+
|       furnished|                   2.0|           (3,[2],[1.0])|
|  semi-furnished|                   0.0|           (3,[0],[1.0])|
|     unfurnished|                   1.0|           (3,[1],[1.0])|
+----------------+----------------------+------------------------+
```

The furnishingstatus column displays the original categorical values of this feature (furnished, semi-furnished, unfurnished). The furnishingstatus_label column shows the label-encoded version of the furnishing status feature. We obtained this version by applying the `StringIndexer` in substep 2.1. Label encoding converts categorical values into numerical labels. For example, furnished is assigned the label 2.0, semi-furnished is assigned the label 0.0, and unfurnished is assigned the label 1.0.

The furnishingstatus_encoded column represents the one-hot encoded version of the furnishingstatus feature. This was obtained by applying the `OneHotEncoder` to the indexed features in substep 2.2. One-hot encoding converts categorical values into binary vectors, where each vector has a 1 in the position corresponding to the category and 0s in all other positions. This format is more suitable for the machine learning algorithm. The notation (3,[2],[1.0]) indicates that the vector has a length of 3 (reflecting the three categories of the furnishing status column), and it has a 1.0 value in the position corresponding to the label 2 (which represents furnished). Similarly, (3,[0],[1.0]) represents semi-furnished, and (3,[1],[1.0]) represents unfurnished.

2.3 Combining all features into a single feature vector

To meet PySpark's requirement, the one-hot encoded categorical features are combined into a single numerical vector. The `VectorAssembler` class is employed for this task. The specified input columns are concatenated to create a new column named features.

The following is an example of a row from the features column, which contains the assembled features:

```
(20,[0,1,2,3,4,5,7,9,11,14,16,19],
[7420.0,4.0,2.0,3.0,2.0,1.0,1.0,1.0,1.0,1.0,1.0,1.0])
```

There are 20 features as indicated by the number before the square brackets. This is the sum of the original numerical features (12) and those that have been one-hot encoded (8). In other words, five original numerical features (area, bedrooms, bathrooms, stories, and parking), seven one-hot encoded features (for the seven categorical columns, mainroad, guestroom, basement, hot water heating, air conditioning, preferred area, and furnishing status, with two unique values each, that is, yes or no), and one additional slot for the furnishing status column with three unique values (furnished, semi-furnished, unfurnished).

There are 12 indices and 12 corresponding values in the first bracket and second bracket, respectively. These indices and values correspond to the 12 features in the dataset, including both original numerical features and one-hot encoded features.

2.4. Splitting data into training and testing sets

The dataset is divided into training and testing sets for model development and evaluation. Using the `randomSplit()` method, the data is partitioned into an 80% training set and a 20% test set. The split is carried out with a specific seed value (42) for reproducibility.

Step 3: Training the random forest model

The training sample is used to train the algorithm. A `RandomForestRegressor` instance is initialized with parameters including the input features column (features) and the target column (price). This model is trained using the training dataset.

We can check the number of decision trees used in the model and the depth of each tree with the following code:

```
[In]: from collections import Counter
[In]: tree_depths = []
[In]: for tree in individual_trees:
          depth = tree.depth
          tree_depths.append(depth)
[In]: depth_counter = Counter(tree_depths)
```

```
[In]: print("Tree Depths and Counts:")
[In]: for depth, count in depth_counter.items():
          print(f"Depth {depth}: {count} trees")
[Out]: Depth 5: 20 trees
```

After importing the Counter class from the collections module, the code initializes an empty list to store tree depths. Next, it iterates through each tree and retrieves its depth. The code then counts the occurrences of each tree depth and finally prints the depths and their counts.

The output indicates that there are 20 trees, each with a depth of 5. These are the default values in PySpark, which differ from those of the Scikit-Learn algorithm: Scikit-Learn uses 100 trees by default and does not have a specific default value for the depth of a tree. These variations can lead to differences in the model's performance.

In tree-based algorithms, the depth of a tree refers to the number of levels or layers of splits it contains. Each split divides the data into subsets based on a specific feature and threshold. Deeper trees can capture more intricate patterns in the data. In our specific example, each tree has been constructed with a depth of 5, indicating relatively shallow trees. This implies that the PySpark model might have lower performance compared to Scikit-Learn's, as the latter's tree depths range from 14 to 20, enabling it to capture more intricate information from the dataset.

We can also print the feature importances of the model as follows:

```
[In]: feature_importances = spark_model.featureImportances.toArray()
[In]: feature_names = feature_cols
[In]: feature_importance_tuples = list(zip(feature_names, feature_importances))
[In]: sorted_feature_importance = sorted(feature_importance_tuples,
      key=lambda x: x[1], reverse=True)
[In]: print("Sorted Feature Importances:")
[In]: for feature_name, importance in sorted_feature_importance:
          print(f"Feature '{feature_name}': {importance:.4f}")
[Out]:
Sorted Feature Importances:
area: 0.3693
bathrooms: 0.2030
stories: 0.0604
bedrooms: 0.0545
parking: 0.0446
```

```
furnishingstatus_encoded: 0.0096
basement_encoded: 0.0094
prefarea_encoded: 0.0087
guestroom_encoded: 0.0080
mainroad_encoded: 0.0072
airconditioning_encoded: 0.0072
hotwaterheating_encoded: 0.0045
```

The code begins by directly printing the feature importances obtained from the random forest model. It then retrieves the column names from the feature_cols list. Afterward, it combines the feature names and importances into a list of tuples. Next, the code sorts the list of tuples in descending order based on their importance values and proceeds to print the sorted feature importances alongside their corresponding names.

The output indicates that the area of the house and the number of bathrooms are the main drivers behind the random forest predictions. Air conditioning and hot water heating are the least important features in the model.

Step 4: Prediction

With the trained model, predictions are made on the test data. The `transform` method is applied to the test dataset using the trained model, yielding predicted values for the target variable.

As we have done with Scikit-Learn, let's print a sample of actual and predicted price values for comparison (Note that differences exist in the first 5 rows of actual prices between the Scikit-Learn and PySpark implementations. These differences can be attributed to variations in data splitting and the subsequent prediction process. Specifically, the Scikit-Learn code uses `train_test_split`, while the PySpark code, which employs randomSplit, utilizes distinct test data subsets for making predictions, resulting in these disparities):

```
[In]: predictions.select('price', 'prediction').show(5)
[Out]:
```

Price	Prediction
6930000	5198444
7070000	4911813
7210000	7193449
7350000	5423289
7455000	3970584

This sample output indicates that the model has underestimated the target price in all cases. In the first observation, for example, the actual price is $6,930,000, while the predicted price is $5,198,444. Similarly, in the last observation, the predicted price is much lower than the actual price ($3,970,584 vs. $7,455,000).

Step 5: Model evaluation

In this final step of the modeling process in PySpark, two evaluation metrics (root mean squared error, RMSE, and R-squared, r2) are used to assess the model's performance. The `RegressionEvaluator` class is utilized to compute these metrics. RMSE measures the average prediction error, while R-squared assesses the model's goodness of fit to the data. The `evaluate` method calculates these metrics based on the predicted and actual target values, providing insights into the model's accuracy and predictive capabilities.

We can print the values of the two metrics with the following code:

```
[In]: print(f"RMSE: {rmse}")
[In]: print(f"R2: {r2}")
[Out]: RMSE: 1237365 R2: 0.50
```

The RMSE is relatively large (1,237,365), while the R2 is relatively small (0.50), indicating a rather modest performance of the PySpark random forest regression model. Nevertheless, when compared to the decision tree regressor from the preceding chapter, the model in this chapter exhibits a noticeable improvement. This is evident in the following table, where the RMSE of the random forest is lower than that of the decision tree (1,237,365 vs. 1,463,512), and its r2 is higher (0.50 vs. 0.32). This improvement is expected due to the ensemble nature of the random forest, which is less prone to overfitting compared to the decision tree.

Model	RMSE	r2
Random forest	1,237,365	0.50
Decision tree	1,463,512	0.32

Bringing It All Together

In the previous sections of this chapter, we have used code snippets to demonstrate each step of the modeling process individually. However, in this section, we aim to consolidate all the relevant code from those steps into a single block. This will enable the data scientist to view how all the code snippets work together and to execute the code as a single unit. Combining the code from each step into one block will also make it easier for the data scientist to modify the code and experiment with different parameters or configurations.

Scikit-Learn

The following is the code to prepare data, build and train the model, and evaluate its performance:

```
# Import necessary libraries
[In]: import pandas as pd
[In]: from sklearn.ensemble import RandomForestRegressor
[In]: from sklearn.model_selection import train_test_split
[In]: from sklearn.metrics import r2_score, mean_squared_error
[In]: from sklearn.preprocessing import OneHotEncoder
[In]: import numpy as np

# Define function for loading data
[In]: def load_housing_data():
          url = ('https://raw.githubusercontent.com/abdelaziztestas/'
                 'spark_book/main/housing.csv')
          return pd.read_csv(url)

# Create a Pandas DataFrame by calling the function:
[In]: pandas_df = load_housing_data()

# Split the data into features and target variable
[In]: X = pandas_df.drop('price', axis=1)
[In]: y = pandas_df['price']
```

```
# Perform one-hot encoding on categorical variables
[In]: cat_cols = ['mainroad', 'guestroom', 'basement', 'hotwaterheating',
      'airconditioning','prefarea','furnishingstatus']
[In]: onehot_encoder = OneHotEncoder(sparse=False)
[In]: X_encoded = onehot_encoder.fit_transform(X[cat_cols])
[In]: X_encoded_df = pd.DataFrame(X_encoded, columns=onehot_encoder.get_
      feature_names_out(cat_cols))
[In]: X.drop(cat_cols, axis=1, inplace=True)
[In]: X = pd.concat([X, X_encoded_df], axis=1)

# Split the data into training and testing sets
[In]: X_train, X_test, y_train, y_test = train_test_split(X, y, test_
      size=0.2, random_state=42)

# Train the Random Forest model
[In]: random_forest_model = RandomForestRegressor()
[In]: random_forest_model.fit(X_train, y_train)

# Print the number of trees in the forest
[In]: print("Number of trees in the Random Forest:", random_forest_model.n_
      estimators)

# Print feature importances
[In]: importances = random_forest_model.feature_importances_
[In]: feature_names = X.columns
[In]: indices = np.argsort(importances)[::-1]
[In]: print('Feature importances:')
[In]: for i in indices:
          print(feature_names[i], ':', importances[i])

# Print the depth of each tree
[In]: tree_depths = [estimator.tree_.max_depth for estimator in random_
      forest_model.estimators_]
[In]: print('Depth of each tree:', tree_depths)

# Make predictions on the testing set
[In]: y_pred = random_forest_model.predict(X_test)
```

```
# Print actual and predicted values
[In]: results_df = pd.DataFrame({'Price': y_test, 'Prediction': y_pred})
[In]: print(results_df.head())

# Calculate evaluation metrics
[In]: r2 = r2_score(y_test, y_pred)
[In]: rmse = np.sqrt(mean_squared_error(y_test, y_pred))

# Print evaluation metrics
[In]: print('R-squared score:', r2)
[In]: print('Root mean squared error:', rmse)
```

PySpark

Provided here is the code to prepare the data, build and train the random forest in PySpark, and evaluate its performance:

```
# Import necessary libraries
[In]: from pyspark.sql import SparkSession
[In]: from pyspark.ml.feature import StringIndexer, OneHotEncoder,
      VectorAssembler
[In]: from pyspark.ml.regression import RandomForestRegressor
[In]: from pyspark.ml.evaluation import RegressionEvaluator
[In]: import pandas as pd
[In]: import numpy as np

# Create a Spark session
[In]: spark = SparkSession.builder.appName("RandomForestRegressorExample").
      getOrCreate()

[In]: def load_housing_data():
          url = ('https://raw.githubusercontent.com/abdelaziztestas/'
                 'spark_book/main/housing.csv')
          return pd.read_csv(url)
# Load data and create Spark DataFrame
[In]: pandas_df = load_housing_data()
[In]: spark_df = spark.createDataFrame(pandas_df)
```

```
# Label encoding categorical columns
[In]: cat_columns = ['mainroad', 'guestroom', 'basement',
      'hotwaterheating', 'airconditioning', 'prefarea', 'furnishingstatus']
[In]: indexers = [StringIndexer(inputCol=col, outputCol=col+'_label',
      handleInvalid='keep') for col in cat_columns]
[In]: for indexer in indexers:
          spark_df = indexer.fit(spark_df).transform(spark_df)

# One-hot encoding for all categorical columns
[In]: encoder = OneHotEncoder(inputCols=[col+'_label' for col in cat_
      columns], outputCols=[col+'_encoded' for col in cat_columns])
[In]: spark_df = encoder.fit(spark_df).transform(spark_df)

# Combining all features into a single feature vector
[In]: feature_cols = ['area', 'bedrooms', 'bathrooms', 'stories',
      'parking'] + [col+'_encoded' for col in cat_columns]
[In]: assembler = VectorAssembler(inputCols=feature_cols,
      outputCol='features')
[In]: spark_df = assembler.transform(spark_df)

# Splitting data into training and test sets
[In]: (training_data, test_data) = spark_df.randomSplit([0.8, 0.2],
      seed=42)

# Training the Random Forest model
[In]: rf = RandomForestRegressor(featuresCol='features', labelCol='price')
[In]: spark_model = rf.fit(training_data)

# Access the individual trees from the RandomForest model
[In]: individual_trees = spark_model.trees

# Initialize an empty list to store tree depths
[In]: tree_depths = []

# Iterate through each tree and retrieve its depth
[In]: for tree in individual_trees:
          depth = tree.depth
          tree_depths.append(depth)
```

```
# Print the tree depths
[In]: print("Tree Depths:", tree_depths)

# Printing feature importances directly from RandomForest model
[In]: feature_importances = spark_model.featureImportances.toArray()

# Get column names from feature_cols
[In]: feature_names = feature_cols

# Combine feature names and importances into a list of tuples
[In]: feature_importance_tuples = list(zip(feature_names, feature_
      importances))

# Sort the list of tuples by importance in descending order
[In]: sorted_feature_importance = sorted(feature_importance_tuples,
      key=lambda x: x[1], reverse=True)

# Printing sorted feature importances along with names
[In]: print("Sorted Feature Importances:")
[In]: for feature_name, importance in sorted_feature_importance:
          print(f"Feature '{feature_name}': {importance:.4f}")

# Making predictions on test data
[In]: predictions = spark_model.transform(test_data)

# Printing actual and predicted values
[In]: predictions.select('price', 'prediction').show(5)

# Evaluating the model using RMSE and R2
[In]: evaluator_rmse = RegressionEvaluator(labelCol='price',
      predictionCol='prediction', metricName='rmse')
[In]: rmse = evaluator_rmse.evaluate(predictions)
[In]: evaluator_r2 = RegressionEvaluator(labelCol='price',
[In]: predictionCol = 'prediction', metricName='r2')
[In]: r2 = evaluator_r2.evaluate(predictions)

# Printing RMSE and R2 values
[In]: print(f"RMSE: {rmse}")
[In]: print(f"R2: {r2}")
```

Summary

This chapter introduced random forest as an alternative to decision tree regression. The strengths and weaknesses of the two algorithms were highlighted. For example, decision trees are simpler but more susceptible to overfitting, a concern mitigated by random forests. Additionally, random forests offer improved accuracy compared to individual decision trees. In terms of interpretability, decision trees are more straightforward to understand since they could be visualized.

The chapter also showed that the steps of data preparation and model development were similar between Scikit-Learn and PySpark. During this process, parallels between the two frameworks were drawn. Subsequently, the hypothesis that random forests outperformed decision trees was assessed by applying the random forest model to the same housing dataset. The hypothesis was accepted as random forests showed higher R2 and lower RMSE values. The chapter also demonstrated that Pandas and PySpark shared similar approaches to reading and exploring data.

In the next chapter, we will continue with tree-based machine learning, hence introducing gradient-boosted tree (GBT) regression. The chapter will demonstrate how to construct a GBT and compare its performance with that of both decision tree and random forest regression.

CHAPTER 6

Gradient-Boosted Tree Regression with Pandas, Scikit-Learn, and PySpark

In this chapter, we continue with supervised learning and tree-based regression. Specifically, we develop a gradient-boosted tree (GBT) regression model using the same housing dataset we used for decision tree and random forest regression in the preceding chapters. This way, we can have a better idea about which tree type performs better by comparing their performance metrics.

There are similarities between random forest and GBT regression models. The most obvious similarity is that they are both ensemble techniques that combine predictions from multiple trees—the difference being that random forests combine predictions from strong trees, while GBTs combine predictions from weaker ones. This characteristic endows GBTs with strong predictive capabilities, as they leverage the strengths of multiple predictive models. This enables effective capturing of complex data relationships, which mitigates overfitting. Similar to random forest, gradient boosting can accommodate nonlinear relationships in data, rendering it adaptable to a wide array of data structures.

Furthermore, similar to other tree-based models, GBTs provide access to feature importances by assigning weights to features based on their contributions to the model's performance. This attribute is invaluable for identifying the driving factors behind predictions. The algorithm's resilience to outliers is another significant advantage, as the decision trees within the ensemble can isolate outliers, thereby reducing their impact.

On the other hand, even though both GBTs and random forests are tree-based ensemble algorithms, they exhibit distinct training processes. More specifically, GBTs progress by training one tree at a time, whereas random forests have the capability to

A. Testas, *Distributed Machine Learning with PySpark*, https://doi.org/10.1007/978-1-4842-9751-3_6

train multiple trees simultaneously. Consequently, GBTs might necessitate more time for training compared to random forests. As a practical consequence, it's frequently advisable to employ smaller (shallower) trees with GBTs rather than with random forests, as training smaller trees consumes less time.

Additionally, it's worth noting that while both random forests and GBTs mitigate overfitting in contrast to decision trees, random forests tend to exhibit a lesser susceptibility to overfitting than GBTs. This means that increasing the number of trees in a random forest diminishes the risk of overfitting, whereas increasing the number of trees in GBTs increases the likelihood of overfitting.

Lastly, random forests offer a more straightforward tuning process, as performance consistently improves alongside the increase of tree count. In contrast, the performance of GBTs can begin to deteriorate if the number of trees becomes excessively large.

In this chapter, we showcase the similarities in modeling steps between Scikit-Learn and PySpark, even in cases where the syntax varies. We utilize both libraries to make predictions on the same housing price dataset. Additionally, we will compare the performance of all tree-based models we have employed so far (decision trees, random forests, and gradient-boosted trees) by evaluating their performance metrics. Furthermore, we will highlight the parallel approach that Pandas and PySpark adopt for data loading and exploration.

The Dataset

For this chapter's project, we will be using the same housing price dataset we used in the previous two chapters for decision tree and random forest modeling. The rationale behind this is that using the same dataset would make it easier for the reader to follow and compare the results between the last three chapters.

Since we have explored this dataset in great detail in the previous chapters, our analysis of the data in this chapter will be brief. The main purpose is only to remind ourselves of the main attributes of the dataset and how Pandas and PySpark can be used interchangeably for exploratory data analysis with PySpark having the advantage of handling much larger volumes due to the nature of its distributed computing.

Similar to what we did in the previous chapters, we can define a Python function to load and read the housing dataset. We then create a Pandas DataFrame and convert it to a PySpark DataFrame. The following code achieves these objectives:

```
[In]: import pandas as pd
[In]: def load_housing_data():
          url = ('https://raw.githubusercontent.com/abdelaziztestas/'
                 'spark_book/main/housing.csv')
          return pd.read_csv(url)
[In]: pandas_df = load_housing_data()
[In]: spark_df = spark.createDataFrame(pandas_df)
```

Recall from the previous chapters that this dataset had 545 rows and 13 columns (12 features and 1 target variable). We can confirm this by using the Pandas shape attribute:

```
[In]: print(pandas_df.shape)
[Out]: (545, 13)
```

In PySpark, we use a combination of the count() and len(columns) methods:

```
[In]: print((spark_df.count(), len(spark_df.columns)))
[Out]: (545, 13)
```

We can list the name of the columns or labels using the columns attribute for both Pandas and PySpark:

Pandas:

```
[In]: print(pandas_df.columns)
[Out]: Index(['price', 'area', 'bedrooms', 'bathrooms', 'stories',
       'mainroad', 'guestroom', 'basement', 'hotwaterheating',
       'airconditioning', 'parking', 'prefarea',
       'furnishingstatus'],      dtype='object')
```

PySpark:

```
[In]: print(spark_df.columns)
[Out]: ['price', 'area', 'bedrooms', 'bathrooms', 'stories', 'mainroad',
       'guestroom', 'basement', 'hotwaterheating', 'airconditioning',
       'parking', 'prefarea', 'furnishingstatus']
```

As we recall from the previous chapters, the dataset is a mixture of numerical and categorical variables. To check which columns are numerical and which are categorical, we can use the dtypes attribute in both Pandas and PySpark:

Pandas:

```
[In]: print(pandas_df.dtypes)
[Out]:
price                  int64
area                   int64
bedrooms               int64
bathrooms              int64
stories                int64
mainroad              object
guestroom             object
basement              object
hotwaterheating       object
airconditioning       object
parking                int64
prefarea              object
furnishingstatus      object
dtype: object
```

PySpark:

```
[In]: for col_name, col_type in spark_df.dtypes:
          print(col_name, col_type)
[Out]:
price bigint
area bigint
bedrooms bigint
bathrooms bigint
stories bigint
mainroad string
guestroom string
basement string
hotwaterheating string
airconditioning string
parking bigint
prefarea string
furnishingstatus string
```

The numerical variables have the *int64* data type in Pandas and *bigint* in PySpark, while the categorical variables have the *object* and *string* data types, respectively. There are, therefore, seven categorical variables that will need to be one-hot encoded (mainroad, guestroom, basement, hotwaterheating, airconditioning, prefarea, furnishingstatus).

Six of these categorical variables have two categories each (yes or no, or vice versa), while the furnishing status column has three categories: furnished, semi-furnished, and unfurnished. We can confirm this with the following code:

Pandas:

```
[In]: for col in ['mainroad', 'guestroom', 'basement', 'hotwaterheating',
      'airconditioning', 'prefarea', 'furnishingstatus']:
          unique_values = pandas_df[col].unique()
          print(f"Unique values in {col}: {unique_values}")
[Out]:
Unique values in mainroad: ['yes' 'no']
Unique values in guestroom: ['no' 'yes']
Unique values in basement: ['no' 'yes']
Unique values in hotwaterheating: ['no' 'yes']
Unique values in airconditioning: ['yes' 'no']
Unique values in prefarea: ['yes' 'no']
Unique values in furnishingstatus: ['furnished' 'semi-furnished'
'unfurnished']
```

PySpark:

```
[In]: from pyspark.sql.functions import col
[In]: for col_name in ['mainroad', 'guestroom', 'basement',
      'hotwaterheating', 'airconditioning', 'prefarea',
      'furnishingstatus']: unique_values=spark_df.select(col(col_name)).
      distinct().rdd.flatMap(lambda x: x).collect()
          print(f"Unique values in {col_name}: {unique_values}")
[Out]:
Unique values in mainroad: ['yes', 'no']
Unique values in guestroom: ['no', 'yes']
Unique values in basement: ['no', 'yes']
Unique values in hotwaterheating: ['no', 'yes']
```

```
Unique values in airconditioning: ['no', 'yes']
Unique values in prefarea: ['no', 'yes']
Unique values in furnishingstatus: ['semi-furnished', 'unfurnished',
'furnished']
```

Now that we have reminded ourselves of the main attributes of the housing dataset, we can proceed to build the gradient-boosted tree (GBT) regression model.

Gradient-Boosted Tree (GBT) Regression

As stated in the introduction to this chapter, gradient boosting is a type of ensemble learning technique that combines multiple weak learning models (decision trees) to create a single strong learning model. The idea behind gradient boosting is to iteratively add new trees to the model, each one correcting the errors of the previous ones. At each iteration, the model fits a new tree to the residuals of the previous iteration. The final prediction is then a weighted sum of the predictions of all the trees in the model. The weights assigned to each tree depend on its performance in the training set and the number of trees in the model. The term "gradient" in GBT refers to the fact that the algorithm uses gradient descent optimization to minimize the loss function of the model, which is typically the mean squared error for regression problems.

In this section, we will use Scikit-Learn and PySpark libraries to build, train, and evaluate a GBT regression model and use it to predict the price of housing. We will be using the 13 features we have already learned about in the previous section.

Similar to decision trees and random forests in the previous chapters, standard scaling is not strictly required for GBT because tree-based algorithms are not affected by the scale of the input features. This is because gradient boosting does not use a distance-based approach to find the optimal split points in the decision trees. Therefore, the scale of the input features does not affect the relative importance of the features or the performance of the algorithm.

Scikit-Learn/PySpark Similarities

Scikit-Learn and PySpark share the same modeling steps when conducting gradient-boosted tree regression. As we'll find out here, they adhere to comparable processes for building, training, and evaluating the gradient-boosted tree regressor:

1. Data preparation:

   Scikit-Learn: In Scikit-Learn, functions such as train_test_split are used to partition data into training and testing sets. Additional functions include StandardScaler for feature scaling, OneHotEncoder for categorical variable encoding, and SimpleImputer for handling missing data.

   PySpark: Within PySpark, similar tasks are accomplished through DataFrames. The randomSplit function is utilized for data splitting, StandardScaler handles feature scaling, OneHotEncoder manages categorical encoding, and Imputer addresses missing data. Furthermore, the VectorAssembler is employed to group feature columns into a single vector column, a step that is not required in Scikit-Learn's methodology.

2. Model training:

   Scikit-Learn: To conduct gradient-boosted tree regression in Scikit-Learn, one instantiates the GradientBoostingRegressor model and employs the fit method. This involves providing the feature matrix (X_train) along with the target variable (y_train).

   PySpark: In PySpark, the GBTRegressor model is used along with its corresponding fit method. A DataFrame (train_data) that holds the features and the target variable is provided. The VectorAssembler step to combine the features in one single vector is required before starting the training process.

3. Model evaluation:

   Scikit-Learn: The evaluation of gradient-boosted tree regression model performance involves two metrics: mean_squared_error and r2_score. These metrics are computed by comparing predicted (y_pred) and actual (y_test) target variable values.

   PySpark: PySpark's RegressionEvaluator class facilitates metric computation, including mean squared error and r2. This is executed by specifying columns containing predicted values (predictions) and actual values (y_test) for evaluation.

4. Prediction:

   Scikit-Learn: Making predictions with the trained gradient-boosted tree regression model involves the use of the predict method. By supplying the feature matrix (X_test), one obtains predicted target variable values (y_pred).

   PySpark: Within the PySpark framework, the transform function of the trained model produces predictions for new data instances. This involves feeding a DataFrame (test_data) containing features, which leads to the creation of a new column (prediction). The VectorAssembler step is required to assemble the features into the appropriate format necessary for accurate prediction.

The preceding comparison indicates that Scikit-Learn and PySpark share largely similar modeling steps, despite instances where the syntax and implementation of the two algorithms differ due to Scikit-Learn's Python-based nature and PySpark being a distributed computing framework built on Apache Spark.

GBT with Scikit-Learn

In the following two subsections, we convert the preceding descriptive modeling steps (data preparation, model creation, model training, prediction, and model evaluation) into Scikit-Learn and PySpark code. Starting with Scikit-Learn, let's briefly remind ourselves of the data we will be using to operationalize the random forest regressor in Scikit-Learn:

```
[In]: import pandas as pd
[In]: def load_housing_data():
          url = ('https://raw.githubusercontent.com/abdelaziztestas/'
                 'spark_book/main/housing.csv')
      return pd.read_csv(url)
[In]: pandas_df = load_housing_data()
[In]: X = pandas_df.drop('price', axis=1)
[In]: y = pandas_df['price']
```

In this code, we define a function named `load_housing_data()` in which the URL pointing to the CSV file location is passed to the Pandas `read_csv()` method. This reads the CSV file and returns a Pandas DataFrame. The pandas_df DataFrame is created once we call the `load_housing_data()` function.

The code then creates two new DataFrames: X containing the 12 features and y holding the target (price) variable. Separating X and y allows the gradient-boosted tree algorithm to predict the y target values based on the X input features.

Having generated the X and y DataFrames, we can now start the modeling process in Scikit-Learn by first importing the necessary libraries.

Step 1: Importing necessary libraries

```
[In]: import pandas as pd
[In]: from sklearn.ensemble import GradientBoostingRegressor
[In]: from sklearn.model_selection import train_test_split
[In]: from sklearn.metrics import r2_score, mean_squared_error
[In]: from sklearn.preprocessing import OneHotEncoder
```

Step 2: Data preparation

2.1. Perform one-hot encoding on categorical variables

```
[In]: cat_cols = ['mainroad', 'guestroom', 'basement', 'hotwaterheating',
      'airconditioning', 'prefarea', 'furnishingstatus']
[In]: onehot_encoder = OneHotEncoder(sparse=False)
[In]: X_encoded = onehot_encoder.fit_transform(X[cat_cols])
[In]: X_encoded_df = pd.DataFrame(X_encoded, columns=onehot_encoder.get_
      feature_names_out(cat_cols))
[In]: X.drop(cat_cols, axis=1, inplace=True)
[In]: X = pd.concat([X, X_encoded_df], axis=1)
```

2.2. Splitting the data into training and testing sets

```
[In]: X_train, X_test, y_train, y_test = train_test_split(X, y,
      test_size=0.2, random_state=42)
```

Step 3: Training the GBT regression model

```
[In]: gbt_regressor = GradientBoostingRegressor()
[In]: gbt_regressor.fit(X_train, y_train)
```

Step 4: Prediction

```
[In]: y_pred = gbt_regressor.predict(X_test)
```

Step 5: Model evaluation

```
[In]: r2 = r2_score(y_test, y_pred)
[In]: rmse = np.sqrt(mean_squared_error(y_test, y_pred))
```

One key observation from the preceding steps is that in less than 20 code lines, we were able to prepare the data, create and train the gradient-boosted tree model, make predictions, and evaluate the model's performance.

Here are the steps in detail:

Step 1: Importing necessary libraries

In this first step of the modeling process, essential libraries are imported to enable various aspects of building, training, and evaluating the GBT regression model. The libraries include Pandas for data manipulation, Scikit-Learn's `GradientBoostingRegressor` class for creating the regression model, the `train_test_split` function for splitting data into training and testing sets, `r2_score` and `mean_squared_error` for model evaluation, and `OneHotEncoder` for encoding categorical variables.

Step 2: Data preparation

2.1. One-hot encoding of categorical variables

Seven categorical columns (mainroad, guestroom, basement, hotwaterheating, airconditioning, prefarea, furnishingstatus) are prepared for modeling. The `OneHotEncoder` is used to transform these categorical variables into numerical representations. The categorical column names are stored in the cat_cols list. An instance of the `OneHotEncoder` is created, specifying that the output should not be sparse. The encoder is fitted to the categorical columns in the feature data X, and the encoded features are stored in X_encoded DataFrame. A DataFrame (X_encoded_df) is created using these encoded features with appropriately labeled columns. The original categorical columns are then dropped from X, and the encoded columns are concatenated back to X to form the fully encoded input data, which is now in the format that the machine learning algorithm understands.

2.2. Data splitting

The data is divided into training and testing sets using the `train_test_split` function. This division is important for training the model on one subset of the data and evaluating its performance on another independent subset. The function takes the feature data (X) and the target variable (y) as inputs and splits them into X_train, X_test, y_train, and y_test. The test size is set to 20%, and a random state of 42 ensures reproducibility.

Step 3: Train the gradient-boosted tree regression model

A `GradientBoostingRegressor` model is initialized without specifying any hyperparameters, meaning that the model is using the default parameters. This model will be trained using the training data (X_train and y_train) and the `fit()` method.

Step 4: Prediction

In this step, the trained GBT regression model is employed to predict the target variable (y_pred) for the test data (X_test) using the `predict()` method. This step enables us to assess how well the model's predictions align with the actual values.

Step 5: Model evaluation

The code in this step calculates two key evaluation metrics: R-squared (r2) and root mean squared error (RMSE). R-squared measures the proportion of the variance in the dependent variable (price in this case) that's predictable from the independent variables. RMSE quantifies the average difference between predicted and actual values, providing insight into the model's prediction accuracy and overall performance.

Before we print the RMSE and r2 of the model, it's worth going through a few points of consideration. First, since we didn't specify the number of estimators (also known as boosting stages), the model will use the default number of trees, which is 100 in Scikit-Learn. We can confirm this with the following code:

```
[In]: print(gbt_regressor.n_estimators_)
[Out]: 100
```

The model also uses the default maximum depth of the individual trees (base learners), which is set to 3. The following code and output confirm this:

```
[In]: print(gbt_regressor.max_depth)
[Out]: 3
```

The depth of a decision tree refers to the number of levels or layers of splits it has in its structure. Each level represents a decision point based on a feature, and the tree's leaves hold the final predictions. In our example, each of the 100 trees will have a maximum depth of 3, meaning that each tree has been split into a series of decisions up to 3 levels deep. It can potentially create a tree structure with 2^3 (8) leaf nodes.

We can customize the number of trees and maximum depth by setting the n_estimators and max_depth parameters, respectively, to different values. In this chapter, we stick to the default parameters, while in Chapter 16, we demonstrate how to fine-tune a model's hyperparameters.

Like decision trees and random forests in the preceding chapters, the gradient-boosted tree algorithm allows us to extract the list of feature importances. We can do this with the following code:

```
[In]: importances = gbt_regressor.feature_importances_
[In]: feature_names = X.columns
[In]: indices = np.argsort(importances)[::-1]
[In]: print('Feature importances:')
[In]: for i in indices:
        print(feature_names[i], ':', importances[i])
[Out]:
area : 0.4607
bathrooms : 0.1673
airconditioning_no : 0.0556
parking : 0.0511
stories : 0.0455
bedrooms : 0.0441
airconditioning_yes : 0.0350
furnishingstatus_unfurnished : 0.0297
basement_no : 0.0222
prefarea_yes : 0.0199
hotwaterheating_no : 0.0134
mainroad_no : 0.0129
basement_yes : 0.0106
guestroom_no : 0.0089
prefarea_no : 0.0069
guestroom_yes : 0.0048
mainroad_yes : 0.0037
furnishingstatus_furnished : 0.0030
furnishingstatus_semi-furnished : 0.0023
hotwaterheating_yes : 0.0023
```

The main takeaway from the output is that the property area and the number of bathrooms are the key drivers of the model's predictions. On the contrary, the semi-furnished status and availability of hot water heating contribute the least.

Step 6: Making predictions on the testing set

The trained random forest model is used to predict the target variable for the testing feature set (X_test). The predicted values are stored in the y_pred array.

We can compare a sample of the actual and predicted target values as follows:

```
[In]: results_df = pd.DataFrame({'Price': y_test, 'Prediction': y_pred})
[In]: print(results_df.head())
[Out]:
```

Price	Prediction
4,060,000	4,502,828
6,650,000	7,301,498
3,710,000	3,697,702
6,440,000	4,415,743
2,800,000	3,722,737

We can tell from the output that the accuracy of the model isn't 100% because there are discrepancies between the actual and predicted prices. In the first row, for example, the actual price is $4,060,000, while the predicted price is $4,502,828, indicating that the model overestimates the price. The model also overestimates the price in the second row (7,301,498 vs. 6,650,000), as well as the fifth row (3,722,737 vs. 2,800,000). In the third and fourth rows, however, the model underestimates the price.

Step 7: Model evaluation

In this final step, the performance of the gradient-boosted tree regression model is evaluated using two evaluation metrics: the coefficient of determination (R-squared, r2) and the root mean squared error (RMSE, rmse). The `r2_score()` function computes the R-squared value, which indicates the proportion of the variance in the target variable that is predictable from the features. The `mean_squared_error()` function computes the mean squared error between the predicted and actual target values, and then the square root is taken to obtain RMSE. These metrics provide insights into how well the model's predictions align with the actual target values, helping to assess the model's performance.

We can print these two performance metrics with the following code:

```
[In]: print('R-squared score:', r2)
[In]: print('Root mean squared error:', rmse)
[Out]: R-squared score: 0.67 Root mean squared error: 1298307
```

Now that we have obtained the values of the evaluation metrics for the gradient-boosted tree regression model, we can compare them with the values from the preceding two chapters for the random forest and decision tree regression models.

Here is the comparison:

Model	RMSE	r2
Gradient-boosted tree	1,298,307	0.67
Random forest	1,388,213	0.62
Decision tree	1,715,691	0.42

In this table, the first column represents the type of algorithm used. The second column contains the values of the RMSE, which measures the average difference between predicted and actual values. A lower RMSE indicates better model performance. The last column contains r2 (R-squared), which is a measure of how well the model's predictions match the variability of the actual data. It ranges from 0 to 1, with higher values indicating a better fit.

We can see that the gradient-boosted tree regression model outperforms both the random forest and decision tree regression models, as it has a lower RMSE and a higher r2. However, this difference in performance could be attributed to variations in default hyperparameters, a topic we examine in detail in Chapter 16. For example, the default maximum depth in a random forest in Scikit-Learn is set to None, allowing the tree to expand to any depth. This contrasts with the default maximum depth of 3 in gradient boosting, indicating that the model can be shallower. As mentioned in the introduction to this chapter, shallower GBTs tend to perform better than deeper ones, often yielding more robust and generalized results.

The differences in performance could also mean that the particular dataset we are working on has patterns and relationships that GBT is particularly good at capturing. GBT has a sequential nature that aims to correct the errors of previous trees, which can be advantageous when dealing with complex relationships in data.

GBT with PySpark

We will now proceed to build the same GBT regression model and train it using the same housing dataset using PySpark. First, let's have a quick reminder of the code that we used to set up PySpark in the "The Dataset" section.

As shown in the following code, we first create a Pandas DataFrame and then convert it to PySpark using the createDataFrame() method.

```
[In]: import pandas as pd
[In]: from pyspark.sql import SparkSession
[In]: spark = SparkSession.builder.appName("GBTRegression").getOrCreate()
[In]: def load_housing_data():
          url = ('https://raw.githubusercontent.com/abdelaziztestas/'
                 'spark_book/main/housing.csv')
          return pd.read_csv(url)
[In]: pandas_df = load_housing_data()
[In]: spark_df = spark.createDataFrame(pandas_df)
```

In this code, several steps are taken to work with the housing data using both the Pandas library and the PySpark framework. The first line imports the Pandas library and names it pd. The second line imports the SparkSession, which is the entry point to utilizing Spark functionalities. The code then establishes a Spark Session using the SparkSession.builder.appName() method. This session serves as the connection to the Spark cluster and provides the environment to perform distributed data processing. The next step defines a function named load_housing_data() that retrieves housing data from a URL using the Pandas read_csv() function. This reads a CSV file from the specified URL and returns a Pandas DataFrame containing the data. Following this, the code reads the housing data into a Pandas DataFrame named pandas_df using the defined load_housing_data() function. Finally, the code converts the Pandas DataFrame pandas_df into a Spark DataFrame named spark_df using the createDataFrame() method. This conversion allows the data to be processed and analyzed using Spark's distributed computing capabilities.

We will now write the PySpark code for the steps to prepare data, build and train the model, and evaluate its performance using the test set. First, let's import the necessary libraries:

Step 1: Importing necessary libraries

```
[In]: from pyspark.ml.feature import StringIndexer, OneHotEncoder,
      VectorAssembler
[In]: from pyspark.ml.regression import GBTRegressor
[In]: from pyspark.ml.evaluation import RegressionEvaluator
[In]: import pandas as pd
```

Step 2: Data preparation

2.1. Label encoding categorical columns

```
[In]: cat_columns = ['mainroad', 'guestroom', 'basement',
      'hotwaterheating', 'airconditioning', 'prefarea','furnishingstatus']
[In]: indexers = [StringIndexer(inputCol=col, outputCol=col+'_label', or
      indexer in indexers:
          spark_df = indexer.fit(spark_df).transform(spark_df)
```

2.2. One-hot encoding for all categorical columns

```
[In]: encoder = OneHotEncoder(inputCols=[col+'_label' for col in cat_
      columns], outputCols=[col+'_encoded' for col in cat_columns])
[In]: spark_df = encoder.fit(spark_df).transform(spark_df)
```

2.3. Combining all features into a single feature vector

```
[In]: feature_cols = ['area', 'bedrooms', 'bathrooms', 'stories',
      'parking'] + [col+'_encoded' for col in cat_columns]
[In]: assembler = VectorAssembler(inputCols=feature_cols,
      outputCol='features')
[In]: spark_df = assembler.transform(spark_df)
```

2.4. Splitting data into training and testing sets

```
[In]: (training_data, test_data) = spark_df.randomSplit([0.8, 0.2],
      seed=42)
```

Step 3: Training the gradient-boosted tree regression model

```
[In]: gbt = GBTRegressor(featuresCol='features', labelCol='price')
[In]: spark_model = gbt.fit(training_data)
```

Step 4: Making predictions on test data

```
[In]: predictions = spark_model.transform(test_data)
```

Step 5: Evaluating the model using RMSE and R2

```
[In]: evaluator_rmse = RegressionEvaluator(labelCol='price',
      predictionCol='prediction', metricName='rmse')
[In]: rmse = evaluator_rmse.evaluate(predictions)
[In]: evaluator_r2 = RegressionEvaluator(labelCol='price',
      predictionCol='prediction', metricName='r2')
[In]: r2 = evaluator_r2.evaluate(predictions)
```

The preceding steps demonstrate how we can efficiently solve a regression problem using just a few lines of PySpark code.

In the following, we provide a detailed explanation for each of the modeling steps outlined previously: preparing data, creating and training a GBT regression model, making predictions on test data, and evaluating the model's performance using RMSE and r2.

Step 1: Importing necessary libraries

In this first step of the modeling process, libraries for working with PySpark are imported. These include `StringIndexer`, `OneHotEncoder`, and `VectorAssembler` for data preparation, `GBTRegressor` for performing regression analysis, and `RegressionEvaluator` for evaluating the model's performance. The `Pandas` library is also imported and named pd.

Step 2: Data preparation

This section focuses on preparing the data for training and evaluation.

2.1. Label encoding of categorical columns

In this substep of the data preparation step, the categorical columns in the dataset are identified. Each categorical column is then label-encoded using `StringIndexer`, generating a corresponding column with _label suffix in the DataFrame.

In the first line, a list named cat_columns is created, containing the names of categorical columns in the dataset that require encoding. These categorical columns represent binary attributes such as mainroad, guestroom, basement, and others. The code then proceeds to create a list named indexers, which will store instances of the `StringIndexer` transformation for each categorical column. The transformation is initialized by specifying the input column (e.g., mainroad) and the desired output column (e.g., mainroad_label) with _label appended to it. This new column will store the numerical labels corresponding to the categorical values.

In a loop, each categorical column is processed using the `StringIndexer` instances stored in the indexers list. The `fit()` method is applied to the indexer, using the current dataset spark_df as input. This step calculates the mapping of categorical values to numerical labels. Then, the `transform()` method is used to apply the calculated transformation to the dataset, generating the new column with numerical labels.

2.2. One-hot encoding

In this substep, one-hot encoding is applied to the label-encoded categorical columns using `OneHotEncoder`. This results in the creation of new columns with _ encoded suffix.

The code starts by creating an instance of the `OneHotEncoder` by specifying two lists: inputCols and outputCols. The inputCols list is generated using a list comprehension that iterates through the categorical columns stored in cat_columns and appends _label to each column name. This aligns with the previously created categorical label columns from the `StringIndexer` step. Similarly, the outputCols list is generated by appending _encoded to each corresponding categorical column name.

Once the encoder instance is initialized, it is applied to the DataFrame spark_df using the `fit()` method. This step calculates the parameters necessary for the encoding process based on the provided data. Subsequently, the `transform()` method is used to apply the calculated encoding to the DataFrame. This generates new columns with binary vector representations of the categorical values. Each new column corresponds to a distinct categorical label, containing binary values that indicate the presence or absence of that specific label for each row in the DataFrame.

2.3. Feature vector assembly

All relevant feature columns, including encoded categorical columns, are combined into a single feature vector using `VectorAssembler`. This step is necessary for model training.

The first line of code defines a list named feature_cols, containing the names of the features that will be used for modeling. These features include numerical attributes like area, bedrooms, bathrooms, stories, and parking. Additionally, the list comprehension adds the names of the encoded categorical columns with _encoded appended to each column name. These encoded columns were generated in the previous step using the `OneHotEncoder` technique.

The second line of code initializes an instance of the `VectorAssembler`. This transformer takes the specified input columns, listed in the feature_cols list, and combines them into a single vector column named features, which is specified in the outputCol parameter.

Finally, the third line of code applies the `VectorAssembler` to the DataFrame spark_df using the `transform()` method. This step generates a new column named features in the DataFrame. Each row in this column contains a vector that encapsulates the values of all selected features for that particular row.

2.4. Data splitting

The dataset is split into training and testing sets with an 80-20 ratio. This split is essential to evaluate the model's performance on unseen data.

Going through the code, the `randomSplit()` method is applied to the DataFrame spark_df. The method takes two arguments: a list [0.8, 0.2] and a seed value of 42. The list defines the proportion of data to be allocated to the training and test datasets, respectively. In this case, 80% of the data is assigned to the training dataset and the remaining 20% to the test dataset. The method randomly shuffles the rows of the DataFrame and separates them according to the specified proportions. The seed parameter ensures reproducibility, allowing the same partitioning to be generated across different executions.

The result of this operation is the creation of two separate DataFrames named training_data and test_data, each containing the subset of rows assigned to the training and test datasets, respectively. These datasets will be used for training and evaluating the GBT regression model.

Step 3: Training the gradient-boosted tree regression model

In this step, a GBT regression model is trained. A `GBTRegressor` is instantiated with specified parameters, including the input feature column (features) and the target label column (price). The model uses the default hyperparameters and is trained on the training data using the `fit()` method.

In the first line, an instance of the `GBTRegressor` class is created and assigned to the variable gbt. The featuresCol parameter is set to features, which corresponds to the column containing the feature vectors that were assembled earlier. The labelCol parameter is set to price, which corresponds to the column containing the target variable (in this case, the price of a property).

In the second line, the `fit()` method is applied to the gbt instance using the training_data DataFrame as input. This method initiates the training process of the GBT regression model. During training, the model learns to predict the target variable (price) based on the provided feature vectors (features) in the training dataset. The result of this operation is the creation of a trained model, stored in the spark_model variable.

With this trained model, we can now access the number of trees and maximum depth from the GBT model:

```
[In]: num_trees = spark_model.getNumTrees
[In]: max_depth = spark_model._java_obj.getMaxDepth()
```

Then print the number of trees and maximum depth as follows:

```
[In]: print("Number of Trees:", num_trees)
[In]: print("Maximum Depth:", max_depth)
[Out]: Number of Trees: 20 Maximum Depth: 5
```

The output indicates that the GBT model is composed of an ensemble of 20 individual decision trees. In GBT, multiple decision trees are sequentially trained, and their predictions are combined to improve the overall model's performance. Having a larger number of trees can potentially enhance the model's ability to capture complex relationships within the data and lead to better predictive accuracy. However, there's a trade-off between model complexity and overfitting, so choosing the right number of trees is crucial. In this project, however, we have chosen the default number of trees, which is 20, for illustration. In Chapter 16, we demonstrate how to tune the hyperparameters of a model to override the default values.

The output also shows the maximum depth of each of the 20 trees. The depth of a tree signifies how many splits or nodes it can have, allowing it to learn intricate patterns in the data. A depth of 5 indicates that each of the 20 trees of the GBT model can have up to 5 levels of splits, potentially capturing relatively detailed features of the data. However, a deeper tree may also lead to overfitting, especially if the dataset is not large enough to support such complexity.

We can also access the feature importances from the trained GBT model:

```
[In]: feature_importances = spark_model.featureImportances.toArray()
[In]: feature_names = feature_cols
```

```
[In]: feature_importance_tuples = list(zip(feature_names, feature_
      importances))
[In]: sorted_feature_importance = sorted(feature_importance_tuples,
      key=lambda x: x[1], reverse=True)
```

Then print the value of each feature importance with the following loop:

```
[In]: print("Sorted Feature Importances:")
[In]: for feature_name, importance in sorted_feature_importance:
          print(f"Feature '{feature_name}': {importance:.4f}")
[Out]:
Sorted Feature Importances:
area: 0.3818
bathrooms: 0.1089
stories: 0.1085
parking: 0.0900
bedrooms: 0.0808
airconditioning_encoded: 0.0343
mainroad_encoded: 0.0240
basement_encoded: 0.0227
furnishingstatus_encoded: 0.0081
guestroom_encoded: 0.0000
hotwaterheating_encoded: 0.0000
prefarea_encoded: 0.0000
```

The output confirms what we already know from the Scikit-Learn GBT regression model of the previous subsection: that the area of a property and the number of bathrooms it has are the key drivers in the model. The output also shows that whether a house has a guest room, hot water heating system, or is located in a preferred area has no importance at all. This suggests that removing these features from the model may boost performance by reducing the model's complexity and its potential for overfitting.

Step 4: Making predictions on test data

The trained model is employed to make predictions on the test dataset. The `transform()` method is utilized, which applies the model to the test data and generates predictions for the target variable (price).

The resulting DataFrame is named predictions, which contains the original columns from the test_data along with an additional column named prediction. Each row in the prediction column contains the model's predicted value for the target variable (price) based on the corresponding feature vector.

We can print the price and prediction columns as follows:

```
[In]: predictions.select('price', 'prediction').show(5)
[Out]:
+---------+----------+
|  Price  | Prediction|
+---------+----------+
| 6930000 | 5365539   |
| 7070000 | 5303199   |
| 7210000 | 7355716   |
| 7350000 | 5057867   |
| 7455000 | 2507188   |
+---------+----------+
only showing top 5 rows
```

This output suggests that the model both overestimates and underestimates the property prices. It tends to underestimate the prices more frequently than overestimating them.

For the first property with an actual price of $6,930,000, the model predicts a value of $5,365,539. This prediction is less than the actual price, indicating an underestimation by the model. Similarly, the second property with an actual price of $7,070,000 receives a prediction of $5,303,199, which is lower than the actual price, also suggesting underestimation. The third property with an actual price of $7,210,000 gets a prediction of $7,355,716, which is slightly higher than the actual price, indicating overestimation. For the fourth property with an actual price of $7,350,000, the model predicts $5,057,867, reflecting underestimation. Finally, the fifth property with an actual price of $7,455,000 receives a prediction of $2,507,188, significantly lower than the actual price, suggesting substantial underestimation.

Step 5: Evaluating the model

This step focuses on evaluating the model's performance using RMSE and R2. In the first line, an instance of the `RegressionEvaluator` class is created and assigned to the variable evaluator_rmse. The labelCol parameter is set to price, which represents the actual target variable, and the predictionCol parameter is set to prediction, which corresponds to the predicted values generated by the model. The metricName parameter is set to rmse, indicating that the evaluator will calculate the RMSE metric.

In the second line, the `evaluate()` method is applied to the evaluator_rmse instance using the predictions DataFrame as input. This method calculates the RMSE between the actual prices (price column) and the predicted values (prediction column) in the predictions DataFrame. The calculated RMSE value is stored in the variable rmse.

The same procedure is repeated for the R-squared metric. In the third line, an instance of the `RegressionEvaluator` class is created and assigned to the variable evaluator_r2. The metricName parameter is set to r2, indicating that the evaluator will calculate the R-squared metric.

In the fourth line, the `evaluate()` method is applied to the evaluator_r2 instance using the predictions DataFrame as input. This method computes the R-squared value, which quantifies how well the model's predictions align with the variability of the actual data. The calculated R-squared value is stored in the variable r2.

We can print the values of the RMSE and r2 as follows:

```
[In]: print(f"RMSE: {rmse}")
[In]: print(f"R2: {r2}")
[Out]: RMSE: 1335830 R2: 0.42
```

With these results, let's compare the performance of the GBT regression model in this chapter with the other tree-based regression algorithms (decision trees and random forests) in the preceding chapters. Based on the evaluation metrics shown in the following table, the random forest model appears to be better compared to the GBT and decision tree models. This conclusion is drawn from the lower RMSE value and the higher R-squared (r2) value shown by the random forest model, as elaborated here:

- Lower RMSE: The random forest model has the lowest root mean squared error (RMSE) of 1,237,365, indicating that its predictions are, on average, closer to the actual property prices in the dataset. A lower RMSE signifies better predictive accuracy, as it measures the magnitude of prediction errors. The fact that the random forest model has the lowest RMSE suggests that it is making predictions that are more aligned with the actual prices.

- Higher R-squared (r2) value: The random forest model also boasts the highest R-squared (r2) value of 0.50. This signifies that approximately 50% of the variance in property prices can be explained by the model. A higher r2 value indicates that the model's predictions better capture the variability present in the dataset. In this case, the random forest model's ability to explain a larger portion of the variance in property prices suggests that it captures underlying patterns in the data more effectively.

On the other hand, even though the random forest model stands out as the best-performing model among the three evaluated, with a lower RMSE and a relatively higher R-squared (r2) value of 0.50, there is still room for improvement in terms of its predictive capability. While the R-squared value of 0.50 indicates that the model explains approximately 50% of the variance in property prices, there remains potential to enhance its ability to capture more nuances and patterns within the data. This suggests that further optimization of the model's hyperparameters and potentially removing the features with the least importances could lead to even more accurate predictions and a higher r2 value.

Model	RMSE	r2
Gradient-boosted tree	1,335,830	0.42
Random forest	1,237,365	0.50
Decision tree	1,463,512	0.32

Bringing It All Together

In the previous section, we discussed the GBT regression methodology by using different individual code snippets to explain each step. However, in this section, we bring all of the code snippets together in one block for each of the Scikit-Learn and PySpark frameworks. This makes it easier to see how the entire code fits together and allows for running it. A copy of this code will also be available on GitHub.

Scikit-Learn

In this subsection, we combine all the Scikit-Learn code snippets we have used to prepare data, build and train the GBT regression model, make predictions, and evaluate the model's performance.

```
# Import necessary libraries
[In]: from sklearn.ensemble import GradientBoostingRegressor
[In]: from sklearn.model_selection import train_test_split
[In]: from sklearn.metrics import r2_score, mean_squared_error
[In]: from sklearn.preprocessing import OneHotEncoder
[In]: import pandas as pd
[In]: import numpy as np

# Define function for loading data
[In]: def load_housing_data():
          url = ('https://raw.githubusercontent.com/abdelaziztestas/'
                 'spark_book/main/housing.csv')
          return pd.read_csv(url)

# Create a Pandas DataFrame by calling the function:
[In]: pandas_df = load_housing_data()

# Split the data into features and target variable
[In]: X = pandas_df.drop('price', axis=1)
[In]: y = pandas_df['price']

# Perform one-hot encoding on categorical variables
[In]: cat_cols = ['mainroad', 'guestroom', 'basement', 'hotwaterheating',
      'airconditioning', 'prefarea', 'furnishingstatus']
[In]: onehot_encoder = OneHotEncoder(sparse=False)
[In]: X_encoded = onehot_encoder.fit_transform(X[cat_cols])
[In]: X_encoded_df = pd.DataFrame(X_encoded, columns=onehot_encoder.get_
      feature_names_out(cat_cols))
[In]: X.drop(cat_cols, axis=1, inplace=True)
[In]: X = pd.concat([X, X_encoded_df], axis=1)
```

```
# Split the data into training and testing sets
[In]: X_train, X_test, y_train, y_test = train_test_split(X, y, test_
      size=0.2, random_state=42)

# Train the Gradient Boosted Tree Regression model
[In]: gbt_regressor = GradientBoostingRegressor()
[In]: gbt_regressor.fit(X_train, y_train)

# Print the number of boosting stages
[In]: print("Number of boosting stages:", gbt_regressor.n_estimators_)

# Print the maximum depth of trees
[In]: print("Maximum tree depth:", gbt_regressor.max_depth)

# Print feature importances
[In]: importances = gbt_regressor.feature_importances_
[In]: feature_names = X.columns
[In]: indices = np.argsort(importances)[::-1]
[In]: print('Feature importances:')
[In]: for i in indices:
          print(feature_names[i], ':', importances[i])

# Make predictions on the testing set
[In]: y_pred = gbt_regressor.predict(X_test)

# Print actual and predicted values
[In]: results_df = pd.DataFrame({'Price': y_test, 'Prediction': y_pred})
[In]: print(results_df.head())

# Calculate evaluation metrics
[In]: r2 = r2_score(y_test, y_pred)
[In]: rmse = np.sqrt(mean_squared_error(y_test, y_pred))

# Print evaluation metrics
[In]: print('R-squared score:', r2)
[In]: print('Root mean squared error:', rmse)
```

PySpark

In this subsection, we do the same with PySpark code: combine all the snippets we have used to prepare the housing data, build and train the GBT regression model, make predictions, and evaluate the model's performance.

```
# Import necessary libraries
[In]: from pyspark.sql import SparkSession
[In]: from pyspark.ml.feature import StringIndexer, OneHotEncoder,
      VectorAssembler
[In]: from pyspark.ml.regression import GBTRegressor
[In]: from pyspark.ml.evaluation import RegressionEvaluator
[In]: import pandas as pd
[In]: import numpy as np
# Create a Spark session
[In]: spark = SparkSession.builder.appName("GBTRegressorExample").
      getOrCreate()

[In]: def load_housing_data():
         url = ('https://raw.githubusercontent.com/abdelaziztestas/'
                'spark_book/main/housing.csv')
         return pd.read_csv(url)

# Load data and create Spark DataFrame
[In]: pandas_df = load_housing_data()
[In]: spark_df = spark.createDataFrame(pandas_df)

# Label encoding categorical columns
[In]: cat_columns = ['mainroad', 'guestroom', 'basement',
      'hotwaterheating', 'airconditioning', 'prefarea', 'furnishingstatus']
[In]: indexers = [StringIndexer(inputCol=col, outputCol=col+'_label',
      handleInvalid='keep') for col in cat_columns]
[In]: for indexer in indexers:
         spark_df = indexer.fit(spark_df).transform(spark_df)
```

```
# One-hot encoding for all categorical columns
[In]: encoder = OneHotEncoder(inputCols=[col+'_label' for col in cat_
      columns], outputCols=[col+'_encoded' for col in cat_columns])
[In]: spark_df = encoder.fit(spark_df).transform(spark_df)

# Combining all features into a single feature vector
[In]: feature_cols = ['area', 'bedrooms', 'bathrooms', 'stories',
      'parking'] + [col+'_encoded' for col in cat_columns]
[In]: assembler = VectorAssembler(inputCols=feature_cols,
      outputCol='features')
[In]: spark_df = assembler.transform(spark_df)
# Splitting data into training and test sets
[In]: (training_data, test_data) = spark_df.randomSplit([0.8, 0.2],
      seed=42)

# Training the Gradient-Boosted Tree Regression model
[In]: gbt = GBTRegressor(featuresCol='features', labelCol='price')
[In]: spark_model = gbt.fit(training_data)

# Access the number of trees and maximum depth from the GBT model
[In]: num_trees = spark_model.getNumTrees
[In]: max_depth = spark_model._java_obj.getMaxDepth()

# Print the number of trees and maximum depth
[In]: print("Number of Trees:", num_trees)
[In]: print("Maximum Depth:", max_depth)

# Access the feature importances from the GBT model
[In]: feature_importances = spark_model.featureImportances.toArray()

# Get column names from feature_cols
[In]: feature_names = feature_cols

# Combine feature names and importances into a list of tuples
[In]: feature_importance_tuples = list(zip(feature_names, feature_
      importances))

# Sort the list of tuples by importance in descending order
[In]: sorted_feature_importance = sorted(feature_importance_tuples,
      key=lambda x: x[1], reverse=True)
```

```
# Printing sorted feature importances along with names
[In]: print("Sorted Feature Importances:")
[In]: for feature_name, importance in sorted_feature_importance:
          print(f"Feature '{feature_name}': {importance:.4f}")

# Making predictions on test data
[In]: predictions = spark_model.transform(test_data)

# Printing actual and predicted values
[In]: predictions.select('price', 'prediction').show(5)

# Evaluating the model using RMSE and R2
[In]: evaluator_rmse = RegressionEvaluator(labelCol='price',
      predictionCol='prediction', metricName='rmse')
[In]: rmse = evaluator_rmse.evaluate(predictions)

[In]: evaluator_r2 = RegressionEvaluator(labelCol='price',
      predictionCol='prediction', metricName='r2')
[In]: r2 = evaluator_r2.evaluate(predictions)

# Printing RMSE and R2 values
[In]: print(f"RMSE: {rmse}")
[In]: print(f"R2: {r2}")
```

Summary

This chapter introduced gradient-boosted tree (GBT) regression and demonstrated the construction of a regression model using the gradient boosting algorithm. The chapter also examined the similarities between Pandas and PySpark in terms of reading and exploring data and showed that Scikit-Learn and PySpark follow similar modeling steps, which include transformation, categorical encoding, data scaling, model construction, prediction, and evaluation.

The chapter also compared the performance of the GBT regression model with the performance of the other two tree-based algorithms (decision trees and random forests). The key takeaway is that even the best-performing model among the three can still benefit from further hyperparameter optimization and dimensionality reduction.

In the next chapter, we focus on classification, a distinct form of supervised learning. Our objective will be to build, train, and evaluate a logistic regression model and then use it to predict the likelihood of diabetes. The aim is to demonstrate that despite variations in syntax and function names, the fundamental procedures for conducting logistic regression in Scikit-Learn and PySpark are comparable. Furthermore, we will demonstrate that Pandas and PySpark share similar methods for handling data loading and manipulation, which simplifies tasks like reading and exploring data.

CHAPTER 7

Logistic Regression with Pandas, Scikit-Learn, and PySpark

This chapter focuses on classification, a distinct form of supervised learning. Our objective is to build, train, and evaluate a logistic regression model and then use it to predict the likelihood of diabetes.

Despite its name suggesting a connection to linear regression, logistic regression is fundamentally distinct in its purpose and methodology. Contrary to linear regression, which predicts continuous numerical values (see Chapter 3), logistic regression predicts the probability of an instance belonging to a particular class—in the case of diabetes, whether a person has the disease (class 1) or doesn't have the disease (class 0).

Logistic regression has both advantages and disadvantages. One of its advantages is simplicity and interpretability. Another advantage is efficiency, especially when dealing with large datasets. This advantage stems from its linear nature and straightforward structure, contributing to its ability to quickly process data. Moreover, logistic regression is known for low variance, reducing the likelihood of overfitting. This attribute is specifically beneficial when the dataset is limited. By maintaining stable performance on smaller datasets, logistic regression becomes a reliable option for various applications.

On the other hand, logistic regression does come with a number of disadvantages. Its fundamental limitation lies in its assumption of a linear decision boundary. This can lead to suboptimal performance when the actual boundary is nonlinear, making more sophisticated algorithms (e.g., tree-based algorithms, support vector machines, neural networks) more suitable for such classification tasks.

A. Testas, *Distributed Machine Learning with PySpark*, https://doi.org/10.1007/978-1-4842-9751-3_7

Another limitation lies in the model's sensitivity to outliers, which can disproportionately affect the estimated coefficients, potentially leading to skewed results. Furthermore, logistic regression assumes that features are independent of one another, and its performance can be compromised when this assumption is violated—in other words, when correlations among features are present. Furthermore, while logistic regression excels in binary classification tasks, it is less versatile in handling multiclass classification. Although extensions like multinomial logistic regression exist, they might not match the performance of algorithms explicitly designed for multiclass scenarios.

In this chapter, we use both Scikit-Learn and PySpark to construct a logistic regression model. We demonstrate that despite variations in syntax and function names, the fundamental procedures in the two frameworks are comparable. Furthermore, we demonstrate that Pandas and PySpark share similar methods for handling data loading and manipulation, which simplifies tasks like reading and exploring data.

For the purpose of this project, we will use the same diabetes dataset that we utilized in Chapter 2 to select algorithms. Before building our logistic regression model and training it on this data, let's first remind ourselves of the main attributes of this dataset.

The Dataset

The dataset used for this project is the Pima Indians Diabetes Database we used in Chapter 2. The dataset contains 768 records, each of which represents a Pima Indian woman and includes various health-related attributes, along with a target variable indicating whether the woman developed diabetes or not.

As indicated in Chapter 2, the source of this dataset is Kaggle—a website designed for data scientists. The contributor's name, approximate upload date, dataset name, site name, and the URL from which we downloaded a copy of the CSV file are given here:

Title: Pima Indians Diabetes Database

Source: Kaggle

URL: `www.kaggle.com/uciml/pima-indians-diabetes-database`

Contributor: UCI Machine Learning

Date: 2016

As in Chapter 2, we begin by defining a function named load_diabetes_data() to load the data from a CSV file named diabetes.csv located on the GitHub repository. The data is loaded using the read_csv() function, which reads the file into a Pandas DataFrame.

```
[In]: import pandas as pd
[In]: def load_diabetes_data():
          url = ('https://raw.githubusercontent.com/abdelaziztestas/'
                 'spark_book/main/diabetes.csv')
          return pd.read_csv(url)
```

We then write a Python function that prints basic information about the pandas_df:

```
[In]: def explore_pandas_df():
          pandas_df = load_diabetes_csv()
          columns = ['Pregnancies', 'Glucose', 'BloodPressure',
                     'BMI', 'DiabetesPedigreeFunction', 'Age']
          selected_df = pandas_df[columns]
          selected_df = selected_df.loc[(pandas_df['Glucose'] != 0)
              & (pandas_df['BloodPressure'] != 0)
              & (pandas_df['BMI'] != 0)]
          print(f'First 5 rows:\n{selected_df.head()}\n')
          print(f'Shape: {selected_df.shape}\n')
          print("Info:")
          selected_df.info()
          print(f'Summary statistics:\n'
              f'{selected_df.describe().round(2)}\n')
```

The explore_pandas_df() function selects a subset of columns from pandas_df by creating a list of column names and then using this list to select only the relevant columns. The selected columns are assigned to a new DataFrame called selected_df.

Next, the function uses the loc[] method to filter the selected_df, selecting only the rows where the Glucose, Blood Pressure, and BMI columns are not equal to zero. This filters out any rows where there is missing or invalid data for these columns.

The reason for this filtering was explained back in Chapter 2, where we used the same dataset to illustrate the concept of k-fold cross validation. We realized there were invalid readings for Glucose, Blood Pressure, Skin Thickness, Insulin, and BMI as they had 0 values. We reasoned that it made sense to exclude Skin Thickness and Insulin as the number of cases with 0 values was too large. We also decided to exclude rows with 0 values for Glucose, Blood Pressure, and BMI. The number of invalid cases was not too large, so excluding rows won't significantly impact the sample size.

The function then prints the first five rows of selected_df using the `head()` method and displays its shape using the `shape` attribute. Moving on, the function displays summary information about selected_df using the `info()` method. This provides a concise summary of the DataFrame's structure, including the number of rows, columns, and non-null values for each column, as well as the data types. In the final step, the `explore_pandas_df()` function displays summary statistics for selected_df using the `describe()` method, which provides descriptive statistics such as the mean, standard deviation, and quartiles for each column. The output is formatted as a string using f-strings and printed.

We can now call the `explore_pandas_df()` function to see the output:

```
{In]: print(explore_pandas_df())
[Out]:
```

```
First five rows:
```

	Pregnancies	Glucose	Blood Pressure	BMI	Diabetes Pedigree Function	Age
0	6	148	72	33.6	0.627	50
1	1	85	66	26.6	0.351	31
2	8	183	64	23.3	0.672	32
3	1	89	66	28.1	0.167	21
4	0	137	40	43.1	2.288	33

```
Shape: (724, 6)

Info:
<class 'pandas.core.frame.DataFrame'>
Int64Index: 724 entries, 0 to 767
Data columns (total 6 columns):
 #   Column                      Non-Null Count  Dtype
---  ------                      --------------  -----
 0   Pregnancies                 724 non-null    int64
 1   Glucose                     724 non-null    int64
 2   BloodPressure               724 non-null    int64
 3   BMI                         724 non-null    float64
```

```
 4   DiabetesPedigreeFunction  724 non-null    float64
 5   Age                       724 non-null    int64
dtypes: float64(2), int64(4)

Summary statistics:
```

	Pregnancies	Glucose	BloodPressure	BMI	Diabetes Pedigree Function	Age
count	724	724	724	724	724	724
mean	3.87	121.88	72.4	32.47	0.47	33.35
std	3.36	30.75	12.38	6.89	0.33	11.77
min	0	44	24	18.2	0.08	21
25%	1	99.75	64	27.5	0.24	24
50%	3	117	72	32.4	0.38	29
75%	6	142	80	36.6	0.63	41
max	17	199	122	67.1	2.42	81

We get a set of four outputs. The first section of the output shows the first five rows of the DataFrame. Each row represents a set of values for the columns Pregnancies, Glucose, Blood Pressure, BMI, Diabetes Pedigree Function, and Age. The Shape section indicates the shape of the DataFrame, showing the number of rows and columns. In this case, the DataFrame has 724 rows and 6 columns. The Info section provides information about the DataFrame, including the data types of each column and the count of non-null values. Finally, the Summary statistics section presents summary statistics for each column in the DataFrame. It provides key statistical measures such as the count, mean, standard deviation, minimum, 25th percentile, median (50th percentile), 75th percentile, and maximum value for each numerical column.

We can achieve the same results with PySpark code:

```
[In]: from pyspark.sql import SparkSession
[In]: from pyspark.sql.functions import col
[In]: def explore_spark_df():
          pandas_df = load_diabetes_csv()
```

```
    spark = SparkSession.builder.getOrCreate()
    spark_df = spark.createDataFrame(pandas_df)
    columns = ['Pregnancies', 'Glucose', 'BloodPressure',
               'BMI', 'DiabetesPedigreeFunction', 'Age']
    spark_df = spark_df.select(columns)
    spark_df = spark_df.filter((col('Glucose') != 0)
                               & (col('BloodPressure') != 0)
                               & (col('BMI') != 0))
    spark_df.show(5)
    print(
     f'Shape: ({spark_df.count()}, {len(spark_df.columns)})\n')
    spark_df.printSchema()
    spark_df.summary().show()
```

The code first imports the necessary classes (SparkSession and col) and defines the `explore_spark_df()` function, which loads the diabetes CSV file into pandas_df DataFrame using the `load_diabetes_csv()` function. It then creates a SparkSession object named spark using the `SparkSession.builder.getOrCreate()` method and converts the pandas_df to spark_df using the `createDataFrame()` method.

Next, the spark_df is filtered by selecting specific columns and excluding rows where the Glucose, Blood Pressure, and BMI values are equal to zero, which are considered invalid readings. The function then displays the first five rows of the filtered Spark DataFrame using the `show(5)` method. It also displays the spark_df's shape using the `count()` and `len()` methods, as well as the schema using the `printSchema()` method. Finally, the function displays a summary that includes statistics for the selected columns using the `summary()` method.

We can look at the output by calling the `explore_spark_df()` function:

```
[In]: print(explore_spark_df())
[Out]:
```

Pregnancies	Glucose	Blood Pressure	BMI	Diabetes Pedigree Function	Age
6	148	72	33.6	0.627	50
1	85	66	26.6	0.351	31
8	183	64	23.3	0.672	32
1	89	66	28.1	0.167	21
0	137	40	43.1	2.288	33

```
Shape: (724, 6)

root
 |-- Pregnancies: long (nullable = true)
 |-- Glucose: long (nullable = true)
 |-- BloodPressure: long (nullable = true)
 |-- BMI: double (nullable = true)
 |-- DiabetesPedigreeFunction: double (nullable = true)
 |-- Age: long (nullable = true)
```

summary	Pregnancies	Glucose	Blood Pressure	BMI	Diabetes Pedigree Function	Age
count	724	724	724	724	724	724
mean	3.87	121.88	72.40	32.47	0.47	33.35
stddev	3.36	30.75	12.38	6.89	0.33	11.77
min	0	44	24	18.2	0.078	21
25%	1	99	64	27.5	0.245	24
50%	3	117	72	32.4	0.378	29
75%	6	142	80	36.6	0.627	41
max	17	199	122	67.1	2.42	81

The output from the Pandas and PySpark code is comparable, with the diabetes dataset having 724 rows and 6 features. These features include a comprehensive spectrum of health-related measures: Pregnancies pertains to the count of pregnancy occurrences; Glucose captures the two-hour plasma glucose concentration obtained during an oral glucose tolerance test; Blood Pressure denotes the diastolic blood pressure, measured in millimeters of mercury (mm Hg); BMI represents the body mass index, an important indicator of body composition; Diabetes Pedigree Function quantifies the likelihood of diabetes based on familial medical history; and finally, Age denotes the individual's age, measured in years.

The target variable, which is not shown in the output, signifies whether a Pima woman has diabetes (assigned a value of 1) or does not have diabetes (assigned a value of 0). This variable is the focus of our prediction.

Logistic Regression

In this section, we build, train, and evaluate a logistic regression model and use it to predict the likelihood of a Pima Indian woman having diabetes based on a set of predictor variables. We use six features: number of pregnancies, blood pressure, BMI, glucose level, diabetes pedigree function, and age. These features are important factors in predicting the presence of diabetes.

The logistic regression model estimates the coefficients of each predictor variable and uses them to calculate the probability of the woman having diabetes. The model outputs a probability score between 0 and 1, with a score closer to 1 indicating a higher likelihood of having diabetes. For example, if a woman has had four pregnancies, high blood pressure, high BMI, high glucose levels, a family history of diabetes, and is 50 years old, the logistic regression model might output a probability score of 0.8, indicating that this woman has a high likelihood of having diabetes. On the other hand, a woman who has had only one pregnancy, normal blood pressure, low BMI, low glucose levels, no family history of diabetes, and is 30 years old might have a probability score of 0.1, indicating a low likelihood of having diabetes.

We will be using both Scikit-Learn and PySpark. While Scikit-Learn is a popular library for building logistic regression models, PySpark offers several advantages for handling big data. By migrating from Scikit-Learn to PySpark, data scientists can take advantage of PySpark's distributed computing capabilities, which allow for parallel processing of large datasets, and seamless integration with other big data tools. While the syntax and functions used in Scikit-Learn and PySpark may differ, the underlying principles of logistic regression remain the same. Therefore, data scientists who are familiar with Scikit-Learn can easily transition to PySpark and build logistic regression models for big data problems.

Before building the logistic regression model, let's begin by examining the parallels between Scikit-Learn and PySpark in terms of their modeling steps and the functions and classes they utilize.

Scikit-Learn and PySpark Similarities for Logistic Regression

Both Scikit-Learn and PySpark share a parallel workflow when implementing logistic regression, including data preparation, model training, model evaluation, and prediction. The following is an outline of the key machine learning functionalities in both frameworks specifically for logistic regression.

1. Data preparation:

   Scikit-Learn leverages functions like train_test_split to split the dataset into training and testing sets, StandardScaler for feature scaling, OneHotEncoder for one-hot encoding categorical variables, and SimpleImputer for handling missing data.

   In PySpark, we can work with DataFrames using operations like randomSplit for splitting data into training and testing sets, StandardScaler for feature scaling, OneHotEncoder for categorical variables, and Imputer for handling missing data. Moreover, the VectorAssembler combines feature columns into a single vector column.

2. Model training:

   Logistic regression models in Scikit-Learn and PySpark are instantiated using LogisticRegression and `LogisticRegression` classes, respectively. The models are trained using the fit() method. In Scikit-Learn, features (X_train) and the binary target variable (y_train) are provided.

   In PySpark, the fit() method accepts a DataFrame (train_data) containing features and a column for the binary target variable. A prior step with the VectorAssembler is necessary in PySpark to consolidate feature columns.

3. Model evaluation:

   Evaluation metrics in Scikit-Learn, such as accuracy (computed with `accuracy_score`), area under the ROC curve (AUC, computed with `roc_auc_score`), precision (computed with `precision_score`), recall (computed with `recall_score`), F1 score (computed with `f1_score`), and the confusion matrix (computed with `confusion_matrix`), are calculated through a direct comparison of predicted and actual binary target variable values (y_pred, y_test).

   In PySpark, the `BinaryClassificationEvaluator` calculates the AUC. This involves specifying the predicted and actual values (predictions, y_test), along with their respective columns. While there are no built-in functions for calculating the other metrics for binary classification, manual calculation is straightforward.

4. Prediction:

   Scikit-Learn models employ the predict() method to generate binary predictions for new data using features (X_test), yielding predicted target variable values (y_pred).

   Similarly, PySpark models utilize the transform() method to generate predictions. Employing a DataFrame (test_data) with features, the method produces predictions stored in a new column (prediction). As with training, the VectorAssembler ensures feature vector uniformity for new data.

These comparisons reveal that despite occasional differences in syntax, functions, and classes between Scikit-Learn and PySpark, the fundamental modeling steps remain largely consistent. In the next subsections, we translate these steps into code. It's crucial to note from the outset that we are constructing models with default hyperparameters. This means that the model probabilities and evaluation metrics may vary due to the differing default parameters between the two frameworks. In Chapter 16, we demonstrate how to fine-tune and align these parameters.

Logistic Regression with Scikit-Learn

Let's start with Scikit-Learn. To predict the likelihood of diabetes using the Pima Indian diabetes dataset, we adhere to the same steps outlined in the preceding subsection: data preparation, model training, model evaluation, and prediction. The code will be encapsulated within functions to enhance the ease of execution and comprehension.

It is common practice to start the modeling process by importing the necessary libraries, which provide functions and classes for data manipulation, preprocessing, model training, and evaluation.

Step 1: Importing necessary packages

```
[In]: import pandas as pd
[In]: from sklearn.model_selection import train_test_split
[In]: from sklearn.linear_model import LogisticRegression
[In]: from sklearn.metrics import accuracy_score, precision_score, recall_
      score, f1_score, confusion_matrix
[In]: from sklearn.preprocessing import StandardScaler
```

Step 2: Loading the diabetes dataset

This step defines a function `load_diabetes_csv(url)` to load the diabetes dataset from the provided URL. The dataset is loaded into a Pandas DataFrame.

```
[In]: def load_diabetes_csv(url):
          """
          Loads the diabetes.csv file from a URL and returns a Pandas
          DataFrame.

          Parameters:
          url (str): The URL of the CSV file.

          Returns:
          pandas.DataFrame: The loaded DataFrame.
          """
          pandas_df = pd.read_csv(url)
          return pandas_df
```

Step 3: Data preparation

3.1. Splitting the data into training and testing sets

In this substep of the data preparation step, the function split_data(X, y, test_size, random_state) is defined to split the feature matrix X and target variable y into training and testing sets.

```
[In]: def split_data(X, y, test_size=0.2, random_state=42):
          """
          Splits data into training and testing sets.

          Parameters:
          X (pd.DataFrame): Feature matrix.
          y (pd.Series): Target variable.
          test_size (float): Proportion of the dataset to include in
          test split.
          random_state (int): Seed for random number generator.

          Returns:
          tuple: X_train, X_test, y_train, y_test.
          """
          return train_test_split(X, y, test_size=test_size,
          random_state=random_state)
```

3.2. Scaling the data

Here, the function scale_data(X_train, X_test) scales the training and testing features using the StandardScaler to ensure that all features have similar scales.

```
[In]: def scale_data(X_train, X_test):
          """
          Scales the features using StandardScaler.

          Parameters:
          X_train (pd.DataFrame): Scaled training features.
          X_test (pd.DataFrame): Scaled testing features.

          Returns:
          tuple: X_train_scaled, X_test_scaled.
          """
          scaler = StandardScaler()
          X_train_scaled = scaler.fit_transform(X_train)
          X_test_scaled = scaler.transform(X_test)
          return X_train_scaled, X_test_scaled
```

Step 4: Model training

In this step, the function `train_lr(X_train, y_train)` trains the logistic regression model using the `LogisticRegression` class from Scikit-Learn.

```
[In]: def train_lr(X_train, y_train):
        """
        Trains a logistic regression model on the scaled training
        data.
        Parameters:
        X_train (pd.DataFrame): Scaled training features.
        y_train (pd.Series): Training target variable.

        Returns:
        sklearn.linear_model.LogisticRegression: Trained logistic
        regression model.
        """
        model = LogisticRegression()
        model.fit(X_train, y_train)
        return model
```

Step 5: Model evaluation

In this evaluation step, the function `evaluate_model(model, X_test, y_test)` evaluates the trained model using various metrics such as accuracy, precision, recall, F1 score, and confusion matrix. The metrics are printed, and a DataFrame containing the actual and predicted values is displayed.

```
[In]: def evaluate_model(model, X_test, y_test):
        """
        Evaluates the model using various metrics and prints the
        results.
        Parameters:
        model (sklearn.linear_model.LogisticRegression): Trained
        logistic regression model.
        X_test (pd.DataFrame): Scaled testing features.
        y_test (pd.Series): Testing target variable.
        """
        y_pred = model.predict(X_test)
```

```
        accuracy = accuracy_score(y_test, y_pred)
        precision = precision_score(y_test, y_pred)
        recall = recall_score(y_test, y_pred)
        f1 = f1_score(y_test, y_pred)
        confusion = confusion_matrix(y_test, y_pred)

        print(f'Accuracy: {accuracy:.2f}')
        print(f'Precision: {precision:.2f}')
        print(f'Recall: {recall:.2f}')
        print(f'F1 Score: {f1:.2f}')
        print(f'Confusion Matrix:\n{confusion}')

        result_df = pd.DataFrame({'Actual': y_test, 'Predicted':
        y_pred})
        print(result_df.head())
```

Step 6: Using the functions

In this final step, we call the functions that were defined previously. There are nine substeps as indicated here:

6.1. Define the URL for the dataset

```
[In]: url = ('https://raw.githubusercontent.com/abdelaziztestas/'
             'spark_book/main/diabetes.csv')
```

6.2. Call the load_diabetes_csv() function to read the dataset

```
[In]: pandas_df = load_diabetes_csv(url)
```

6.3. Specify the columns to be used from the dataset

```
[In]: columns = ['Pregnancies', 'Glucose', 'BloodPressure',
                 'BMI', 'DiabetesPedigreeFunction', 'Age', 'Outcome']
```

6.4. Select only the specified columns and filter out rows with zeros in key features

```
[In]: pandas_df = pandas_df[columns]
[In]: pandas_df = pandas_df.loc[(pandas_df['Glucose'] != 0)
                           & (pandas_df['BloodPressure'] != 0)
                           & (pandas_df['BMI'] != 0)]
```

6.5. Separate the features (X) and target variable (y)

```
[In]: X = pandas_df.drop('Outcome', axis=1)
[In]: y = pandas_df['Outcome']
```

6.6. Split the data into training and testing sets

```
[In]: X_train, X_test, y_train, y_test = split_data(X, y)
```

6.7. Scale the features in the training and testing sets

```
[In]: X_train_scaled, X_test_scaled = scale_data(X_train, X_test)
```

6.8. Train a logistic regression model on the scaled training data

```
[In]: model = train_lr(X_train_scaled, y_train)
```

6.9. Evaluate the model on the scaled testing data

```
[In]: evaluate_model(model, X_test_scaled, y_test)
```

```
[Out]:
Accuracy: 0.79
Precision: 0.66
Recall: 0.63
F1 Score: 0.64
Confusion Matrix:
[[88 14]
[16 27]]
Actual  Predicted
  0         0
  0         0
  1         1
  0         1
  1         1
```

After executing the nine substeps, we get the output shown previously. Before explaining this output, let's first go through the Scikit-Learn code step by step.

Step 1: Importing necessary libraries

In this first step, the required libraries are imported to use their functions and classes for data manipulation, model training, evaluation, and preprocessing:

- `pandas` for data manipulation
- `train_test_split` for splitting data into training and testing sets
- `LogisticRegression` for model training
- `accuracy_score`, `precision_score`, `recall_score`, `f1_score, and confusion_matrix` for model evaluation
- `StandardScaler` for feature scaling

Step 2: Loading the diabetes dataset

In this step, the `load_diabetes_csv(url)` function is defined. This takes a single parameter, url, which represents the URL of the dataset in CSV format. Inside the function, the line `pandas_df = pd.read_csv(url)` uses the Pandas `read_csv()` function to fetch the data from the specified URL and convert it into a Pandas DataFrame. The function returns the resulting DataFrame.

Step 3: Data preparation

This step has two substeps. The first substep defines a function named `split_data()` that splits the dataset into training and testing sets. The function takes several parameters:

- X: Matrix containing the independent variables used for prediction.
- y: Dependent variable associated with each data point.
- test_size: Specifies the proportion of the dataset that is allocated to the testing set. It defaults to 0.2, which corresponds to a 20% allocation.
- random_state: Sets the seed for the random number generator, ensuring reproducibility of the split. It defaults to 42.

Inside the `split_data()` function, the splitting operation is performed using the `train_test_split()` function. This takes the feature matrix X and the target variable y, along with the specified test_size and random_state values. It returns four separate datasets:

- X_train: The feature matrix for training
- X_test: The feature matrix for testing
- y_train: The target variable corresponding to the training set
- y_test: The target variable corresponding to the testing set

Finally, the function returns these four datasets as a tuple.

The second substep defines the `scale_data()` function. This standardizes the features of the dataset, ensuring that they have similar scales. This is particularly important in machine learning, where models like logistic regression can be sensitive to the varying magnitudes of features. The function takes two parameters:

- X_train: A feature matrix representing the training data. This is a Pandas DataFrame containing the independent variables.
- X_test: A feature matrix representing the testing data. It contains the independent variables of the testing set.

Inside the function, a `StandardScaler` object is created. This object is a tool from the Scikit-Learn library that facilitates the scaling of features. It's used to ensure that each feature has a mean of 0 and a standard deviation of 1.

The next step involves applying the scaling transformation to the training data using the line `X_train_scaled=scaler.fit_transform(X_train)`. Here, the `fit_transform()` method of the scaler object is used to compute the scaling parameters from the training data and simultaneously transform the training features accordingly. The scaled training features are stored in the variable X_train_scaled.

Next, the same scaling transformation is applied to the testing data using the line `X_test_scaled=scaler.transform(X_test)`. However, in this case, the `transform()` method is used because the scaling parameters were already computed using the training data. The scaled testing features are stored in the variable X_test_scaled.

Finally, the function returns the scaled training and testing feature matrices.

Step 4: Model training

In this step, a function called `train_lr` is defined. This function takes two input arguments: X_train and y_train. Its purpose is to train a logistic regression model using the provided training data and then return the trained model.

Within the function, a `LogisticRegression` object is created and assigned to the variable model. `LogisticRegression` is a class provided by Scikit-Learn that implements the logistic regression algorithm for classification tasks.

Moving to the next step, the `fit` method of the `LogisticRegression` model is called with the training data X_train and the corresponding labels y_train. This method trains the logistic regression model using the provided data by adjusting its parameters to fit the given training examples. Once the model has been trained, it is returned from the function as the output.

Step 5: Model evaluation

In this step, the code defines a function called `evaluate_model` that takes three input arguments: the trained model, X_test, and y_test. This function is used to evaluate the performance of the logistic regression model on the given test data using various metrics and print the results.

Inside the function, the `predict` method of the classification model is called with the test features X_test to make predictions on the test data. The predicted labels are assigned to the variable y_pred.

Next, the following key evaluation metrics of the model are computed:

- The accuracy metric measures the proportion of correctly predicted instances (both true positives and true negatives) out of the total number of instances in the dataset. In other words, accuracy tells us how often the model's predictions match the actual labels. This metric is calculated by comparing the predicted labels y_pred with the true labels y_test.
- Precision is calculated as a measure of the model's ability to correctly predict positive instances out of all instances it predicted as positive.
- Recall, also known as sensitivity or true positive rate, measures the model's ability to correctly predict positive instances out of all actual positive instances.
- F1 score is the weighted mean of precision and recall and provides a balanced measure between the two.
- The confusion matrix is a table showing the count of true positive, true negative, false positive, and false negative predictions.

The print statements are then used to display the calculated metrics and the confusion matrix. After this, a Pandas DataFrame is created to store the actual test labels (y_test) and the predicted labels (y_pred), while the last line of code prints the first five rows of the DataFrame to show a comparison between the actual and predicted labels.

Step 6: Using the functions

In this final step, the functions that were defined to prepare the data, train and evaluate the model, and make predictions are called. The order of using these functions is as follows:

- Define the URL for the dataset.
- Call the `load_diabetes_csv()` function to read the dataset.
- Specify the columns to be used from the dataset.

- Select only the specified columns and filter out rows with zeros in key features.
- Separate the features (X) and target variable (y).
- Split the data into training and testing sets by calling the `split_data(X, y)` function.
- Scale the features in the training and testing sets by using the `scale_data(X_train, X_test)` function.
- Train the model on the scaled training data using the `train_lr(X_train_scaled, y_train)` function.
- Evaluate the model on the scaled testing data using the `evaluate_model(model, X_test_scaled, y_test)` function.

After running the code in these steps, we get the output from evaluating the classification model using various metrics, including accuracy, precision, recall, F1 score, and the confusion matrix. The following is an interpretation of the output:

- Accuracy (0.79): This means that the model's predictions were correct for approximately 79% of the instances in the dataset. It's the ratio of the number of correctly predicted instances to the total number of instances.
- Precision (0.66): Precision is the ratio of correctly predicted positive instances (true positives) to all instances predicted as positive (true positives + false positives). In this example, the model's precision is approximately 66%, meaning that when it predicts a positive outcome, it is correct about 66% of the time.
- Recall (0.63): Recall, also known as sensitivity or true positive rate, is the ratio of correctly predicted positive instances (true positives) to all actual positive instances (true positives + false negatives). Here, the model's recall is about 63%, indicating that it correctly identified about 63% of the actual positive instances.
- F1 score (0.64): The F1 score is the weighted mean of precision and recall. It provides a balance between precision and recall. In this example, the F1 score is about 64%.

- Confusion matrix: This is a table that shows the counts of true positives (top-left), true negatives (bottom-right), false positives (top-right), and false negatives (bottom-left). In this example, there are 88 true negatives (instances correctly predicted as negative), 27 true positives (instances correctly predicted as positive), 14 false positives (instances incorrectly predicted as positive), and 16 false negatives (instances incorrectly predicted as negative).
- Actual vs. predicted: The table shows a comparison of actual labels and predicted labels for the top five instances. In four out of the five instances, or 80%, the model's predictions matched the actual labels (0 and 1). This is slightly higher than the overall accuracy of 79%, which takes into account all the samples, not just the first five rows.

Logistic Regression with PySpark

In this subsection, we will be developing PySpark code to prepare data, train and evaluate the logistic regression model, and make predictions. Similar to the preceding subsection, where we used Scikit-Learn, we encapsulate the code within functions to facilitate smooth execution and comprehension.

It is important to note that unlike Scikit-Learn, PySpark doesn't have built-in functions to calculate evaluation metrics such as accuracy, precision, recall, F1 score, and the confusion matrix for binary classification. However, we can easily compute these metrics manually, as will be demonstrated in the evaluation step. Additionally, when working with PySpark, there are two extra initial steps compared to Scikit-Learn: creating a Spark Session, which establishes the connection to the Spark cluster, and assembling features into a single column using the PySpark `VectorAssembler` class.

Just like in Scikit-Learn, we start by importing the necessary packages.

Step 1: Import necessary libraries

In this step, we import essential libraries, including those for SparkSession, LogisticRegression, VectorAssembler, StandardScaler, col, and Pandas.

```
[In]: from pyspark.sql import SparkSession
[In]: from pyspark.ml.classification import LogisticRegression
[In]: from pyspark.ml.feature import VectorAssembler, StandardScaler
[In]: from pyspark.sql.functions import col
[In]: import pandas as pd
```

Step 2: Create a Spark Session

This step creates a Spark Session named "DiabetesPrediction" using SparkSession.builder.

```
[In]: spark = SparkSession.builder.appName("DiabetesPrediction").
    getOrCreate()
```

Step 3: Read the data from a CSV file

This step defines a function to read data from a URL and convert it into a Spark DataFrame using pd.read_csv and spark.createDataFrame.

```
[In]: def read_data(url):
          """
          Read data from a given URL and convert it to a Spark
          DataFrame.

          Parameters:
          url (str): The URL to the CSV file.

          Returns:
          pyspark.sql.DataFrame: A Spark DataFrame containing the
          data.
          """
          pandas_df = pd.read_csv(url)
          spark_df = spark.createDataFrame(pandas_df)
          return spark_df
```

Step 4: Data preparation

This step has three substeps:

4.1. Create a function that splits the data into training and testing datasets using randomSplit

```
[In]: def split_data(spark_df):
          """
          Split data into training and testing datasets.

          Parameters:
          spark_df (pyspark.sql.DataFrame): The input Spark DataFrame.
```

```
        Returns:
        tuple: A tuple containing the training and testing Spark
        DataFrames.
        """
        return spark_df.randomSplit([0.8, 0.2], seed=42)
```

4.2. Define a function to convert feature columns into a single vector column using VectorAssembler

```
[In]: def vectorize_features(train_data, test_data, feature_cols):
        """
        Convert feature columns into a single vector column.

        Parameters:
        train_data (pyspark.sql.DataFrame): The training Spark
        DataFrame.
        test_data (pyspark.sql.DataFrame): The testing Spark
        DataFrame.
        feature_cols (list): List of feature column names.

        Returns:
        tuple: A tuple containing the vectorized training and
        testing Spark DataFrames.
        """
        assembler = VectorAssembler(inputCols=feature_cols,
        outputCol='features')
        return assembler.transform(train_data),
        assembler.transform(test_data)
```

4.3. Create a function to scale the features using StandardScaler

```
[In]: def scale_features(train_data, test_data):
        """
        Scale the features in the dataset.

        Parameters:
        train_data (pyspark.sql.DataFrame): The training Spark
        DataFrame.
```

```
        test_data (pyspark.sql.DataFrame): The testing Spark
        DataFrame.

        Returns:
        tuple: A tuple containing the scaled training and testing
        Spark DataFrames.
        """
        scaler = StandardScaler(inputCol='features',
        outputCol='scaled_features')
        scaler_model = scaler.fit(train_data)
        return scaler_model.transform(train_data),
        scaler_model.transform(test_data)
```

Step 5: Model training

In this step, we define a function that trains a logistic regression model on the training dataset using LogisticRegression.

```
[In]: def train_model(train_data):
        """
        Train a logistic regression model on the training dataset.
        Set a seed for reproducibility.

        Parameters:
        train_data (pyspark.sql.DataFrame): The training Spark
        DataFrame.

        Returns:
        pyspark.ml.classification.LogisticRegressionModel: The
        trained logistic regression model.
        """
        spark.conf.set("spark.seed", "42")

        lr = LogisticRegression(labelCol='Outcome',
        featuresCol='scaled_features')
        return lr.fit(train_data)
```

Step 6: Model evaluation

This step creates a function that evaluates the trained model on the testing dataset. This function calculates evaluation metrics such as accuracy, precision, recall, and F1 score and generates a confusion matrix.

```
[In]: def evaluate_model(lr_model, test_data):
        """
        Evaluate the trained model on the testing dataset.

        Parameters:
        lr_model
        (pyspark.ml.classification.LogisticRegressionModel): The
        trained logistic regression model.
        test_data (pyspark.sql.DataFrame): The testing Spark
        DataFrame.
        """
        predictions = lr_model.transform(test_data)
        predictions = predictions.withColumn('prediction',
        col('prediction').cast('int'))
        predictions_and_labels = predictions.select(['Outcome',
        'prediction'])
        tp = predictions_and_labels[(predictions_and_labels.Outcome
        == 1) & (predictions_and_labels.prediction == 1)].count()
        tn = predictions_and_labels[(predictions_and_labels.Outcome
        == 0) & (predictions_and_labels.prediction == 0)].count()
        fp = predictions_and_labels[(predictions_and_labels.Outcome
        == 0) & (predictions_and_labels.prediction == 1)].count()
        fn = predictions_and_labels[(predictions_and_labels.Outcome
        == 1) & (predictions_and_labels.prediction == 0)].count()
        acc = (tp + tn) / predictions_and_labels.count()
        prec = tp / (tp + fp)
        rec = tp / (tp + fn)
        f1 = 2 * prec * rec / (prec + rec)
        confusion = [[tp, fp], [fn, tn]]
        print(f'Accuracy: {acc:.2f}')
        print(f'Precision: {prec:.2f}')
```

```
    print(f'Recall: {rec:.2f}')
    print(f'F1 Score: {f1:.2f}')
    print(f'Confusion Matrix:\n{confusion}')
    predictions.select('Outcome', 'prediction').show(5)
```

Step 7: Using the functions

This step calls the functions defined in the previous steps. It also provides the URL location.

7.1. Provide the URL location

```
[In]: url = ('https://raw.githubusercontent.com/abdelaziztestas/'
             'spark_book/main/diabetes.csv')
```

7.2. Call the read_data() function to read dataset and create Spark DataFrame

```
[In]: spark_df = read_data(url)
columns = ['Pregnancies', 'Glucose', 'BloodPressure',
         'BMI', 'DiabetesPedigreeFunction', 'Age', 'Outcome']
[In]: spark_df = spark_df.select(columns)
[In]: spark_df = spark_df.filter((col('Glucose') != 0)
                        & (col('BloodPressure') != 0)
                        & (col('BMI') != 0))
```

7.3. Call the split_data() function to split data into training and testing sets

```
[In]: train_data, test_data = split_data(spark_df)
```

7.4. Define the list of feature columns

```
[In]: feature_cols = ['Pregnancies', 'Glucose', 'BloodPressure', 'BMI',
    'DiabetesPedigreeFunction', 'Age']
```

7.5. Call the vectorize_features() function to convert features into a single vector

```
[In]: train_data, test_data = vectorize_features(train_data, test_data,
    feature_cols)
```

7.6. Call the scale_features() function to scale features

```
[In]: train_data, test_data = scale_features(train_data, test_data)
```

7.7. Call the train_model() function to train the logistic regression model

```
[In]: lr_model = train_model(train_data)
```

7.8. Call the evaluate_model() function to evaluate the model on the test set

```
[In]: evaluate_model(lr_model, test_data)

[Out]:

Accuracy: 0.78
Precision: 0.62
Recall: 0.56
F1 Score: 0.59
Confusion Matrix:
[[23, 14],
[18, 91]]
+-------+----------+
|Outcome|prediction|
+-------+----------+
|      0|         0|
|      0|         1|
|      0|         0|
|      0|         0|
|      0|         0|
+-------+----------+
only showing top 5 rows
```

Before interpreting the output, let's take a moment to go through the code we have used to generate it. After importing the necessary libraries required for different aspects of working with PySpark and performing the machine learning task, we created a Spark Session to allow us to interact with the PySpark API. We then defined a function (read_data) that reads data from the given URL pointing to a CSV file and converts it into a Spark DataFrame. The function takes a URL as input and returns a Spark DataFrame. Inside the function, the Pandas library is utilized to read the data from the URL into a Pandas DataFrame named pandas_df. The `spark.createDataFrame(pandas_df)` method is used to convert the Pandas DataFrame into a Spark DataFrame named spark_df. The function then returns the resulting Spark DataFrame.

What follows are the steps to prepare data, train and evaluate the model, and make predictions:

Data Preparation (Step 4)

This step has three substeps. In the first substep, a function named `split_data` is defined, which takes a Spark DataFrame named spark_df as an argument. The purpose of this function is to partition the input DataFrame into two subsets: one for training and the other for testing.

The function employs the PySpark `randomSplit` method on the spark_df DataFrame. It divides the DataFrame into two segments, based on user-specified weights, which are set as [0.8, 0.2]. This signifies an allocation of approximately 80% of the data for training and 20% for testing. To ensure consistent results between repeated runs, the seed=42 parameter is utilized. When the function is executed, it yields a tuple with two separate DataFrames: one designed for training and the other for testing.

In the second substep, the code defines a function named `vectorize_features` that takes three arguments:

- train_data: A Spark DataFrame containing the training data
- test_data: A Spark DataFrame containing the test data
- feature_cols: A list of column names representing the features that we want to combine into a single feature vector

The purpose of this function is to convert the specified feature columns in both the training and test data into a single vector column. This is done to prepare the data for the machine learning algorithm, which requires input features to be in vector format.

Inside the function, the code first creates an instance of the `VectorAssembler` class from PySpark's pyspark.ml.feature module. This class is used to assemble feature columns into a single vector column. It takes two key arguments:

- inputCols: This is set to the list of column names provided in the feature_cols argument. It specifies the feature columns that should be combined.
- outputCol: This is set to features, which is the name of the new vector column that will be created.

The `transform()` method is then applied to the assembler instance for both the train_data and test_data DataFrames. This transformation takes the specified feature columns and combines them into a new vector column named features.

The function returns a tuple of two DataFrames: one is the train_data DataFrame with the newly created features column, and the other is the test_data DataFrame with the same transformation applied.

In the final substep of the data preparation step, the code defines a function named `scale_features` that takes two arguments:

- train_data: A Spark DataFrame containing the training data with a features column
- test_data: A Spark DataFrame containing the test data with a features column

The purpose of this function is to scale the features within the features column using the `StandardScaler` transformation provided by PySpark's machine learning library.

The function starts by creating an instance of the `StandardScaler` class, which is used for feature scaling. It takes two key arguments:

- inputCol: This is set to features, indicating the name of the column containing the features that need to be scaled.
- outputCol: This is set to scaled_features, which is the name of the new column where the scaled features will be stored.

The `scaler.fit(train_data)` line inside the function fits the scaler model using the training data. This calculates the mean and standard deviation needed for scaling the features. The `transform()` method is then applied to the scaler_model for both the train_data and test_data DataFrames. This transformation scales the features in the features column and stores the scaled values in the scaled_features column.

The function returns a tuple of two DataFrames: one is the train_data DataFrame with the newly created scaled_features column containing scaled features, and the other is the test_data DataFrame with the same scaled features applied.

Model Training (Step 5)

The code in this step defines a function named `train_model` that takes one argument:

- train_data: A Spark DataFrame containing the training data with scaled features in a column named scaled_features

The function's purpose is to train the logistic regression model using the training data. Before proceeding with training the model, the code sets a seed for reproducibility using `spark.conf.set("spark.seed", "42")`. This ensures that random operations performed during model training, like initialization or shuffling, will yield consistent results across different runs.

Next, an instance of the `LogisticRegression` class is created. This is used to define and train the logistic regression model. It takes two key arguments:

- labelCol: This is set to Outcome, indicating the column that contains the binary labels (0 or 1) for the logistic regression.
- featuresCol: This is set to scaled_features, indicating the column that contains the scaled features that will be used as input to the model.

The logistic regression model is trained using the `fit()` method on the lr instance. The train_data DataFrame, which contains both the scaled features and the target labels, is provided as the training dataset. The trained logistic regression model is then returned from the function.

Model Evaluation (Step 6)

In this step of the modeling process, the code defines a function named `evaluate_model` that takes two arguments:

- lr_model: A trained logistic regression model
- test_data: A Spark DataFrame containing the test data with scaled features

The function's purpose is to evaluate the performance of the trained logistic regression model using various evaluation metrics.

The function starts by making predictions on the test_data using the `transform` method of the trained logistic regression model (lr_model). The predictions are added as a new column in the resulting DataFrame called prediction.

The second line inside the function casts the predicted values to integers, as they are typically in floating-point format. In the next line, the function creates a new DataFrame named predictions_and_labels that selects only the Outcome and prediction columns from the predictions DataFrame.

The next code lines inside the `evaluate_model(lr_model, test_data)` function calculate various evaluation metrics. The following is an explanation for each of these metrics:

- True positives (tp): The count of instances where the actual outcome is positive (1) and the model's prediction is also positive (1)
- True negatives (tn): The count of instances where the actual outcome is negative (0) and the model's prediction is also negative (0)
- False positives (fp): The count of instances where the actual outcome is negative (0) but the model's prediction is positive (1)
- False negatives (fn): The count of instances where the actual outcome is positive (1) but the model's prediction is negative (0)
- Accuracy (acc): The ratio of correctly predicted instances to the total number of instances
- Precision (prec): The ratio of true positives to the sum of true positives and false positives
- Recall (rec): The ratio of true positives to the sum of true positives and false negatives
- F1 score (f1): The weighted mean of precision and recall
- Confusion matrix (confusion): A 2×2 matrix representing true positive, false positive, false negative, and true negative counts

The function then prints these performance metrics and displays the Outcome and prediction columns for the first five instances in the predictions DataFrame.

The final step (step 7) pertains to calling the functions, which involves the following eight substeps:

- Provide the URL location.
- Read and prepare data using read_data, filter rows, and select columns.
- Split data into training and testing sets using split_data.
- Define feature columns list.
- Vectorize features using vectorize_features.

- Scale features using scale_features.
- Train the logistic regression model using train_model.
- Evaluate the model on test set using evaluate_model, calculating and printing evaluation metrics.

Upon execution of these steps, we get the following results:

- Accuracy (0.78): This indicates that the logistic regression model achieved an accuracy of 78%, meaning that approximately 78% of the predictions made by the model were correct.
- Precision (0.62): The precision score is 0.62, which implies that out of all the instances the model predicted as positive, 62% were truly positive. Precision is a measure of the model's ability to correctly identify positive instances without many false positives.
- Recall (0.56): The recall score is 0.56, indicating that the model successfully captured 56% of the actual positive instances. Recall (also known as sensitivity or true positive rate) measures the model's ability to identify all positive instances without missing many.
- F1 score (0.59): The F1 score is 0.59, which is the weighted mean of precision and recall. It provides a balance between the two metrics and is often used when precision and recall need to be considered together.

The confusion matrix is presented as a 2×2 array:

- True positives (TP): 23 instances were truly positive and were correctly predicted as positive.
- False positives (FP): 14 instances were actually negative but were incorrectly predicted as positive.
- False negatives (FN): 18 instances were actually positive but were incorrectly predicted as negative.
- True negatives (TN): 91 instances were truly negative and were correctly predicted as negative.

The output also shows the Outcome (true label) and prediction (predicted label) columns for the first five instances in the test data. This provides a quick overview of how the model's predictions compare to the true outcomes for these instances. The results indicate that four out of the five predictions listed, or 80%, match the actual labels. This level of accuracy is slightly higher than the overall accuracy of 78%, which takes into account the entire samples.

Putting It All Together

In the previous sections of this chapter, we presented code snippets that demonstrated each step of the modeling process in isolation. However, in this section, we aim to integrate all the relevant code from those steps into a single code block. This will provide data scientists with a comprehensive view of how the code snippets work together and allow them to execute the code as a cohesive unit. Combining the code from each step into one block will also simplify the task of modifying the code and facilitate experimentation with different parameters and configurations.

Scikit-Learn

In this subsection, the entire Scikit-Learn code for preparing data, constructing and training the logistic regression model, and evaluating it on test data using the Pima Indian diabetes dataset is provided:

```
[In]: import pandas as pd
[In]: from sklearn.model_selection import train_test_split
[In]: from sklearn.linear_model import LogisticRegression
[In]: from sklearn.metrics import accuracy_score, precision_score, recall_
      score, f1_score, confusion_matrix
[In]: from sklearn.preprocessing import StandardScaler

[In]: def load_diabetes_csv(url):
          """
          Loads the diabetes.csv file from a URL and returns a Pandas
          DataFrame.
```

```
        Parameters:
        url (str): The URL of the CSV file.

        Returns:
        pandas.DataFrame: The loaded DataFrame.
        """
        pandas_df = pd.read_csv(url)
        return pandas_df

[In]: def split_data(X, y, test_size=0.2, random_state=42):
        """
        Splits data into training and testing sets.

        Parameters:
        X (pd.DataFrame): Feature matrix.
        y (pd.Series): Target variable.
        test_size (float): Proportion of the dataset to include in
        the test split.
        random_state (int): Seed for random number generator.

        Returns:
        tuple: X_train, X_test, y_train, y_test.
        """
        return train_test_split(X, y, test_size=test_size,
        random_state=random_state)

[In]: def scale_data(X_train, X_test):
        """
        Scales the features using StandardScaler.

        Parameters:
        X_train (pd.DataFrame): Scaled training features.
        X_test (pd.DataFrame): Scaled testing features.

        Returns:
        tuple: X_train_scaled, X_test_scaled.
        """
        scaler = StandardScaler()
        X_train_scaled = scaler.fit_transform(X_train)
```

```
        X_test_scaled = scaler.transform(X_test)
        return X_train_scaled, X_test_scaled

[In]: def train_lr(X_train, y_train):
        """
        Trains a logistic regression model on the scaled training
        data.

        Parameters:
        X_train (pd.DataFrame): Scaled training features.
        y_train (pd.Series): Training target variable.

        Returns:
        sklearn.linear_model.LogisticRegression: Trained logistic
        regression model.
        """
        model = LogisticRegression()
        model.fit(X_train, y_train)
        return model

[In]: def evaluate_model(model, X_test, y_test):
        """
        Evaluates the model using various metrics and prints the
        results.

        Parameters:
        model (sklearn.linear_model.LogisticRegression): Trained
        logistic regression model.
        X_test (pd.DataFrame): Scaled testing features.
        y_test (pd.Series): Testing target variable.
        """
        y_pred = model.predict(X_test)

        accuracy = accuracy_score(y_test, y_pred)
        precision = precision_score(y_test, y_pred)
        recall = recall_score(y_test, y_pred)
        f1 = f1_score(y_test, y_pred)
        confusion = confusion_matrix(y_test, y_pred)
```

```
        print(f'Accuracy: {accuracy:.2f}')
        print(f'Precision: {precision:.2f}')
        print(f'Recall: {recall:.2f}')
        print(f'F1 Score: {f1:.2f}')
        print(f'Confusion Matrix:\n{confusion}')

        result_df = pd.DataFrame({'Actual': y_test, 'Predicted':
        y_pred})
        print(result_df.head())
# Use the functions
# Define the URL for the dataset
[In]: url = ('https://raw.githubusercontent.com/abdelaziztestas/'
           'spark_book/main/diabetes.csv')

# Call the load_diabetes_csv() function to read the dataset
[In]: pandas_df = load_diabetes_csv(url)

# Specify the columns of interest
[In]: columns = ['Pregnancies', 'Glucose', 'BloodPressure',
               'BMI', 'DiabetesPedigreeFunction', 'Age', 'Outcome']

# Select only the specified columns and filter out rows with zeros in key
features
[In]: pandas_df = pandas_df[columns]
[In]: pandas_df = pandas_df.loc[(pandas_df['Glucose'] != 0)
                         & (pandas_df['BloodPressure'] != 0)
                         & (pandas_df['BMI'] != 0)]

# Separate the features (X) and target variable (y)
[In]: X = pandas_df.drop('Outcome', axis=1)
[In]: y = pandas_df['Outcome']

# Split the data into training and testing sets
[In]: X_train, X_test, y_train, y_test = split_data(X, y)

# Scale the features in the training and testing sets
[In]: X_train_scaled, X_test_scaled = scale_data(X_train, X_test)
```

```
# Train a logistic regression model on the scaled training data
[In]: model = train_lr(X_train_scaled, y_train)

# Evaluate the model on the scaled testing data
[In]: evaluate_model(model, X_test_scaled, y_test)
```

PySpark

In this subsection, we provide the PySpark equivalent of the Scikit-Learn code for preparing the data, building and training the logistic regression model, and evaluating it using the Pima Indian diabetes dataset.

```
[In]: from pyspark.sql import SparkSession
[In]: from pyspark.ml.classification import LogisticRegression
[In]: from pyspark.ml.feature import VectorAssembler, StandardScaler
[In]: from pyspark.sql.functions import col
[In]: import pandas as pd

[In]: spark = SparkSession.builder.appName("DiabetesPrediction").
      getOrCreate()

[In]: def read_data(url):
          """
          Read data from a given URL and convert it to a Spark
          DataFrame.

          Parameters:
          url (str): The URL to the CSV file.

          Returns:
          pyspark.sql.DataFrame: A Spark DataFrame containing the
          data.
          """
          pandas_df = pd.read_csv(url)
          spark_df = spark.createDataFrame(pandas_df)
          return spark_df
```

```
[In]: def split_data(spark_df):
        """
        Split data into training and testing datasets.

        Parameters:
        spark_df (pyspark.sql.DataFrame): The input Spark DataFrame.

        Returns:
        tuple: A tuple containing the training and testing Spark
        DataFrames.
        """
        return spark_df.randomSplit([0.8, 0.2], seed=42)
[In]: def vectorize_features(train_data, test_data, feature_cols):
        """
        Convert feature columns into a single vector column.

        Parameters:
        train_data (pyspark.sql.DataFrame): The training Spark
        DataFrame.
        test_data (pyspark.sql.DataFrame): The testing Spark
        DataFrame.
        feature_cols (list): List of feature column names.

        Returns:
        tuple: A tuple containing the vectorized training and
        testing Spark DataFrames.
        """
        assembler = VectorAssembler(inputCols=feature_cols,
        outputCol='features')
        return assembler.transform(train_data),
        assembler.transform(test_data)
[In]: def scale_features(train_data, test_data):
        """
        Scale the features in the dataset.
```

```
        Parameters:
        train_data (pyspark.sql.DataFrame): The training Spark
        DataFrame.

        test_data (pyspark.sql.DataFrame): The testing Spark
        DataFrame.

        Returns:
        tuple: A tuple containing the scaled training and testing
        Spark DataFrames.
        """
        scaler = StandardScaler(inputCol='features',
        outputCol='scaled_features')
        scaler_model = scaler.fit(train_data)
        return scaler_model.transform(train_data),
        scaler_model.transform(test_data)

[In]: def train_model(train_data):
        """
        Train a logistic regression model on the training dataset.

        Parameters:
        train_data (pyspark.sql.DataFrame): The training Spark
        DataFrame.

        Returns:
        pyspark.ml.classification.LogisticRegressionModel: The
        trained logistic regression model.
        """
        # Set a seed for reproducibility
        spark.conf.set("spark.seed", "42")

        lr = LogisticRegression(labelCol='Outcome',
        featuresCol='scaled_features')
        return lr.fit(train_data)

[In]: def evaluate_model(lr_model, test_data):
        """
        Evaluate the trained model on the testing dataset.
```

```
    Parameters:
    lr_model
    (pyspark.ml.classification.LogisticRegressionModel): The
    trained logistic regression model.
    test_data (pyspark.sql.DataFrame): The testing Spark
    DataFrame.
    """
    predictions = lr_model.transform(test_data)
    predictions = predictions.withColumn('prediction',
    col('prediction').cast('int'))
    predictions_and_labels = predictions.select(['Outcome',
    'prediction'])
    tp = predictions_and_labels[(predictions_and_labels.Outcome
    == 1) & (predictions_and_labels.prediction == 1)].count()
    tn = predictions_and_labels[(predictions_and_labels.Outcome
    == 0) & (predictions_and_labels.prediction == 0)].count()
    fp = predictions_and_labels[(predictions_and_labels.Outcome
    == 0) & (predictions_and_labels.prediction == 1)].count()
    fn = predictions_and_labels[(predictions_and_labels.Outcome
    == 1) & (predictions_and_labels.prediction == 0)].count()
    acc = (tp + tn) / predictions_and_labels.count()
    prec = tp / (tp + fp)
    rec = tp / (tp + fn)
    f1 = 2 * prec * rec / (prec + rec)
    confusion = [[tp, fp], [fn, tn]]
    # Print evaluation metrics
    print(f'Accuracy: {acc:.2f}')
    print(f'Precision: {prec:.2f}')
    print(f'Recall: {rec:.2f}')
    print(f'F1 Score: {f1:.2f}')
    print(f'Confusion Matrix:\n{confusion}')
    predictions.select('Outcome', 'prediction').show(5)

[In]: url = ('https://raw.githubusercontent.com/abdelaziztestas/'
             'spark_book/main/diabetes.csv')

[In]: spark_df = read_data(url)
```

```
[In]: columns = ['Pregnancies', 'Glucose', 'BloodPressure',
                 'BMI', 'DiabetesPedigreeFunction', 'Age', 'Outcome']
[In]: spark_df = spark_df.select(columns)
[In]: spark_df = spark_df.filter((col('Glucose') != 0)
                     & (col('BloodPressure') != 0)
                     & (col('BMI') != 0))

[In]: train_data, test_data = split_data(spark_df)

[In]: feature_cols = ['Pregnancies', 'Glucose', 'BloodPressure', 'BMI',
      'DiabetesPedigreeFunction', 'Age']

[In]: train_data, test_data = vectorize_features(train_data, test_data,
      feature_cols)

[In]: train_data, test_data = scale_features(train_data, test_data)

[In]: lr_model = train_model(train_data)

[In]: evaluate_model(lr_model, test_data)
```

Summary

This chapter introduced logistic regression and demonstrated the construction of a classification model using the logistic regression algorithm in both Scikit-Learn and PySpark. Additionally, the chapter discussed the advantages and disadvantages of logistic regression and delved into standard evaluation metrics including accuracy, precision, recall, F1 score, and the confusion matrix. Pandas and PySpark code were also used to compare and contrast how the two libraries handled data manipulation and exploration.

In the next chapter, we will continue our journey with supervised learning for classification algorithms. Specifically, we will introduce an alternative classification model to logistic regression based on decision trees. We will demonstrate how to prepare data, build, train, and evaluate the model, and make predictions on new, unseen data. Both Scikit-Learn and PySpark will be used. Pandas and PySpark will also be compared.

CHAPTER 8

Decision Tree Classification with Pandas, Scikit-Learn, and PySpark

In this chapter, we will continue with classification as a form of supervised learning. Our objective is to develop, train, and evaluate a decision tree classification model for predicting the species of an Iris flower based on its feature measurements. We will leverage the well-known Iris dataset, which consists of measurements of four features (sepal length, sepal width, petal length, and petal width) from three distinct species of Iris flowers (setosa, versicolor, and virginica).

Decision tree classifiers have several advantages. One key advantage is interpretability. A decision tree classifier is easy to visualize and interpret since the tree structure represents a series of simple decisions based on features, making it understandable for non-experts. Another advantage is that decision tree classifiers can handle nonlinearity. Their ability to model nonlinear relationships between features and the target variable makes them suitable for complex datasets. Moreover, decision tree classifiers make no assumptions; they do not assume a specific distribution of data, which can be beneficial when the data distribution is unknown or nonstandard. Finally, there is the advantage of feature importance. Decision trees provide information about the importance of features in the classification process, aiding in feature selection and understanding the model's behavior.

Where decision tree classifiers excel over the logistic regression models we built in the preceding chapter is in capturing nonlinear relationships. While logistic regression assumes a linear relationship between features and the log-odds of the target, decision

A. Testas, *Distributed Machine Learning with PySpark*, https://doi.org/10.1007/978-1-4842-9751-3_8

trees can effectively model intricate nonlinear interactions. This makes decision trees a powerful choice when dealing with data that exhibits complex dependencies that cannot be adequately captured by a linear model.

On the other hand, decision tree classifiers aren't without shortcomings. First, they are prone to overfitting as they tend to learn the training data too well, leading to poor performance on new, unseen data. There is also the issue of instability, as small changes in the data can lead to significant changes in the structure of the tree, causing instability in the model. Finally, decision tree classifiers tend to exhibit bias toward dominant classes. In imbalanced datasets, decision trees can be biased toward the majority class, resulting in suboptimal performance on minority classes. However, as will be seen in the next chapter, these issues can be effectively mitigated by employing a technique known as random forests.

In this chapter, we will utilize two powerful libraries to build a decision tree classifier: Scikit-Learn and PySpark. By comparing their Python code, we will highlight the similarities between them, making it easier for data scientists to switch from Scikit-Learn to PySpark and leverage the advantages offered by PySpark's distributed computing capabilities.

In addition, we will demonstrate the comparability of Pandas with PySpark, particularly in terms of data loading and exploration. By demonstrating how Pandas performs similar tasks to PySpark, we aim to provide data scientists with the flexibility to choose between the two libraries based on their specific requirements and data sizes.

By the end of this chapter, you will have a thorough understanding of classification using decision tree models. You will be able to utilize both Scikit-Learn and PySpark, enabling a seamless transition and allowing you to take full advantage of PySpark's distributed computing capabilities. Furthermore, you will gain insights into how PySpark can serve as an alternative to Pandas for larger datasets, offering flexibility in your data analysis workflows.

The Dataset

For the project presented in this chapter, we are utilizing the well-known Iris dataset to construct a decision tree classifier. The dataset contains measurements of four distinct attributes—sepal length, sepal width, petal length, and petal width—pertaining to three different species of iris flowers: setosa, versicolor, and virginica. The dataset, which is pre-installed with Scikit-Learn by default, contains 150 records, corresponding to 50 samples from each of the three species.

We can access the Iris dataset by writing a few lines of code. In the following code, we define a Python function named `create_pandas_dataframe()`, which creates a Pandas DataFrame from the dataset using the Scikit-Learn library and Pandas:

```
[In]: from sklearn.datasets import load_iris
[In]: import pandas as pd
[In]: def create_pandas_dataframe():
          """
          Creates a pandas DataFrame from the Iris dataset.

          Returns:
          pandas.DataFrame: A DataFrame containing feature data,
          target labels, and target names.
          """
          iris = load_iris()
          X = iris.data
          y = iris.target
          feature_names = iris.feature_names
          target_names = iris.target_names

          pandas_df = pd.DataFrame(X, columns=feature_names)
          pandas_df['target'] = y
          pandas_df['target_names'] = pandas_df['target'].map(
          dict(zip(range(len(target_names)), target_names)))

          return pandas_df
```

The first two lines of code import the necessary libraries. `load_iris` is a function that allows us to load the Iris dataset, and pd is an alias for the Pandas library. Next, the code defines the function `create_pandas_dataframe()`, which doesn't take any arguments. Inside the function, the `load_iris()` function loads the Iris dataset, which contains feature data, target labels, feature names, and target names. The code then proceeds to extract the feature data (X) and target labels (y) from the loaded dataset. It also extracts the feature names and target names.

Next, a Pandas DataFrame named pandas_df is created with the extracted feature data (X) and feature names as columns. Moving on, the target labels (y) are added as a new column named target to the DataFrame. Additionally, a column target_names is created by mapping the target labels to their corresponding target names using a dictionary created with the zip() function. Finally, the create_pandas_dataframe() function returns the created DataFrame containing feature data, target labels, and target names.

We can call the function to create the pandas_df DataFrame with the following line:

```
[In]: pandas_df = create_pandas_dataframe()
```

We can convert the pandas_df to a PySpark DataFrame with the following code:

```
[In]: from pyspark.sql import SparkSession
[In]: def convert_to_spark_df(pandas_df):
          """
          Converts a pandas DataFrame to a PySpark DataFrame.

          Args:
          pandas_df (pandas.DataFrame): The pandas DataFrame to be
          converted.

          Returns:
          pyspark.sql.DataFrame: A PySpark DataFrame.
          """
          spark =
          SparkSession.builder.appName("PandasToSpark").getOrCreate()
          spark_df = spark.createDataFrame(pandas_df)

          return spark_df
```

The code first imports the SparkSession class, which is the entry point for using Spark functionality. It then defines the convert_to_spark_df() function. This takes one argument, pandas_df, which is the pandas DataFrame that we want to convert to a PySpark DataFrame. Inside the function, a Spark Session with the app name "PandasToSpark" is created using the SparkSession.builder.appName() method. The createDataFrame() method of the Spark Session is then used to convert the provided Pandas DataFrame (pandas_df) to a PySpark DataFrame (spark_df). Finally, the function returns the converted PySpark DataFrame (spark_df).

We can call the convert_to_spark_df() function to convert the pandas_df to the spark_df as follows:

```
[In]: spark_df = convert_to_spark_df(pandas_df)
```

We can explore the two DataFrames (pandas_df and spark_df) to gain a deeper understanding of the dataset's structure, contents, and characteristics. We begin by showing the top five rows of the DataFrames.

In Pandas, we can use the head() method:

```
[In]: print(pandas_df.head())
[Out]:
```

	sepal length (cm)	sepal width (cm)	petal length (cm)	petal width (cm)	target	target_names
0	5.1	3.5	1.4	0.2	0	setosa
1	4.9	3	1.4	0.2	0	setosa
2	4.7	3.2	1.3	0.2	0	setosa
3	4.6	3.1	1.5	0.2	0	setosa
4	5	3.6	1.4	0.2	0	setosa

We can observe from the output that the label has already been converted to a numerical format. Therefore, there will be no need for us to perform label encoding during the modeling steps. Additionally, all features are presented in centimeters, implying that standard scaling is unnecessary, as the features are already on a uniform scale. It's worth noting, however, that while tree-based algorithms do not strictly require scaling, some practitioners do recommend its application.

In PySpark, we can use the show(5) method to display the top five rows:

```
[In]: spark_df.show(5)
[Out]:
```

sepal length (cm)	sepal width (cm)	petal length (cm)	petal width (cm)	target	target_names
5.1	3.5	1.4	0.2	0	setosa
4.9	3	1.4	0.2	0	setosa
4.7	3.2	1.3	0.2	0	setosa
4.6	3.1	1.5	0.2	0	setosa
5	3.6	1.4	0.2	0	setosa

The output from pandas_df and spark_df is identical, except for the absence of the index in the PySpark output. The absence of the index in the PySpark output compared to the Pandas DataFrame is due to the differing approaches in how these frameworks handle data. In Pandas, the index serves as a unique identifier for rows, allowing for quick access and manipulation. However, PySpark DataFrames are built for distributed processing, where individual row indexing is less relevant.

We can print the shape of the `pandas_df` (i.e., number of rows and columns) using the Pandas `shape` attribute:

```
[In]: print(pandas_df.shape)
[Out]: (150, 6)
```

We can achieve the same result with the following PySpark code:

```
[In]: print((spark_df.count(), len(spark_df.columns)))
[Out]: (150, 6)
```

Since PySpark doesn't have the `shape` attribute, we combined two methods: `count()` and `len()` to get the number of rows and columns, respectively.

To explore the dataset further, we can print summary statistics of the features. In Pandas, we can use the `describe()` method:

```
[In]: pandas_df.drop(columns=['target']).describe()
[Out]:
```

	Sepal length (cm)	Sepal width (cm)	Petal length (cm)	Petal width (cm)
count	150	150	150	150
mean	5.84	3.06	3.76	1.20
std	0.83	0.44	1.77	0.76
min	4.3	2	1	0.1
25%	5.1	2.8	1.6	0.3
50%	5.8	3	4.3	1.3
75%	6.4	3.3	5.1	1.8
max	7.9	4.4	6.9	2.5

The use of the Pandas drop() method, in conjunction with the describe() method, effectively excluded the target variable as its inclusion doesn't contribute further information. Recall that, earlier, when we employed the head() method to inspect the initial five rows, there were two target variables: target and target_names. Notably, the former is numerical, while the latter is not. Pandas automatically omits non-numerical variables; hence, it was unnecessary to incorporate the target_names in the drop()/ describe() code.

The PySpark equivalent to Pandas describe() is the summary() method. PySpark, however, doesn't automatically exclude non-numerical variables from the calculation, so we need to explicitly exclude the target_names variable from the calculation:

```
[In]: from pyspark.sql.functions import col
[In]: numeric_columns = [col(column) for column in spark_df.columns if
      column not in ['target', 'target_names']]
[In]: spark_df.select(numeric_columns).summary().show()
[Out]:
```

summary	sepal length (cm)	sepal width (cm)	petal length (cm)	petal width (cm)
count	150	150	150	150
mean	5.84	3.06	3.76	1.20
stddev	0.83	0.44	1.77	0.76
min	4.3	2	1	0.1
25%	5.1	2.8	1.6	0.3
50%	5.8	3	4.3	1.3
75%	6.4	3.3	5.1	1.8
max	7.9	4.4	6.9	2.5

By importing the col function, we create a list of column references named numeric_ columns. This list includes all columns from the DataFrame except for the target and target_names columns, which are excluded using a conditional statement. The select() method is then employed to choose only the specified numeric columns. Next, the summary() method is applied to generate the summary statistics, including count, mean, standard deviation, min, 25th percentile, median, 75th percentile, and max values for each of the selected numeric columns. The show() method is utilized to display these calculated statistics.

We can observe that the outputs from Pandas and PySpark are nearly identical, differing only in two cosmetic aspects. In PySpark, the statistics column is labeled as summary, whereas in Pandas, it remains blank. Additionally, PySpark labels the standard deviation as stddev, while Pandas uses std.

In fact, PySpark does have a describe() method, similar to Pandas. However, it excludes quartiles (25th, 50th, and 75th percentiles) from its output. Let's execute the same code as before, but this time using the describe() method instead of the summary() method:

```
[In]: from pyspark.sql.functions import col
[In]: numeric_columns = [col(column) for column in spark_df.columns if
column not in ['target', 'target_names']]
[In]: spark_df.select(numeric_columns).describe().show()
[Out]:
```

summary	sepal length (cm)	sepal width (cm)	petal length (cm)	petal width (cm)
count	150	150	150	150
mean	5.84	3.06	3.76	1.20
stddev	0.83	0.44	1.77	0.76
min	4.3	2	1	0.1
max	7.9	4.4	6.9	2.5

The describe() method may come handy when the user doesn't need the quartiles.

To dig deeper into the dataset, we can check for null values in both pandas_df and spark_df DataFrames.

Starting with Pandas, we can use the isnull() and sum() functions:

```
[In]: print(pandas_df.isnull().sum())
[Out]:
sepal length (cm)    0
sepal width (cm)     0
petal length (cm)    0
```

```
petal width (cm)    0
target              0
target_names        0
dtype: int64
```

In PySpark, we can use the isNull() and sum() functions:

```
[In]: from pyspark.sql.functions import col, sum
[In]: null_counts = spark_df.select([sum(col(column).isNull().
      cast("integer")).alias(column) for column in spark_df.columns])
[In]: null_counts.show()
[Out]:
```

sepal length (cm)	sepal width (cm)	petal length (cm)	petal width (cm)	target	target_ names
0	0	0	0	0	0

The PySpark code begins by utilizing the `select()` method on the spark_df DataFrame. Within this method, a list comprehension iterates through each column in the DataFrame, accessed through the `spark_df.columns` attribute. For each column, a series of functions is applied: the `col()` function references the current column, followed by the `isNull()` function to determine null values. To convert resulting Booleans into integers (1 for True, 0 for False), `cast("integer")` is used. The integers are then summed up using the `sum()` function. The `alias(column)` function assigns names to columns. Finally, the `show()` method prints the results.

Both Pandas and PySpark outputs indicate that there are no missing records within the Iris dataset.

Decision Tree Classification

In this section, we aim to build, train, and evaluate a decision tree classifier using the Iris dataset. This contains four features (length and width of sepals and petals) of 50 samples of three species of Iris (setosa, virginica, and versicolor). Since the features are in the same unit of measurement (cm), scaling is not required. Decision trees perform equally well with or without scaling.

Using the Iris dataset, a decision tree classifier will make predictions by following a set of rules based on the features of the input data. It starts at the root node and evaluates a feature to determine which branch to follow. For example, if the petal length is less than 2.5 cm, it may follow the left branch, while if it is greater than or equal to 2.5 cm, it may follow the right branch. This process continues until a leaf node is reached, where the classifier assigns a class label based on the majority class of training samples that reached that leaf node. This assigned class label serves as the prediction for the given input.

We will be using Scikit-Learn and PySpark to construct and train the decision tree classifier and evaluate its performance using the Iris dataset. This comparison is aimed to help data scientists already familiar with the Scikit-Learn library transition smoothly to PySpark in order to take advantage of the distributed ML environment that it provides. This has the ability to process much larger volumes of data and increase the speed of training and evaluation.

Before we begin the process of model building, let's see how Scikit-Learn and PySpark compare in terms of the modeling steps and the functions and classes they use.

Scikit-Learn and PySpark Similarities

The following is a table comparing the classes and functions used in Scikit-Learn and their equivalent counterparts in PySpark for model construction in the context of multiclass decision tree classification:

Task	Scikit-Learn	PySpark
Data preparation		
Splitting data	train_test_split	randomSplit
Feature scaling	StandardScaler	StandardScaler
Categorical encoding	OneHotEncoder	OneHotEncoder
Missing data handling	SimpleImputer	Imputer
Feature vectorization	Not applicable	VectorAssembler
Model training		

(*continued*)

Task	Scikit-Learn	PySpark
Model class	DecisionTreeClassifier	DecisionTreeClassifier
Training	fit()	fit()
Model evaluation		
Accuracy	accuracy_score	MultiClassificationEvaluator
Precision	precision_score	
Recall	recall_score	
F1 score	f1_score	
Prediction		
Generate predictions	predict	transform

These comparisons reveal that both Scikit-Learn and PySpark adhere to the same modeling steps of data preparation, model training, model evaluation, and prediction. In some cases, even the names of functions or classes are the same in both platforms. These include StandardScaler, used for standardizing features by removing the mean and scaling to unit variance; OneHotEncoder, employed to convert categorical data into one-hot encoded features; DecisionTreeClassifier, which represents a decision tree classifier model; and the fit() method, which is used to train the machine learning model.

There are, however, notable differences. For example, Scikit-Learn doesn't require the VectorAssembler step to combine the features into a single vector. The reason for this is that this step isn't necessary for nondistributed systems like Scikit-Learn. Similarly, PySpark's evaluation class MultiClassificationEvaluator is designed to handle all performance metrics, unlike Scikit-Learn where separate functions exist for evaluation metrics such as accuracy_score, precision_score, recall_score, and f1_score.

Additionally, Scikit-Learn provides a dedicated function named confusion_matrix for calculating the confusion matrix, while in PySpark, this calculation is usually performed manually. There is, however, an RDD class (MulticlassMetrics) that can automatically calculate the confusion matrix, along with all the other evaluation metrics (precision, recall, F1 score) as shown in the following syntax:

```
[In]: from pyspark.mllib.evaluation import MulticlassMetrics
[In]: metrics = MulticlassMetrics(predictions)
[In]: precision = metrics.precision()
```

```
[In]: recall = metrics.recall()
[In]: f1_score = metrics.fMeasure()
[In]: confusion_matrix = metrics.confusionMatrix().toArray()
```

In this book, however, our focus is on the ML library instead of RDD's mllib, as RDD is less commonly used and is likely to be deprecated at some point.

In the next subsections of this section, we translate these modeling steps into code. It's worth noting that we are building models with default hyperparameters, meaning that the results between Scikit-Learn are likely to differ. We demonstrate how to fine-tune the hyperparameters of a model in Chapter 16.

Decision Tree Classification with Scikit-Learn

In this subsection, we develop the Scikit-Learn code to build, train, and evaluate the performance of a decision tree classification model and use it to predict the species of an Iris flower based on its feature measurements. The steps we follow are the same steps described earlier: data preparation, model training, evaluation, and prediction. Prior to these steps, however, we need to import the necessary libraries and create a DataFrame with the Iris flower data.

Step 1: Importing necessary libraries

In this step, we import the necessary libraries to load the Iris data, split the data into training and testing sets, build the decision tree classifier, and calculate the accuracy score:

```
[In]: from sklearn.datasets import load_iris
[In]: from sklearn.model_selection import train_test_split
[In]: from sklearn.tree import DecisionTreeClassifier
[In]: from sklearn.metrics import accuracy_score
```

Step 2: Reading data

In this step, we define a function to load and return the Iris dataset features (X) and labels (y).

```
[In]: def load_iris_dataset():
          iris = load_iris()
          X = iris.data
          y = iris.target
          return X, y
```

Step 3: Data preparation

In this step, we define a function to split data into training and testing sets.

```
[In]: def split_data(X, y, test_size=0.2, random_state=42):
        X_train, X_test, y_train, y_test = train_test_split(X, y,
        test_size=test_size, random_state=random_state)
        return X_train, X_test, y_train, y_test
```

Step 4: Model training

In this step, we define a function to build and return a decision tree classifier using the provided features (X) and labels (y).

```
[In]: def build_decision_tree(X, y):
        clf = DecisionTreeClassifier()
        clf.fit(X, y)
        return clf
```

Step 5: Model evaluation

The code in this step defines a function that evaluates and returns the accuracy score of the decision tree classifier (clf) on the test features (X_test) and labels (y_test).

```
[In]: def evaluate_model(clf, X_test, y_test):
        y_pred = clf.predict(X_test)
        accuracy = accuracy_score(y_test, y_pred)
        return accuracy
```

Step 6: Using the functions

This step includes a number of substeps to call the functions previously defined:

6.1. Load the Iris dataset

```
[In]: X, y = load_iris_dataset()
```

6.2. Split the dataset into training and testing sets using the split_data function

```
[In]: X_train, X_test, y_train, y_test = split_data(X, y)
```

6.3. Build the decision tree model using the training data

```
[In]: clf = build_decision_tree(X_train, y_train)
```

6.4. Evaluate the model on the test data

```
[In]: accuracy = evaluate_model(clf, X_test, y_test)
```

6.5. Print the accuracy

```
[In]: print("Accuracy:", accuracy)
[Out]: 1.00
```

Let's go through the code step by step:

Step 1: Importing necessary libraries

In this step, the code imports several modules and classes to prepare the data, build and train the model, evaluate its performance, and make predictions on test data.

The code begins by importing the `load_iris` function to load the Iris dataset. It then brings in the `train_test_split` function, which is used to split the dataset into training and testing subsets. This is to ensure that the model's performance is evaluated on data it hasn't seen during training. The next import statement brings in the `DecisionTreeClassifier` class, which represents the decision tree classifier. The final import statement brings in the `accuracy_score` function, which calculates the accuracy of the classification model's predictions. It compares the predicted labels with the actual labels and returns the fraction of correctly predicted instances.

Step 2: Reading data

In this step, the code defines a function to load and return the Iris dataset. Inside the function, the code uses the `load_iris` function to load the Iris dataset, including both the features and target labels. It then assigns the feature data and target labels of the Iris dataset to the variables X and y, respectively. The function finally returns the feature matrix X and the target array y as a tuple.

Step 3: Data preparation

In this step, the code defines a function named `split_data` to split the dataset into training and testing sets. It takes the following arguments:

- X: The feature data (matrix) of the dataset.
- y: The target labels (array) of the dataset.
- test_size: The proportion of the dataset to include in the test split (default is 0.2, or 20%).
- random_state: Seed for the random number generator (default is 42). It ensures reproducibility of the split.

The line inside the function uses the `train_test_split` function to split the data into training and testing sets. It takes the feature matrix X and target labels y as inputs. The test_size argument determines the proportion of the data that should be allocated for testing, in this case, 0.2 or 20%. The random_state argument sets the seed for random shuffling of the data during the split.

The function returns four variables:

- X_train: The feature data for the training set
- X_test: The feature data for the testing set
- y_train: The target labels for the training set
- y_test: The target labels for the testing set

Finally, the function returns the four sets (feature data and labels for training and testing) as a tuple.

Step 4: Model training

The code in this step defines a function named `build_decision_tree` to build and return a decision tree classifier using the features X and labels y. The function takes two arguments:

- X: The feature data (matrix) of the dataset
- y: The target labels (array) of the dataset

The first line inside the function creates an instance of the `DecisionTreeClassifier` class. This represents a decision tree-based classification model. The classifier is initialized with its default settings, which we can customize its behavior by passing various parameters to this constructor. We show how to customize a model's parameters in Chapter 16.

The next line trains the decision tree classifier on the provided feature matrix X and target labels y. The `fit` method learns the patterns and relationships in the data, enabling the classifier to make predictions based on the learned knowledge.

Finally, the function returns the trained decision tree classifier (clf) as the output of the function. This allows us to use the trained classifier for making predictions on new, unseen data.

Having trained the model, we can check which features are driving the model's predictions:

```
[In]: feature_names = ['sepal length', 'sepal width', 'petal length',
      'petal width']
[In]: feature_importances = clf.feature_importances_
[In]: importances_with_names = list(zip(feature_names, feature_
      importances))
[In]: importances_with_names_sorted = sorted(importances_with_names,
      key=lambda x: x[1], reverse=True)
[In]: for name, importance in importances_with_names_sorted:
          print(f"{name}: {importance:.4f}")
[Out]:
petal length: 0.9061
petal width: 0.0772
sepal length: 0.0167
sepal width: 0.0000
```

We can print the feature importances with just one line:

```
[In]: clf.feature_importances_
```

However, we need to add the feature names and sort the importances by descending order. The code begins by defining a list feature_names, which contains the names of the four features in the Iris dataset. It then extracts the feature importances calculated by the trained decision tree classifier clf. The attribute `feature_importances_` provides importance scores assigned to each feature by the classifier.

Moving to the next step, the code combines the feature_names list and the feature_importances array into a list of tuples using the `zip` function. The code then sorts the list of tuples (importances_with_names) in descending order using the `sorted` function with the reverse=True parameter based on the importance scores (where x[1] refers to the second element of each tuple, i.e., the importance score).

The following step involves a loop that iterates through the sorted list of tuples, importances_with_names_sorted, and unpacks each tuple into the variables name and importance. Within the loop, a line prints the name of a feature (name) along with its corresponding importance score (importance).

The output indicates the relative influence of each feature on the classifier's predictions. Features with higher importance scores contribute more significantly to the decision-making process of the model, while features with lower scores have relatively less impact. In our model, the petal length and width play a more crucial role in the decisions made by the decision tree classifier for the Iris dataset. The sepal width is redundant as its score is 0, meaning that it makes no contribution to the predictions.

We can also take a look at a sample comparison between the actual and predicted values:

```
[In]: import pandas as pd
[In]: y_pred_test = clf.predict(X_test)
[In]: comparison_df = pd.DataFrame({'Actual': y_test[:5], 'Predicted': y_
      pred_test[:5]})
[In]: print(comparison_df)
[Out]:
```

	Actual	Predicted
0	1	1
1	0	0
2	2	2
3	1	1
4	1	1

The output indicates that the model has correctly classified each species, as all five predicted values match the actual label values. This results in an accuracy of 100%. However, to obtain a more comprehensive understanding of the model's performance, it's essential to calculate the overall accuracy using the entire set of sample points.

Step 5: Model evaluation

The code in this step defines a function named `evaluate_model` to evaluate the performance of the trained classifier model on the test dataset and calculate its accuracy. The function takes three arguments:

- clf: The trained classifier model to be evaluated
- X_test: The feature data (matrix) of the test dataset
- y_test: The true target labels (array) of the test dataset

The first line inside the function uses the classifier clf to predict labels for the test dataset's feature matrix X_test. The predict method of the classifier takes the feature data as input and returns the predicted labels for the given data.

The next line calculates the accuracy of the predicted labels (y_pred) when compared to the true labels (y_test) using the accuracy_score function. The `accuracy_score` function measures the proportion of correct predictions in the predicted labels compared to the true labels.

Finally, the function returns the calculated accuracy as the output of the function. The accuracy score indicates how well the trained classifier performs on the test data.

Step 6: Using functions

In this step, we basically utilize the functions we defined previously. In the code sequence of this step, the Iris dataset is loaded using the `load_iris_dataset` function. The dataset is then split into training and testing sets using the `split_data function`, with the default test size of 20%. A decision tree classifier is built using the training data with the `build_decision_tree` function. The classifier's accuracy is evaluated on the test data using the `evaluate_model` function, which calculates the accuracy using `accuracy_score`. Finally, the accuracy of the model on the test data is printed, yielding an accuracy of 1.00 (or 100%), indicating a perfect fit for this specific execution.

Decision Tree Classification with PySpark

In this subsection, we will provide the PySpark code that is equivalent to Scikit-Learn's code in the preceding subsection. We will prepare the Iris dataset and use it to build, train, and evaluate a decision tree classifier, which will be used to predict the species of an Iris flower based on its feature measurements. We start with importing the necessary libraries.

Step 1: Importing necessary libraries

```
[In]: import pandas as pd
[In]: from pyspark.sql import SparkSession
[In]: from pyspark.ml.feature import VectorAssembler
[In]: from pyspark.ml.classification import DecisionTreeClassifier
[In]: from pyspark.ml.evaluation import MulticlassClassificationEvaluator
[In]: from sklearn.datasets import load_iris
```

Step 2: Creating a PySpark DataFrame

This step has two substeps: loading the dataset and converting it to PySpark.

2.1. Loading the Iris dataset and returning it as a Pandas DataFrame

```
[In]: def load_iris_dataset():
          iris = load_iris()
          X = iris.data
          y = iris.target
          feature_names = iris.feature_names
          target_names = iris.target_names
          pandas_df = pd.DataFrame(X, columns=feature_names)
          pandas_df['label'] = y
          return pandas_df
```

2.2. Converting the Pandas DataFrame to a Spark DataFrame using the SparkSession (spark)

```
[In]: def convert_to_spark_df(pandas_df, spark):
          return spark.createDataFrame(pandas_df)
```

Step 3: Data preparation

This step has two substeps: combining the features into a single vector and splitting the data into training and testing sets

3.1. Preparing the feature vector

```
[In]: def prepare_feature_vector(spark_df):
          assembler = VectorAssembler(inputCols=spark_df.columns[:-1],
          outputCol="features")
          transformed_data = assembler.transform(spark_df)
          return transformed_data
```

3.2. Splitting the dataset into training and testing sets

```
[In]: def split_data(transformed_data, train_ratio=0.8, seed=42):
          train_data, test_data =
          transformed_data.randomSplit([train_ratio, 1 - train_ratio],
          seed=seed)
          return train_data, test_data
```

Step 4: Model training

In this step, we build and train a decision tree classifier using the provided training data (Spark DataFrame).

```
[In]: def build_decision_tree(train_data):
          dt = DecisionTreeClassifier(labelCol="label",
          featuresCol="features")
          model = dt.fit(train_data)
          return model
```

Step 5: Model evaluation

In this step, we evaluate the DecisionTreeClassifier using the test data.

```
[In]: def evaluate_model(model, test_data):
          predictions = model.transform(test_data)
          evaluator =
          MulticlassClassificationEvaluator(labelCol="label",
          predictionCol="prediction", metricName="accuracy")
          accuracy = evaluator.evaluate(predictions)
          return accuracy, predictions
```

Step 6: Using the functions

In this step, we use the functions we defined previously. We begin with the creation of a Spark Session.

6.1. Create a SparkSession

```
[In]: spark = SparkSession.builder.appName("IrisDecisionTree").
      getOrCreate()
```

6.2. Load the Iris dataset and convert it to a Pandas DataFrame

```
[In]: pandas_df = load_iris_dataset()
```

6.3. Convert the Pandas DataFrame to PySpark DataFrame

```
[In]: spark_df = convert_to_spark_df(pandas_df, spark)
```

6.4. Prepare the feature vector

```
[In]: transformed_data = prepare_feature_vector(spark_df)
```

6.5. Split the dataset into training and testing sets

```
[In]: train_data, test_data = split_data(transformed_data)
```

6.6. Build the decision tree model

```
[In]: model = build_decision_tree(train_data)
```

6.7. Evaluate the model and get predictions

```
[In]: accuracy, predictions = evaluate_model(model, test_data)
```

6.8. Print the accuracy

```
[In]: print("Accuracy:", accuracy)
[Out]: 0.94
```

Let's summarize what the code does at each step:

Step 1: Importing necessary libraries

In this step, we import the required libraries and modules that will be used throughout the code. These libraries include Pandas, PySpark for distributed data processing, and specific modules for machine learning tasks such as creating feature vectors and training a decision tree classifier. We also import the `load_iris` function from Scikit-Learn to load the Iris dataset.

Step 2: Creating a PySpark DataFrame

This step involves two substeps:

2.1. Loading the Iris dataset and returning it as a Pandas DataFrame

Here, a custom function `load_iris_dataset` is defined. This function loads the Iris dataset using Scikit-Learn's `load_iris` function and extracts the feature data (X), target labels (y), and other relevant information like feature names and target names. Then, it creates a Pandas DataFrame with the data and labels and returns it.

2.2. Converting the Pandas DataFrame to a Spark DataFrame using the SparkSession (spark)

In this substep, another custom function, `convert_to_spark_df`, is defined. It takes the Pandas DataFrame as an input argument along with a SparkSession (spark) and converts the Pandas DataFrame to a Spark DataFrame using the `spark.createDataFrame` method. The resulting Spark DataFrame will be used for subsequent processing in PySpark.

Step 3: Data preparation

This step involves two substeps:

3.1. Preparing the feature vector

Here, a custom function `prepare_feature_vector` is defined. This function takes a Spark DataFrame as input (spark_df) and prepares the feature vector by assembling the feature columns into a single vector column using the `VectorAssembler` from PySpark. The result is a new DataFrame with a feature vector column.

3.2. Splitting the dataset into training and testing sets

In this substep, a custom function `split_data` is defined. It takes the transformed data (the DataFrame with feature vectors) as input along with arguments for the train ratio and seed. This function splits the data into training and testing sets using the `randomSplit` method provided by PySpark.

Step 4: Model training

In this step, a custom function `build_decision_tree` is defined. It takes the training data as input and builds a decision tree classifier using the `DecisionTreeClassifier` from PySpark's machine learning library (pyspark.ml). The trained model is returned.

One important feature of decision trees is that once the model is trained, we can extract the feature importances to identify the features that drive the predictions and those that are redundant. We can use the `featureImportances` attribute to achieve this task.

In the following code, we add a few more lines to this attribute to append the feature names to the feature importances and sort them in descending order:

```
[In]: feature_importances = model.featureImportances.toArray()
[In]: feature_names = spark_df.columns[:-1]
[In]: feature_importance_tuples = [(feature_name, importance) for feature_
      name, importance in zip(feature_names, feature_importances)]
[In]: sorted_feature_importance_tuples = sorted(feature_importance_tuples,
      key=lambda x: x[1], reverse=True)
[In]: for feature_name, importance in sorted_feature_importance_tuples:
          print(f"{feature_name}: {importance}")

[Out]:
petal length (cm): 0.52
petal width (cm): 0.50
sepal length (cm): 0.0
sepal width (cm): 0.0
```

In the preceding code, we first obtain the feature importances and their corresponding names, creating a list of tuples that pair each feature with its importance score. Next, we sort this list in descending order based on importance scores, enabling us to identify the most influential features. Finally, we print the sorted feature importances.

The output indicates that the petal length and width of the flowers are the driving force behind the model's predictions. In the PySpark model, the sepal length and width are redundant as their score is 0, respectively.

Step 5: Model evaluation

In this step, a custom function `evaluate_model` is defined. It takes the trained model and the test data as input. The function makes predictions using the model and evaluates its performance using the `MulticlassClassificationEvaluator` from PySpark. It returns the accuracy of the model and the predictions.

After calling the functions in step 6, we get an output indicating an accuracy of 94%. This indicates that the decision tree model is quite effective at making correct predictions on the test data. It correctly classifies the target variable for a large majority of instances, which is generally a good sign of the model's predictive power.

It's worth noting that the accuracy of the PySpark model is lower than that of Scikit-Learn, which was 100%. This difference arises because we are building models with default hyperparameters, and the two frameworks have distinct default settings. We will explore how to customize algorithm hyperparameters in Chapter 16.

Bringing It All Together

In the preceding sections of this chapter, we provided code that demonstrated each modeling step independently. However, in this section, our goal is to combine all the relevant code from those steps into a single code block. This allows data scientists to execute the code as a unified entity.

Scikit-Learn

```
[In]: from sklearn.datasets import load_iris
[In]: from sklearn.model_selection import train_test_split
[In]: from sklearn.tree import DecisionTreeClassifier
[In]: from sklearn.metrics import accuracy_score
```

```
[In]: def load_iris_dataset():
        """
        Load and return the Iris dataset features (X) and labels
        (y).

        Returns:
        X (numpy.ndarray): Feature matrix of the Iris dataset.
        y (numpy.ndarray): Target labels of the Iris dataset.
        """

        iris = load_iris()
        X = iris.data
        y = iris.target
        return X, y
[In]: def split_data(X, y, test_size=0.2, random_state=42):
        """
        Split data into training and test sets.

        Args:
        X (numpy.ndarray): Feature matrix of the dataset.
        y (numpy.ndarray): Target labels of the dataset.
        test_size (float): Proportion of data to include in the test
        set (default=0.2).
        random_state (int): Seed for random number generation (
        default=42).

        Returns:
        X_train (numpy.ndarray): Feature matrix of the training set.
        X_test (numpy.ndarray): Feature matrix of the test set.
        y_train (numpy.ndarray): Target labels of the training set.
        y_test (numpy.ndarray): Target labels of the test set.
        """

        X_train, X_test, y_train, y_test = train_test_split(X, y,
        test_size=test_size, random_state=random_state)
        return X_train, X_test, y_train, y_test
[In]: def build_decision_tree(X, y):
        """
```

```
        Build and return a decision tree classifier using the
        provided features (X) and labels (y).

        Args:
        X (numpy.ndarray): Feature matrix of the dataset.
        y (numpy.ndarray): Target labels of the dataset.

        Returns:
        clf (DecisionTreeClassifier): Trained decision tree
        classifier.
        """

        clf = DecisionTreeClassifier()
        clf.fit(X, y)
        return clf
[In]: def evaluate_model(clf, X_test, y_test):
        """
        Evaluate and return the accuracy score of the provided
        classifier (clf) on the test features (X_test) and labels
        (y_test).

        Args:
        clf (DecisionTreeClassifier): Trained decision tree
        classifier.
        X_test (numpy.ndarray): Feature matrix of the test set.
        y_test (numpy.ndarray): Target labels of the test set.

        Returns:
        accuracy (float): Accuracy score of the classifier on the
        test data.
        """

        y_pred = clf.predict(X_test)
        accuracy = accuracy_score(y_test, y_pred)
        return accuracy
[In]: X, y = load_iris_dataset()
[In]: X_train, X_test, y_train, y_test = split_data(X, y)
[In]: clf = build_decision_tree(X_train, y_train)
```

```
[In]: accuracy = evaluate_model(clf, X_test, y_test)
[In]: print("Accuracy:", accuracy)
```

PySpark

```
[In]: import pandas as pd
[In]: from pyspark.sql import SparkSession
[In]: from pyspark.ml.feature import VectorAssembler
[In]: from pyspark.ml.classification import DecisionTreeClassifier
[In]: from pyspark.ml.evaluation import MulticlassClassificationEvaluator
[In]: from sklearn.datasets import load_iris
[In]: def load_iris_dataset():
        """
        Load the Iris dataset and return it as a Pandas DataFrame.

        Returns:
        pandas.DataFrame: The Iris dataset as a Pandas DataFrame.
        """

        iris = load_iris()
        X = iris.data
        y = iris.target
        feature_names = iris.feature_names
        target_names = iris.target_names
        pandas_df = pd.DataFrame(X, columns=feature_names)
        pandas_df['label'] = y
        return pandas_df
[In]: def convert_to_spark_df(pandas_df, spark):
        """
        Convert a pandas DataFrame to a Spark DataFrame using the
        provided SparkSession.

        Args:
        pandas_df (pandas.DataFrame): The input pandas DataFrame.
        spark (pyspark.sql.SparkSession): The SparkSession.
```

```
        Returns:
        pyspark.sql.DataFrame: The input data as a Spark DataFrame.
        """
        return spark.createDataFrame(pandas_df)
[In]: def prepare_feature_vector(spark_df):
        """
        Prepare the feature vector for a Spark DataFrame.

        Args:
        spark_df (pyspark.sql.DataFrame): The input Spark DataFrame.

        Returns:
        pyspark.sql.DataFrame: The Spark DataFrame with a feature
        vector column.
        """
        assembler = VectorAssembler(inputCols=spark_df.columns[:-1],
        outputCol="features")
        transformed_data = assembler.transform(spark_df)
        return transformed_data
[In]: def split_data(transformed_data, train_ratio=0.8, seed=42):
        """
        Split the dataset into training and test sets.

        Args:
        transformed_data (pyspark.sql.DataFrame): The Spark
        DataFrame with feature vectors.
        train_ratio (float, optional): The ratio of data to use for
        training. Default is 0.8.
        seed (int, optional): The random seed for reproducibility.
        Default is 42.

        Returns:
        Tuple[pyspark.sql.DataFrame, pyspark.sql.DataFrame]: A tuple
        containing training and test Spark DataFrames.
        """
        train_data, test_data =
        transformed_data.randomSplit([train_ratio, 1 - train_ratio],
```

```
        seed=seed)
        return train_data, test_data
[In]: def build_decision_tree(train_data):
        """
        Build a decision tree classifier using the provided training
        data.

        Args:
        train_data (pyspark.sql.DataFrame): The training data as a
        Spark DataFrame.

        Returns:
        pyspark.ml.classification.DecisionTreeClassificationModel:
        The trained decision tree model.
        """

        dt = DecisionTreeClassifier(labelCol="label",
        featuresCol="features")
        model = dt.fit(train_data)
        return model
[In]: def evaluate_model(model, test_data):
        """
        Evaluate the provided decision tree model using the provided
        test data.

        Args:
        model
        (pyspark.ml.classification.DecisionTreeClassificationModel):
        The trained decision tree model.
        test_data (pyspark.sql.DataFrame): The test data as a Spark
        DataFrame.

        Returns:
        Tuple[float, pyspark.sql.DataFrame]: A tuple containing
        accuracy and prediction results.
        """
        predictions = model.transform(test_data)
        evaluator =
```

```
        MulticlassClassificationEvaluator(labelCol="label",
        predictionCol="prediction", metricName="accuracy")
        accuracy = evaluator.evaluate(predictions)
        return accuracy, predictionsc
[In]: spark = SparkSession.builder.appName("IrisDecisionTree").
      getOrCreate()
[In]: pandas_df = load_iris_dataset()
[In]: spark_df = convert_to_spark_df(pandas_df, spark)
[In]: transformed_data = prepare_feature_vector(spark_df)
[In]: train_data, test_data = split_data(transformed_data)
[In]: model = build_decision_tree(train_data)
[In]: accuracy, predictions = evaluate_model(model, test_data)
[In]: print("Accuracy:", accuracy)
```

Summary

In this chapter, we continued with classification as a form of supervised learning. We used Scikit-Learn and PySpark machine learning libraries to demonstrate how to create, train, and evaluate a decision tree classification model for predicting Iris flower species based on feature measurements. We also compared Pandas and PySpark in how they can be used to load and explore the data.

We also explored the advantages and limitations of decision tree classifiers. These models excel in capturing nonlinear relationships, making them valuable for complex datasets compared to linear models like logistic regression. However, they tend to overfit, are sensitive to data changes, and may exhibit class bias in imbalanced datasets. In the next chapter, we will introduce the concept of random forests to address these issues.

CHAPTER 9

Random Forest Classification with Scikit-Learn and PySpark

In this chapter, we continue with supervised learning tree-based classification, specifically random forests. We proceed by building, training, and evaluating a random forest classifier to classify the species of an Iris flower using the same dataset employed in the previous chapter. Previously, we emphasized that decision trees are powerful machine learning algorithms adept at classification tasks. Nonetheless, they can be susceptible to overfitting, especially when the tree grows excessively deep. To address this concern, ensemble methods such as random forests have gained popularity. These methods combine multiple decision trees to enhance classification performance. By executing the code presented in this chapter, we can compare the accuracy with that obtained from the decision tree. This helps determine how much random forests enhance the performance of tree-based models.

As in the previous chapter, we will employ two libraries: Scikit-Learn and PySpark. By comparing their Python code, we will highlight their similarities, facilitating the transition from Scikit-Learn to PySpark and harnessing the benefits of PySpark's distributed computing capabilities.

We explored the characteristics of the Iris dataset in the previous chapter. In summary, the dataset consists of 150 rows and 6 columns, encompassing four features (sepal length, sepal width, petal length, and petal width), a target variable represented both numerically (target) and in text form (target_names), and no null values.

A. Testas, *Distributed Machine Learning with PySpark*, https://doi.org/10.1007/978-1-4842-9751-3_9

Random Forest Classification

A random forest classifier classifies the species of an Iris flower based on the combination of multiple decision trees. Each decision tree in the random forest is trained on a different subset of the data, created through a process called bootstrap aggregating or bagging. During training, each decision tree in the random forest is exposed to a random subset of the features and a random subset of the training samples. This randomness introduces diversity among the decision trees, reducing overfitting and enhancing the model's performance.

When making predictions for a new Iris flower, each decision tree in the random forest independently predicts the species based on the features of the flower. The final prediction is determined through a majority vote or averaging of the individual predictions from all the decision trees in the forest.

Both Scikit-Learn and PySpark provide tools for building, training, and evaluating random forest algorithms. To start, let's compare these tools before we create models in each of the two libraries.

Scikit-Learn and PySpark Similarities for Random Forests

Here is a table that illustrates the classes and functions employed in Scikit-Learn alongside their corresponding equivalents in PySpark for constructing random forest classifiers:

Task	Scikit-Learn	PySpark
Data preparation		
Splitting data	train_test_split	randomSplit
Feature scaling	StandardScaler	StandardScaler
Categorical encoding	OneHotEncoder	OneHotEncoder
Missing data handling	SimpleImputer	Imputer
Feature vectorization	Not applicable	VectorAssembler
Model training		
Model class	RandomForestClassifier	RandomForestClassifier
Training	fit( )	fit( )

(*continued*)

Task	Scikit-Learn	PySpark
Model evaluation		
Accuracy	accuracy_score	MultiClassificationEvaluator
Precision	precision_score	
Recall	recall_score	
F1 score	f1_score	
Prediction		
Generate predictions	predict	transform

These comparisons show that both Scikit-Learn and PySpark follow the same modeling steps of data preparation, model training, model evaluation, and prediction. Interestingly enough, in some cases, even the names of functions or classes are similar. Examples include `StandardScaler` (used for standardizing features by removing the mean and scaling to unit variance), `OneHotEncoder` (used to convert categorical data into one-hot encoded features), `RandomForestClassifier` (which represents the random forest algorithm), and the `fit()` method (which is used to train the classifier).

In the next subsections, we translate these modeling steps into Scikit-Learn and PySpark code. It is important to note that we are building models with default hyperparameters, meaning that the results between the two platforms can differ. We demonstrate how to fine-tune the hyperparameters of the random forest model in Chapter 16.

Random Forests with Scikit-Learn

We start by using Scikit-Learn to demonstrate how to build, train, and evaluate a random forest model for the purpose of predicting the species of an Iris flower based on its feature measurements. We first import the necessary libraries and then proceed to define essential functions to ensure an intuitive execution of the Random Forest classification.

Step 1. Import necessary libraries

The following code imports the necessary functions and classes, including the load_iris function to load the Iris dataset, the train_test_split function to split data into training and testing sets, the RandomForestClassifier class to build the random forest model, and the accuracy_score function to evaluate the accuracy of predictions:

```
[In]: from sklearn.datasets import load_iris
[In]: from sklearn.model_selection import train_test_split
[In]: from sklearn.ensemble import RandomForestClassifier
[In]: from sklearn.metrics import accuracy_score
```

Step 2. The load_iris_dataset() function

This function loads and returns the features (X) and labels (y) of the Iris dataset. It uses the `load_iris()` function from Scikit-Learn to obtain the dataset and assigns the features to X and the labels to y:

```
[In]: def load_iris_dataset():
        iris = load_iris()
        X = iris.data
        y = iris.target
        return X, y
```

Step 3. The build_random_forest() function

This function builds and returns a random forest classifier using the features (X) and labels (y). It initializes a RandomForestClassifier object, assigns it to rf_classifier, and fits the classifier to the training data using the fit() method:

```
[In]: def build_random_forest(X, y):
        rf_classifier = RandomForestClassifier()
        rf_classifier.fit(X, y)
        return rf_classifier
```

Step 4. The evaluate_model() function

This function evaluates and returns the accuracy score of the random forest classifier (`rf_classifier`) on the test features (X_test) and labels (y_test). It predicts the labels for the test data using the `predict()` method of the classifier, compares the predicted labels with the true labels, and calculates the accuracy score using the `accuracy_score()` function:

```
[In]: def evaluate_model(rf_classifier, X_test, y_test):
        y_pred = rf_classifier.predict(X_test)
        accuracy = accuracy_score(y_test, y_pred)
        return accuracy
```

Step 5. Using the functions

We can now call the preceding functions using the actual Iris dataset:

5.1. Load the Iris dataset

```
[In]: X, y = load_iris_dataset()
```

5.2. Split the dataset into training and testing sets

```
[In]: X_train, X_test, y_train, y_test = train_test_split(X, y, test_
      size=0.2, random_state=42)
```

5.3. Build the random forest model

```
[In]: rf_classifier = build_random_forest(X_train, y_train)
```

5.4. Evaluate the model

```
[In]: accuracy = evaluate_model(rf_classifier, X_test, y_test)
```

5.5. Print the accuracy

```
[In]: print("Accuracy:", accuracy)
[Out]: 1.00
```

The output shows that the model has accurately predicted every case, yielding an accuracy rate of 100%. A comparison between the actual and predicted values for the first five rows demonstrates this perfect match between the predicted and actual values:

```
[In]: y_pred = rf_classifier.predict(X_test)
[In]: print("Predicted\tActual")
[In]: for i in range(5):
          print(f"{y_pred[i]}\t\t{y_test[i]}")
[Out]:
```

Predicted	Actual
1	1
0	0
2	2
1	1
1	1

An accuracy of 100% indicates that the Scikit-Learn random forest algorithm performed as well as the decision tree algorithm used in the preceding chapter, which also achieved 100% accuracy. Given that 100% is the maximum achievable accuracy, we cannot expect an improvement between the two model types. This, however, is not the case with PySpark as we will see in the next subsection.

Finally, we can use the `feature_importances_` attribute in Scikit-Learn to identify the primary drivers of the model's predictions. Running the following code reveals that the flower's petal length and width are the main contributors to the model's predictions:

```
# Get the feature importances and their corresponding names
feature_importances = rf_classifier.feature_importances_
feature_names = load_iris().feature_names

# Create a list of (feature_name, importance) tuples
feature_importance_tuples = [(feature_name, importance) for feature_name,
importance in zip(feature_names, feature_importances)]

# Sort the list by importance in descending order
sorted_feature_importance_tuples = sorted(feature_importance_tuples,
key=lambda x: x[1], reverse=True)

# Print the sorted feature importances
for feature_name, importance in sorted_feature_importance_tuples:
    print(f"{feature_name}: {importance}")

[Out]:
petal length (cm): 0.46
petal width (cm): 0.42
sepal length (cm): 0.09
sepal width (cm): 0.03
```

Random Forests with PySpark

In this subsection, we develop the PySpark code equivalent to the Scikit-Learn code provided in the preceding subsection. The modeling steps are largely the same: data preparation, model training, model evaluation, and prediction. Similar to Scikit-Learn, we start the model building process by importing the necessary libraries:

Step 1. Importing necessary libraries

```
[In]: from sklearn.datasets import load_iris
[In]: import pandas as pd
[In]: from pyspark.sql import SparkSession
[In]: from pyspark.ml.feature import VectorAssembler
[In]: from pyspark.ml.classification import RandomForestClassifier
[In]: from pyspark.ml.evaluation import MulticlassClassificationEvaluator
```

Step 2. The convert_to_spark_df() function

To be able to work with PySpark, we need to convert the Pandas DataFrame containing the Iris dataset to a PySpark DataFrame. The convert_to_spark_df() function takes a Pandas DataFrame (pandas_df) and a SparkSession (spark) as input. It converts the Pandas DataFrame to a Spark DataFrame using the createDataFrame() method of the SparkSession. The resulting Spark DataFrame is then returned:

```
[In]: def convert_to_spark_df(pandas_df, spark):
          return spark.createDataFrame(pandas_df)
```

Step 3. The build_random_forest() function

The build_random_forest() function takes a Spark DataFrame (data) as input. It creates a RandomForestClassifier() object rf with the label column set to label and the features column set to features. The function then fits the random forest model to the provided data using the fit method, resulting in a trained model. Finally, the trained model is returned:

```
[In]: def build_random_forest(data):
          rf = RandomForestClassifier(labelCol="label",
          featuresCol="features")
          model = rf.fit(data)
          return model
```

Step 4. The evaluate_model() function

The evaluate_model() function takes the trained random forest model (model) and a Spark DataFrame (data) as input. It uses the trained model to make predictions on the provided data using the transform() method. The predictions are stored in new DataFrame predictions. The function then creates a MulticlassClassificationEvaluator object evaluator with the label column set to label, the prediction column set to prediction, and the metric set to accuracy. It evaluates the accuracy of the predictions using the evaluate() method of the evaluator, resulting in an accuracy score. The function returns both the accuracy score and the predictions DataFrame:

```
[In]: def evaluate_model(model, data):
          predictions = model.transform(data)
          evaluator =
          MulticlassClassificationEvaluator(labelCol="label",
          predictionCol="prediction", metricName="accuracy")
          accuracy = evaluator.evaluate(predictions)
          return accuracy, predictions
```

Step 5. Using the functions

We are now in a position to call the previous functions with the aim of obtaining the accuracy metric to assess the model's performance:

5.1. Load the Iris dataset

```
[In]: iris = load_iris()
```

5.2. Access the feature data (X) and target labels (y)

```
[In]: X = iris.data
[In]: y = iris.target
```

5.3. Define column names

```
[In]: feature_names = iris.feature_names
[In]: target_names = iris.target_names
```

5.4. Create a DataFrame with column names

```
[In]: df = pd.DataFrame(X, columns=feature_names)
[In]: df['label'] = y
```

5.5. Create a SparkSession

```
[In]: spark = SparkSession.builder.appName("IrisRandomForest").
getOrCreate()
```

5.6. Convert the Pandas DataFrame to a PySpark DataFrame

```
[In]: data = convert_to_spark_df(df, spark)
```

5.7. Prepare the feature vector

```
[In]: assembler = VectorAssembler(inputCols=data.columns[:-1],
      outputCol="features")
[In]: transformed_data = assembler.transform(data)
```

5.8. Split the dataset into training and testing sets

```
[In]: train_data, test_data = transformed_data.randomSplit([0.8, 0.2],
      seed=42)
```

5.9. Build the random forest model

```
[In]: model = build_random_forest(train_data)
```

5.10. Evaluate the model and get predictions

```
[In]: accuracy, predictions = evaluate_model(model, test_data)
```

At this stage, we are already able to compare a sample of predicted and actual data to have some idea of how the model is doing:

```
[In]: def print_actual_vs_predicted(predictions, n_rows=5):
          results = predictions.select("label",
          "prediction").limit(n_rows).toPandas()
          print(results)
[In]: print_actual_vs_predicted(predictions, n_rows=5)
[Out]:
```

Label	Prediction
0	0
0	0
0	0
0	0
0	0

The output indicates that for this sample of five records of the Iris setosa, the model successfully matched each prediction with the actual label. However, as shown in the next output, the overall accuracy is less than 100% because the accuracy metric is based on the entire sample of data points.

5.11. Print the accuracy

```
[In]: print(accuracy)
[Out]: 0.97
```

The output indicates that taking the entire sample of data points, the random forest model achieves a good accuracy of approximately 97%. As a point of reference, in the previous chapter, the decision tree classifier achieved a lower accuracy of 94%. Therefore, it was accurate to hypothesize in the introduction to this chapter that random forests can outperform decision tree algorithms by combining multiple decision trees to improve prediction accuracy.

Finally, we can determine the features that have contributed the most to achieving this accuracy by utilizing the featureImportances attribute in PySpark.

```
# Get the feature importances and their corresponding names
[In]: feature_importances = model.featureImportances.toArray()
[In]: feature_names = data.columns[:-1]

# Create a list of (feature_name, importance) tuples
[In]: feature_importance_tuples = [(feature_name, importance) for feature_
      name, importance in zip(feature_names, feature_importances)]

# Sort the list by importance in descending order
[In]: sorted_feature_importance_tuples = sorted(feature_importance_tuples,
      key=lambda x: x[1], reverse=True)

# Print the sorted feature importances
[In]: for feature_name, importance in sorted_feature_importance_tuples:
          print(f"{feature_name}: {importance}")
.[Out]:

petal width (cm): 0.52
petal length (cm): 0.39
sepal length (cm): 0.07
sepal width (cm): 0.02
```

The output indicates that the petal width and length contributed the most to the model's predictions as they have the most significant importance scores.

Bringing It All Together

In previous sections of this chapter, we presented code examples that illustrated each modeling step individually. However, in this section, our objective is to consolidate all the relevant code from those steps into a single code block. This enables data scientists to execute the code as a cohesive unit.

Scikit-Learn

```
[In]: from sklearn.datasets import load_iris
[In]: from sklearn.model_selection import train_test_split
[In]: from sklearn.ensemble import RandomForestClassifier
[In]: from sklearn.metrics import accuracy_score

[In]: def load_iris_dataset():
          """
        Load and return the Iris dataset features and labels.

        Returns:
            X (array-like): Features (attributes) of the Iris
            dataset.
            y (array-like): Target labels of the Iris dataset.
        """

        iris = load_iris()
        X = iris.data
        y = iris.target
        return X, y

[In]: def build_random_forest(X, y):

        """
        Build and return a Random Forest classifier using the
        provided features and labels.
```

```
        Args:
            X (array-like): Features (attributes) of the dataset.
            y (array-like): Target labels of the dataset.

        Returns:
            RandomForestClassifier: A trained Random Forest
            classifier.
        """

        rf_classifier = RandomForestClassifier()
        rf_classifier.fit(X, y)
        return rf_classifier

[In]: def evaluate_model(rf_classifier, X_test, y_test):
        """
        Evaluate and return the accuracy score of the provided
        classifier on test data.

        Args:
            rf_classifier (RandomForestClassifier): A trained Random
            Forest classifier.
            X_test (array-like): Test features (attributes).
            y_test (array-like): True labels for the test data.

        Returns:
            float: Accuracy score of the classifier on the test
            data.
        """

        y_pred = rf_classifier.predict(X_test)
        accuracy = accuracy_score(y_test, y_pred)
        return accuracy

# Load the Iris dataset
[In]: X, y = load_iris_dataset()

# Split the dataset into training and test sets
[In]: X_train, X_test, y_train, y_test = train_test_split(X, y,
      test_size=0.2, random_state=42)
```

```
# Build the random forest model
[In]: rf = build_random_forest(X_train, y_train)

# Evaluate the model
[In]: accuracy = evaluate_model(rf_classifier, X_test, y_test)

# Print the accuracy
[In]: print("Accuracy:", accuracy)

# Print the first 5 rows of actual and predicted labels
[In]: y_pred = rf_classifier.predict(X_test)
[In]: print("Predicted\tActual")
          for i in range(5):
          print(f"{y_pred[i]}\t\t{y_test[i]}")
```

PySpark

```
[In]: from sklearn.datasets import load_iris
[In]: import pandas as pd
[In]: from pyspark.sql import SparkSession
[In]: from pyspark.ml.feature import VectorAssembler
[In]: from pyspark.ml.classification import RandomForestClassifier
[In]: from pyspark.ml.evaluation import MulticlassClassificationEvaluator
[In]: def convert_to_spark_df(pandas_df, spark):
          """
          Convert a pandas DataFrame to a Spark DataFrame.

          Parameters:
              pandas_df (pd.DataFrame): The pandas DataFrame to be
              converted.
              spark (SparkSession): The SparkSession object to create
              a Spark DataFrame.

          Returns:
              DataFrame: A Spark DataFrame created from the provided
              pandas DataFrame.
              """

          return spark.createDataFrame(pandas_df)
```

```
[In]: def build_random_forest(data):
        """
        Build and train a Random Forest classifier.

        Parameters:
            data (DataFrame): The Spark DataFrame containing the
            training data.

        Returns:
            RandomForestClassificationModel: The trained Random
            Forest model.
        """

        rf = RandomForestClassifier(labelCol="label",
             featuresCol="features")
        model = rf.fit(data)
        return model

[In]: def evaluate_model(model, data):
        """
        Evaluate the performance of a Random Forest classifier on
        the provided data.

        Parameters:
          model (RandomForestClassificationModel): The trained
          Random Forest model.
          data (DataFrame): The Spark DataFrame containing the data
          to be evaluated.

        Returns:
            tuple: A tuple containing the accuracy score and a
            DataFrame with predictions.
        """
        predictions = model.transform(data)
        evaluator =
           MulticlassClassificationEvaluator(labelCol="label",
           predictionCol="prediction", metricName="accuracy")
```

```
        accuracy = evaluator.evaluate(predictions)
        return accuracy, predictions

[In]: def print_actual_vs_predicted(predictions, n_rows=5):
        """
        Print a comparison of actual labels vs. predicted labels
        from a DataFrame of predictions.

        Parameters:
            predictions (DataFrame): The DataFrame containing the
            actual and predicted labels.
            n_rows (int): The number of rows to display. Default is
            5.

        Returns:
            None
        """
        results = predictions.select("label",
                "prediction").limit(n_rows).toPandas()
        print(results)

# Load the Iris dataset
[In]: iris = load_iris()

# Access the feature data (X) and target labels (y)
[In]: X = iris.data
[In]: y = iris.target
# Define column names
[In]: feature_names = iris.feature_names
[In]: target_names = iris.target_names

# Create a DataFrame with column names
[In]: df = pd.DataFrame(X, columns=feature_names)
[In]: df['label'] = y

# Create a SparkSession
[In]: spark = SparkSession.builder.appName("IrisRandomForest").
      getOrCreate()
```

```
# Convert the Pandas DataFrame to PySpark DataFrame
[In]: data = convert_to_spark_df(df, spark)

# Prepare the feature vector
[In]: assembler = VectorAssembler(inputCols=data.columns[:-1],
      outputCol="features")
[In]: transformed_data = assembler.transform(data)

# Split the dataset into training and test sets
[In]: train_data, test_data = transformed_data.randomSplit([0.8, 0.2],
      seed=42)

# Build the random forest model
[In]: model = build_random_forest(train_data)

# Evaluate the model and get predictions
[In]: accuracy, predictions = evaluate_model(model, test_data)

# Print the accuracy
[In]: print("Accuracy:", accuracy)

# Print the first 5 rows of actual vs predicted labels
[In]: print_actual_vs_predicted(predictions, n_rows=5)
```

Summary

This chapter introduced classification using the random forest algorithm on Iris data. It employed the Pandas, Scikit-Learn, and PySpark libraries for data preprocessing and model construction. The chapter showed that Scikit-Learn and PySpark are consistent in terms of the modeling steps, even though syntax may differ. Similar to the logistic regression and decision tree classification chapters, this chapter demonstrated building a classification model in a modularized format. The code was organized into functions for intuitive execution and follow-through.

The chapter also highlighted random forest's advantage over decision trees, particularly in reducing overfitting. We observed that random forests indeed improved classification compared to decision tree classifiers. In the next chapter, we will continue with classification, introducing a new algorithm: support vector machines.

CHAPTER 10

Support Vector Machine Classification with Pandas, Scikit-Learn, and PySpark

This chapter explores support vector machines (SVMs), widely employed supervised learning algorithms recognized for their effectiveness in binary classification tasks. SVMs aim to find an optimal hyperplane (a decision plane that separates objects with different class memberships) that maximizes the margin between data points of different classes. The hyperplane acts as a decision boundary, with one class on each side. The margin represents the perpendicular distance between the hyperplane and the closest points of each class. A larger margin indicates a better separation, while a smaller margin suggests a less optimal decision boundary.

One advantage of SVMs is their capability to handle both linearly separable and nonlinearly separable data. In the case of nonlinear data, SVMs utilize the "kernel trick" to transform the data into a higher-dimensional space where linear separation becomes feasible. This transformation involves increasing the number of dimensions or features beyond the original feature space. Typically, this is achieved using a kernel function, which implicitly applies a nonlinear transformation to the data without the need to explicitly compute and store the transformed feature vectors. In essence, the kernel trick simplifies data transformation, saving both memory and computational resources.

However, SVMs heavily rely on the choice of the kernel function and its associated parameters. Selecting an inappropriate kernel or misconfiguring the parameters can result in suboptimal outcomes. Additionally, training an SVM, particularly with

A. Testas, *Distributed Machine Learning with PySpark*, https://doi.org/10.1007/978-1-4842-9751-3_10

nonlinear kernels or large datasets, can be computationally intensive and time-consuming. As the dataset size grows, so does the training time. SVMs are often considered "black-box" algorithms due to their limited interpretability. Furthermore, while binary classification with SVMs is well established, extending them to multiclass classification can be challenging.

In this chapter, we build, train, and evaluate an SVM classifier and use it to predict whether a tumor is malignant or benign. This is based on the Breast Cancer Wisconsin (Diagnostic) dataset, which is widely used for exploring and evaluating classification algorithms. It offers a real-world scenario where machine learning techniques can aid in distinguishing between benign and malignant tumors based on measurable characteristics.

We will make use of two libraries: Scikit-Learn and PySpark. Through a comparison of their Python code, we aim to emphasize their similarities, making it easier for data scientists to transition from Scikit-Learn to PySpark and take advantage of PySpark's distributed computing capabilities. Additionally, we will conduct a comparison between Pandas and PySpark to demonstrate their similarity, thereby highlighting PySpark's suitability for handling large-scale data projects.

The Dataset

The cancer dataset, also known as the Breast Cancer Wisconsin (Diagnostic) dataset, contains measurements of various features related to breast cancer cell characteristics. The goal is to predict whether a tumor is malignant (cancerous) or benign (noncancerous) based on these features.

The dataset provides information on 30 different attributes, including mean radius, mean texture, and mean smoothness. These features are computed from digitized images of fine needle aspirates (FNA) of breast mass. For each instance, there is a corresponding target label indicating the class: 0 for benign and 1 for malignant.

The dataset is available for download on the UCI Machine Learning website. The following are the contributor's name, donation date, dataset name, site name, and the URL from which the dataset can be downloaded:

Title: Breast Cancer Wisconsin (Diagnostic)
Source: UCI Machine Learning
URL: `https://archive.ics.uci.edu/dataset/17/breast+cancer+wisconsin+diagnostic`

Contributor: William Wolberg et al.

Date: 1995

The dataset is also available in Scikit-Learn. The following code demonstrates how to access it:

Step 1: Import the necessary libraries

```
[In]: import pandas as pd
[In]: from sklearn.datasets import load_breast_cancer
```

Step 2: Load the Breast Cancer Wisconsin dataset

```
[In]: data = load_breast_cancer()
```

Step 3: Create a Pandas DataFrame for the features

```
[In]: pandas_df = pd.DataFrame(data.data, columns=data.feature_names)
```

Step 4: Add the target variable as a column in the DataFrame

```
[In]: pandas_df['target'] = data.target
```

We can now explore the dataset. First, we can use the Pandas shape attribute to retrieve the dimensions (number of rows and columns) of the DataFrame:

```
[In]: pandas_df.shape
[Out]: (569, 31)
```

The dataset has 569 rows and 31 columns (30 features plus the target variable).

We can use the Pandas columns attribute to access the column labels or names of the DataFrame:

```
[In]: print(pandas_df.columns)
[Out]:
Index(['mean radius', 'mean texture', 'mean perimeter', 'mean area', 'mean
smoothness', 'mean compactness', 'mean concavity', 'mean concave points',
'mean symmetry', 'mean fractal dimension', 'radius error', 'texture error',
'perimeter error', 'area error', 'smoothness error', 'compactness error',
'concavity error', 'concave points error', 'symmetry error', 'fractal
dimension error','worst radius', 'worst texture', 'worst perimeter',
'worst area', 'worst smoothness', 'worst compactness', 'worst concavity',
'worst concave points', 'worst symmetry', 'worst fractal dimension',
'target'], dtype='object')
```

Next, we can select and print a subset of rows and columns:

```
[In]: selected_columns = pandas_df.iloc[0:5, 0:4]
[In]: print(selected_columns)
[Out]:
```

	mean radius	mean texture	mean perimeter	mean area
0	17.99	10.38	122.8	1001
1	20.57	17.77	132.9	1326
2	19.69	21.25	130	1203
3	11.42	20.38	77.58	386.1
4	20.29	14.34	135.1	1297

We can also output an array containing the unique values of the target column:

```
[In]: pandas_df['target'].unique()
[Out]: [0 1]
```

The output indicates that the target variable is either 0 (tumor is benign) or 1 (tumor is malignant or cancerous).

To perform exploratory data analysis using PySpark, we need to first create a PySpark DataFrame. The easiest way to do this is to convert the pandas_df DataFrame to PySpark using the PySpark `createDataFrame()` method:

```
[In]: from pyspark.sql import SparkSession
[In]: spark = SparkSession.builder.getOrCreate()
[In]: spark_df = spark.createDataFrame(pandas_df)
```

PySpark does not have a built-in shape attribute like Pandas to retrieve the dimensions (number of rows and columns) of the spark_df. Therefore, we need to perform two operations: `count()` and `len()` to return the number of rows and columns, respectively:

```
[In]: spark_df_shape = (spark_df.count(), len(spark_df.columns))
[In]: print(spark_df_shape)
[Out]: (569, 31)
```

Similar to Pandas, we can use the PySpark columns attribute to print the column names of the DataFrame:

```
[In]: print(spark_df.columns)
[Out]:
['mean radius', 'mean texture', 'mean perimeter', 'mean area', 'mean
smoothness', 'mean compactness', 'mean concavity', 'mean concave points',
'mean symmetry', 'mean fractal dimension', 'radius error', 'texture error',
'perimeter error', 'area error', 'smoothness error', 'compactness error',
'concavity error', 'concave points error', 'symmetry error', 'fractal
dimension error', 'worst radius', 'worst texture', 'worst perimeter', 'worst
area', 'worst smoothness', 'worst compactness', 'worst concavity', 'worst
concave points', 'worst symmetry', 'worst fractal dimension', 'target']
```

We can select and print a subset of rows and columns:

```
[In]: selected_columns = spark_df.columns[0:4]
[In]: spark_df.select(selected_columns).show(5)
[Out]:
```

mean radius	mean texture	mean perimeter	mean area
17.99	10.38	122.8	1001
20.57	17.77	132.9	1326
19.69	21.25	130	1203
11.42	20.38	77.58	386.1
20.29	14.34	135.1	1297

Finally, we can use the distinct() function to get the unique values of the target column:

```
[In]: spark_df.select('target').distinct().show()
[Out]:
```

target
0
1

There are two values for the target variable: 0 indicating a benign tumor and 1 indicating a cancerous one.

Support Vector Machine Classification

The exploratory data analysis done in the previous section was helpful as it helped us learn about the structure and characteristics of the data.

This section introduces the implementation of an SVM classifier using Scikit-Learn and PySpark libraries for the classification of breast cancer. The code utilizes the Breast Cancer Wisconsin dataset, which is loaded and split into training and testing sets. The features are then scaled using `StandardScaler` to ensure consistent scaling across the dataset. In SVMs, feature scaling is not strictly required but is recommended, as it can improve model performance and convergence speed. SVM tries to find the optimal hyperplane that separates the data points of different classes with the maximum margin.

An SVM model with a linear kernel is trained on the scaled training data. The trained model is used to make predictions on the test set. Various evaluation metrics, including accuracy, precision, recall, F1 score, and area under the receiver operating characteristic (ROC) curve, are calculated to assess the performance of the model. Finally, the evaluation metrics are printed.

Linear SVM with Scikit-Learn

We start by building, training, and evaluating the linear SVM classifier using the Scikit-Learn library. We follow the following steps:

Step 1: Import the necessary libraries

```
[In]: from sklearn.datasets import load_breast_cancer
[In]: from sklearn.model_selection import train_test_split
[In]: from sklearn.svm import SVC
```

```
[In]: from sklearn.metrics import accuracy_score, precision_score,
      recall_score, f1_score, confusion_matrix, roc_auc_score
[In]: from sklearn.preprocessing import StandardScaler
```

These imports provide the necessary functions and tools to load the Breast Cancer Wisconsin dataset, split it into training and testing sets, apply feature scaling, train an SVM model, and evaluate its performance using various metrics.

The `load_breast_cancer` function is employed to load the Breast Cancer Wisconsin dataset. The `train_test_split` function is used to split the dataset into training and testing subsets, which is essential for evaluating the performance of the SVM classifier. The SVC (support vector classifier) class is an implementation of the linear SVM algorithm for classification tasks. The `accuracy_score`, `precision_score`, `recall_score`, `f1_score`, `confusion_matrix`, and `roc_auc_score` functions are used to evaluate the performance of the SVM model. Finally, the `StandardScaler` class is used to standardize the features by removing the mean and scaling to unit variance. This preprocessing step ensures that the features have similar scales.

Step 2: Load the Breast Cancer Wisconsin dataset

```
[In]: data = load_breast_cancer()
[In]: X = data.data
[In]: y = data.target
```

This code loads the breast cancer dataset, separates the input features into X, and assigns the corresponding target values to y.

Step 3: Split the data into training and testing sets

```
[In]: X_train, X_test, y_train, y_test = train_test_split(X, y,
      test_size=0.2, random_state=42)
```

In this code, the `train_test_split()` function is used to split the data into training and testing datasets. This function takes the input features X and the target values y as arguments and splits them into four separate datasets: X_train, X_test, y_train, and y_test. The `test_size` parameter is set to 0.2, which means that 20% of the data will be allocated for testing, and the remaining 80% will be used for training the model. The `random_state` parameter is set to 42. This parameter ensures reproducibility by fixing the random seed. It means that each time you run the code, you will get the same split of data into training and testing sets.

Step 4: Scale the features using StandardScaler

```
[In]: scaler = StandardScaler()
[In]: X_train_scaled = scaler.fit_transform(X_train)
[In]: X_test_scaled = scaler.transform(X_test)
```

In this code, the `StandardScaler()` class from the Scikit-Learn library is used to standardize the feature data. First, an instance of the `StandardScaler` class is created and assigned to the variable scaler. Next, the `fit_transform()` method is called on the scaler object with X_train as the argument. This method computes the mean and standard deviation of each feature in the X_train dataset and then applies the transformation to standardize the data. The standardized feature data is assigned to the variable X_train_scaled.

After standardizing the training data, the `transform()` method is called on the scaler object with X_test as the argument. This method applies the same transformation to the test data using the mean and standard deviation computed from the training data. The standardized test data is assigned to the variable X_test_scaled.

Step 5: Train a linear SVM model

```
[In]: svm = SVC(kernel='linear', random_state=42)
[In]: svm.fit(X_train_scaled, y_train)
```

In this code, an SVM classifier is created and trained on the standardized training data. First, an instance of the SVC class from the Scikit-Learn library is created and assigned to the variable svm. The SVC class represents the SVM classifier. The kernel parameter is set to linear, indicating that a linear kernel function will be used for the SVM. The `random_state` parameter is set to 42, which ensures reproducibility by fixing the random seed for the SVM.

Next, the `fit()` method is called on the svm object, with the standardized training data (X_train_scaled) and the corresponding target values (y_train) as arguments. This method trains the SVM classifier on the provided data.

Step 6: Make predictions on the test set

```
[In]: y_pred = svm.predict(X_test_scaled)
```

In this code, the trained SVM classifier is used to predict the target values for the standardized test data. The `predict()` method is called on the SVM object (svm) with the standardized test data (X_test_scaled) as the argument. This method applies the trained SVM model to the test data and returns the predicted target values.

Step 7: Calculate evaluation metrics

```
[In]: accuracy = accuracy_score(y_test, y_pred)
[In]: precision = precision_score(y_test, y_pred)
[In]: recall = recall_score(y_test, y_pred)
[In]: f1 = f1_score(y_test, y_pred)
[In]: cm = confusion_matrix(y_test, y_pred)
[In]: auc = roc_auc_score(y_test, y_pred)
```

In this code, several performance metrics are computed to evaluate the predictions made by the SVM classifier (svm) on the test data:

- Accuracy: The `accuracy_score()` function is used to calculate the accuracy of the predictions. It compares the predicted labels (y_pred) with the true labels (y_test) and returns the fraction of correctly predicted instances.
- Precision: The `precision_score()` function is used to compute the precision of the predictions. Precision measures the proportion of correctly predicted positive instances (true positives) out of all instances predicted as positive (true positives + false positives). It provides an indication of the classifier's ability to avoid false positive predictions.
- Recall: The `recall_score()` function is used to calculate the recall of the predictions. Recall, also known as sensitivity or true positive rate, measures the proportion of correctly predicted positive instances (true positives) out of all actual positive instances (true positives + false negatives). It indicates the classifier's ability to identify positive instances.
- F1 score: The `f1_score()` function is used to compute the F1 score, which is the weighted mean of precision and recall. It provides a single metric that balances both precision and recall, giving an overall performance measure.
- Confusion matrix: The `confusion_matrix()` function is used to create a confusion matrix, which is a table that summarizes the performance of the classification model. It shows the counts of true positive, true negative, false positive, and false negative predictions.

- Area under the ROC curve (AUC): The `roc_auc_score()` function is used to calculate the area under the receiver operating characteristic (ROC) curve. The ROC curve represents the trade-off between the true positive rate and the false positive rate. A higher AUC indicates better classification performance.

Step 8: Print the evaluation metrics

```
[In]: print("Area Under ROC:", auc)
[In]: print("Accuracy:", accuracy)
[In]: print("Precision:", precision)
[In]: print("Recall:", recall)
[In]: print("F1 Score:", f1)
[In]: print("Confusion Matrix:")
[In]: print(cm)
[Out]:
Area Under ROC: 0.96
Accuracy: 0.96
Precision: 0.97
Recall: 0.96
F1 Score: 0.96
Confusion Matrix:
[[41  2]
 [ 3 68]]
```

The interpretation of this output is as follows:

- Area Under ROC (AUC): A value of approximately 0.96 indicates that the SVM classifier performs well in distinguishing between the positive and negative classes. The higher the AUC, the better the classifier's ability to correctly classify instances.
- Accuracy: An accuracy of 0.96 suggests that the SVM classifier correctly predicts the class labels for approximately 96% of the instances in the test data.
- Precision: With a precision score of 0.97, the SVM classifier has a high proportion of true positive predictions (correctly predicted positive instances) compared to the total instances it classified as positive. This indicates a low false positive rate.

- Recall: The recall score of 0.96 indicates that the SVM classifier identifies approximately 96% of the actual positive instances correctly. It has a low false negative rate.
- F1 Score: The F1 score of 0.96 combines precision and recall into a single metric. It provides a balance between precision and recall, indicating a good overall performance of the SVM classifier.
- Confusion Matrix: The confusion matrix is presented in the following format:

  [[True Negative False Positive]

  [False Negative True Positive]]

In this case, the SVM classifier predicted 41 true negatives, 2 false positives, 3 false negatives, and 68 true positives.

Linear SVM with PySpark

We can now write PySpark code to build, train, and evaluate a random forest classifier, similar to the one we constructed with Scikit-Learn using the same cancer dataset. Just as we did in Scikit-Learn, the model is constructed with the default hyperparameters. Since the two platforms use different default hyperparameters, the results between them may vary. In Chapter 16, we demonstrate how to fine-tune the hyperparameters of an algorithm.

As with Scikit-Learn, importing the necessary libraries is our starting point:

Step 1: Import the necessary libraries and modules

```
[In]: from pyspark.ml.feature import StandardScaler
[In]: from pyspark.ml.classification import LinearSVC
[In]: from pyspark.ml.evaluation import BinaryClassificationEvaluator
[In]: from pyspark.ml.linalg import Vectors
[In]: from pyspark.sql import functions as F
[In]: from pyspark.sql import SparkSession
```

By importing these libraries and modules, we can access the necessary functionalities and classes to perform tasks related to data preprocessing, machine learning algorithm, evaluation, and data manipulation within the PySpark framework.

The StandardScaler class is a transformer that will be used for standardizing features in our cancer dataset. The LinearSVC (Linear Support Vector Classifier) is the machine learning algorithm used for classification. The BinaryClassificationEvaluator is used to evaluate the performance of the model. The Vectors module provides data structures for handling vector data, while F provides a collection of built-in functions that can be applied to Spark DataFrames or columns for data manipulation and transformation.

Step 2: Create a SparkSession

```
[In]: spark = SparkSession.builder.getOrCreate()
```

Step 3: Load the Breast Cancer Wisconsin dataset

```
[In]: data = load_breast_cancer()
```

Step 4: Convert the NumPy arrays to lists

```
[In]: X = [Vectors.dense(x) for x in data.data.tolist()]
[In]: y = data.target.tolist()
```

These two lines of code transform the feature data and target labels from the breast cancer dataset into a suitable format (X and y) that can be used for training the SVM classifier.

Step 5: Convert the lists to Spark DataFrames

```
[In]: spark_df = spark.createDataFrame(zip(X, y), ["features", "label"])
```

This code creates a Spark DataFrame from the feature data (X) and target labels (y). zip(X, y) combines the feature data X and target labels y into a list of tuples, where each tuple contains a feature vector and its corresponding label. The createDataFrame() method creates a Spark DataFrame (spark_df) by taking two arguments: the zipped list of tuples and the column names ["features", "label"].

Step 6: Split the data into training and testing sets

```
[In]: train_data, test_data = spark_df.randomSplit([0.8, 0.2], seed=42)
```

This code randomly splits the spark_df into two subsets: a training set (train_data) and a testing set (test_data). The randomSplit() method splits the data based on the provided weights: 80% will be allocated to the training set (train_data), and 20% will be allocated to the testing set (test_data). The seed=42 sets the random seed to ensure that the same split is obtained every time the code is run.

Step 7: Scale the features using `StandardScaler`

```
[In]: scaler = StandardScaler(inputCol="features",
      outputCol="scaledFeatures")
[In]: scaler_model = scaler.fit(train_data)
[In]: train_data_scaled = scaler_model.transform(train_data)
[In]: test_data_scaled = scaler_model.transform(test_data)
```

The `StandardScaler` scales the features of the cancer dataset. This process helps to normalize the features and bring them to a similar scale. The inputCol="features" specifies the name of the input column containing the features in the DataFrame, while the outputCol="scaledFeatures" specifies the name of the output column that will contain the scaled features in the transformed DataFrame. The transformed DataFrame will have a new column named "scaledFeatures" that will store the scaled values.

The next line of code, `scaler_model = scaler.fit(train_data)`, fits the `StandardScaler` object to the training data. It computes the mean and standard deviation of the feature columns in the training set, which will be used to scale the features.

Finally, `train_data_scaled = scaler_model.transform(train_data)` and `test_data_scaled = scaler_model.transform(test_data)` apply the scaling transformation to the training and testing data, respectively. It uses the trained scaler_model to transform the feature columns in the train_data and test_data DataFrames. The transformed DataFrames, train_data_scaled and test_data_scaled, will contain an additional column named "scaledFeatures" with the scaled values of the original features.

Step 8: Train an SVM model

```
[In]: svm = LinearSVC()
[In]: svm_model = svm.fit(train_data_scaled)
```

The `LinearSVC` is the classifier algorithm in PySpark that implements a linear SVM for binary classification. It seeks to find the hyperplane that best separates the data points of different classes by maximizing the margin between the classes. svm is an instance of the `LinearSVC` classifier.

The next line of code, `svm_model = svm.fit(train_data_scaled)`, trains the `LinearSVC` model using the training data that has been scaled using the `StandardScaler`. The `fit(train_data_scaled)` is a method of the `LinearSVC` object that fits the model to the training data. It takes the training data (train_data_scaled) as input and learns the parameters of the SVM model to best fit the data. The fitted model is stored in the svm_model variable, which can be used later for making predictions on new, unseen data.

Step 9: Make predictions on the test set

```
[In]: predictions = svm_model.transform(test_data_scaled)
```

This code generates predictions for the test data using the trained SVM model (svm_model) in PySpark. The svm_model is the trained `LinearSVC` model that was obtained by fitting the model to the scaled training data. The `transform(test_data_scaled)` is a method of the trained model that applies the learned model to the test data (test_data_scaled) to generate predictions. The output of the `transform()` method is a DataFrame (predictions) that contains the original columns of the test data along with additional columns representing the predicted labels and other relevant information.

Step 10: Evaluate the model

To evaluate the performance of the linear SVM binary classifier, we calculate several evaluation metrics, beginning with the area under the ROC curve:

- Calculate the AUC (Area Under the Curve):

  ```
  [In]: evaluator = BinaryClassificationEvaluator(labelCol="label",
        rawPredictionCol="rawPrediction")
  [In]: area_under_roc = evaluator.evaluate(predictions)
  ```

  The `BinaryClassificationEvaluator` is an evaluator in PySpark that is specifically designed for binary classification problems. This suits our case as we are predicting whether the tumor is benign (0) or malignant (1). The labelCol="label" specifies the name of the column in the predictions DataFrame that contains the true labels of the test data. The rawPredictionCol="rawPrediction" specifies the name of the column in the predictions DataFrame that contains the raw predictions or scores produced by the model.

  The next line of code, area_under_roc = evaluator.evaluate(predictions), calculates the area under the receiver operating characteristic (ROC) curve (AUC) using the `BinaryClassificationEvaluator`.

- Calculate the other evaluation metrics:

```
[In]: tp = predictions.filter("label = 1 AND prediction = 1").count()
[In]: fp = predictions.filter("label = 0 AND prediction = 1").count()
[In]: tn = predictions.filter("label = 0 AND prediction = 0").count()
[In]: fn = predictions.filter("label = 1 AND prediction = 0").count()
[In]: accuracy = (tp + tn) / (tp + tn + fp + fn)
[In]: precision = tp / (tp + fp)
[In]: recall = tp / (tp + fn)
[In]: f1 = 2 * (precision * recall) / (precision + recall)
[In]: confusion_matrix = [[tn, fp], [fn, tp]]
```

The `BinaryClassificationEvaluator` in PySpark ML calculates the AUC but lacks built-in functions for computing other metrics such as accuracy, precision, recall, F1 score, and the confusion matrix. Consequently, we calculate these metrics manually using the provided code. This code computes various additional evaluation metrics based on the predictions made by the SVM model on the test data. Let's break it down step by step:

- `tp = predictions.filter("label = 1 AND prediction = 1").count()` counts the number of true positive (tp) predictions. It filters the predictions DataFrame to select the rows where the true label (label) is 1 and the predicted label (prediction) is also 1, and then counts the number of such rows.
- `fp = predictions.filter("label= 0 AND prediction = 1").count()` counts the number of false positive (fp) predictions. It filters the predictions DataFrame to select the rows where the true label is 0 and the predicted label is 1, and then counts the number of such rows.
- `tn = predictions.filter("label = 0 AND prediction = 0").count()` counts the number of true negative (tn) predictions. It filters the predictions DataFrame to select the rows where the true label is 0 and the predicted label is also 0, and then counts the number of such rows.

- `fn = predictions.filter("label = 1 AND prediction = 0").count()` counts the number of false negative (fn) predictions. It filters the predictions DataFrame to select the rows where the true label is 1 and the predicted label is 0, and then counts the number of such rows.

These calculations allow us to compute the four values needed for evaluation metrics: true positives, false positives, true negatives, and false negatives.

- `accuracy = (tp + tn) / (tp + tn + fp + fn)` calculates the accuracy of the model. It divides the sum of true positives and true negatives by the total number of predictions, which includes true positives, true negatives, false positives, and false negatives.
- `precision = tp / (tp + fp)` calculates the precision of the model. It divides the true positives by the sum of true positives and false positives. Precision measures the proportion of correctly predicted positive instances out of the total predicted positive instances.
- `recall = tp / (tp + fn)` calculates the recall (also known as sensitivity or true positive rate) of the model. It divides the true positives by the sum of true positives and false negatives. Recall measures the proportion of correctly predicted positive instances out of the total actual positive instances.
- `f1 = 2 * (precision * recall) / (precision + recall)` calculates the F1 score, which is the weighted mean of precision and recall. It provides a single metric that balances precision and recall.
- `confusion_matrix = [[tn, fp], [fn, tp]]` creates a confusion matrix as a list of lists. The first list represents the true negative (tn) and false positive (fp) counts, and the second list represents the false negative (fn) and true positive (tp) counts.

These evaluation metrics and the confusion matrix help assess the performance of the SVM model by quantifying its accuracy, precision, recall, and F1 score, as well as providing a breakdown of the predictions into different categories based on true and predicted labels.

Step 11: Print the evaluation metrics

```
[In]: print("Area Under ROC:", area_under_roc)
[In]: print("Accuracy:", accuracy)
[In]: print("Precision:", precision)
[In]: print("Recall:", recall)
[In]: print("F1 Score:", f1)
[In]: print("Confusion Matrix:")
[In]: for row in confusion_matrix:
          print(row)
[Out]:
Area Under ROC: 0.9976
Accuracy: 0.9603
Precision: 0.9605
Recall: 0.9733
F1 Score: 0.9669
Confusion Matrix:
[48, 3]
[2, 73]
```

This output can be interpreted as follows:

- Area Under ROC (0.99): The area under the receiver operating characteristic (ROC) curve is a measure of the model's ability to discriminate between the two classes. It ranges from 0 to 1, where a higher value indicates better performance. In this case, the SVM model has achieved a very high AUC close to 1, suggesting that it has excellent discriminatory power.
- Accuracy (0.96): Accuracy represents the proportion of correct predictions (both true positives and true negatives) out of all predictions made by the model. The accuracy of 0.96 indicates that the model is able to correctly predict the class for approximately 96% of the instances in the test data.
- Precision (0.96): Precision measures the proportion of correctly predicted positive instances (true positives) out of all instances predicted as positive (true positives + false positives). The precision of 0.96 indicates that 96% of the instances predicted as positive by the model are actually positive.

- Recall (0.97): Recall, also known as sensitivity or true positive rate, represents the proportion of correctly predicted positive instances (true positives) out of all actual positive instances (true positives + false negatives). The recall value of 0.97 indicates that the model is able to correctly identify 97% of the positive instances.
- F1 Score (0.97): The F1 score is the weighted mean of precision and recall, providing a single metric that balances both measures. It ranges from 0 to 1, where a higher value indicates better balance between precision and recall. The F1 score of 0.97 suggests a good balance between precision and recall for the SVM model.
- Confusion Matrix: The confusion matrix provides a tabular representation of the performance of the model. It summarizes the predictions made by the model in terms of true positives, false positives, true negatives, and false negatives. The confusion matrix of the model is as follows:

  [48, 3]

  [2, 73]

The top-left value (48) represents the number of true negatives (TN), meaning the instances correctly predicted as negative. The top-right value (3) represents the number of false positives (FP), meaning the instances incorrectly predicted as positive. The bottom-left value (2) represents the number of false negatives (FN), meaning the instances incorrectly predicted as negative. The bottom-right value (73) represents the number of true positives (TP), meaning the instances correctly predicted as positive.

Bringing It All Together

In previous sections of this chapter, we presented code examples that illustrated each modeling step individually. However, in this section, our objective is to consolidate all the relevant code from those steps into a single code block. This enables the reader to execute the code as a cohesive unit.

Scikit-Learn

Step 1: Import the necessary libraries

```
[In]: from sklearn.datasets import load_breast_cancer
[In]: from sklearn.model_selection import train_test_split
[In]: from sklearn.svm import SVC
[In]: from sklearn.metrics import accuracy_score, precision_score, [In]:
      recall_score, f1_score, confusion_matrix, roc_auc_score
[In]: from sklearn.preprocessing import StandardScaler
```

Step 2: Load the Breast Cancer Wisconsin dataset

```
[In]: data = load_breast_cancer()
[In]: X = data.data
[In]: y = data.target
```

Step 3: Split the data into training and testing sets

```
[In]: X_train, X_test, y_train, y_test = train_test_split(X, y, test_
      size=0.2, random_state=42)
```

Step 4: Scale the features using StandardScaler

```
[In]: scaler = StandardScaler()
[In]: X_train_scaled = scaler.fit_transform(X_train)
[In]: X_test_scaled = scaler.transform(X_test)
```

Step 5: Train an SVM model

```
[In]: svm = SVC(kernel='linear', random_state=42)
[In]: svm.fit(X_train_scaled, y_train)
```

Step 6: Make predictions on the test set

```
[In]: y_pred = svm.predict(X_test_scaled)
```

Step 7: Calculate evaluation metrics

```
[In]: accuracy = accuracy_score(y_test, y_pred)
[In]: precision = precision_score(y_test, y_pred)
[In]: recall = recall_score(y_test, y_pred)
[In]: f1 = f1_score(y_test, y_pred)
[In]: cm = confusion_matrix(y_test, y_pred)
[In]: auc = roc_auc_score(y_test, y_pred)
```

Step 8: Print the evaluation metrics

```
[In]: print("Area Under ROC:", auc)
[In]: print("Accuracy:", accuracy)
[In]: print("Precision:", precision)
[In]: print("Recall:", recall)
[In]: print("F1 Score:", f1)
[In]: print("Confusion Matrix:")
[In]: print(cm)
```

PySpark

Step 1: Import the necessary libraries

```
[In]: from pyspark.ml.feature import StandardScaler
[In]: from pyspark.ml.classification import LinearSVC
[In]: from pyspark.ml.evaluation import BinaryClassificationEvaluator
[In]: from pyspark.ml.linalg import Vectors
[In]: from pyspark.sql import functions as F
[In]: from pyspark.sql import SparkSession
```

Step 2: Create a SparkSession

```
[In]: spark = SparkSession.builder.getOrCreate()
```

Step 3: Load the Breast Cancer Wisconsin dataset

```
[In]: data = load_breast_cancer()
```

Step 4: Convert the NumPy arrays to lists

```
[In]: X = [Vectors.dense(x) for x in data.data.tolist()]
[In]: y = data.target.tolist()
```

Step 5: Convert the lists to Spark DataFrames

```
[In]: spark_df = spark.createDataFrame(zip(X, y), ["features", "label"])
```

Step 6: Split the data into training and testing sets

```
[In]: train_data, test_data = spark_df.randomSplit([0.8, 0.2], seed=42)
```

Step 7: Scale the features using StandardScaler

```
[In]: scaler = StandardScaler(inputCol="features",
      outputCol="scaledFeatures")
[In]: scaler_model = scaler.fit(train_data)
[In]: train_data_scaled = scaler_model.transform(train_data)
[In]: test_data_scaled = scaler_model.transform(test_data)
```

Step 8: Train an SVM model

```
[In]: svm = LinearSVC()
[In]: svm_model = svm.fit(train_data_scaled)
```

Step 9: Make predictions on the test set

```
[In]: predictions = svm_model.transform(test_data_scaled)
```

Step 10: Evaluate the model

```
[In]: evaluator = BinaryClassificationEvaluator(labelCol="label",
      rawPredictionCol="rawPrediction")
[In]: area_under_roc = evaluator.evaluate(predictions)
```

Step 11: Calculate evaluation metrics

```
[In]: tp = predictions.filter("label = 1 AND prediction = 1").count()
[In]: fp = predictions.filter("label = 0 AND prediction = 1").count()
[In]: tn = predictions.filter("label = 0 AND prediction = 0").count()
[In]: fn = predictions.filter("label = 1 AND prediction = 0").count()
[In]: accuracy = (tp + tn) / (tp + tn + fp + fn)
```

```
[In]: precision = tp / (tp + fp)
[In]: recall = tp / (tp + fn)
[In]: f1 = 2 * (precision * recall) / (precision + recall)
[In]: confusion_matrix = [[tn, fp], [fn, tp]]
```

Step 12: Print the evaluation metrics

```
[In]: print("Area Under ROC:", area_under_roc)
[In]: print("Accuracy:", accuracy)
[In]: print("Precision:", precision)
[In]: print("Recall:", recall)
[In]: print("F1 Score:", f1)
[In]: print("Confusion Matrix:")
[In]: for row in confusion_matrix:
          print(row)
```

Summary

This chapter introduced support vector machines (SVMs) using the Breast Cancer dataset. It used Pandas, Scikit-Learn, and PySpark for data processing, exploration, and machine learning. The chapter discussed the advantages and disadvantages of SVMs, as well as the kernel trick for handling nonlinearly separable data. We constructed, trained, and evaluated the linear SVM classifier, using it to predict whether a tumor is malignant or benign. This was fitting since SVMs excel in binary classification, and the breast cancer data consists of binary labels: 0 (tumor is benign) and 1 (tumor is cancerous). The model achieved a high accuracy rate.

In the next chapter, we will introduce another supervised classification algorithm: Naive Bayes. Naive Bayes is a popular and simple classification algorithm based on Bayes' theorem. It finds applications in various classification tasks, including text classification, spam detection, and sentiment analysis.

CHAPTER 11

Naive Bayes Classification with Pandas, Scikit-Learn, and PySpark

This chapter focuses on the development, training, and evaluation of a Naive Bayes algorithm. Naive Bayes classification is a well-known supervised machine learning technique widely recognized for its simplicity and ease of implementation in classification tasks. It is computationally efficient, making it suitable for large datasets and real-time applications. It can work well with relatively small datasets because it relies on simple probability calculations. The model's probabilistic nature makes it easy to interpret the results and understand why a particular classification decision was made.

The algorithm is called naive because it assumes feature independence. This means that irrelevant features do not significantly affect the classification performance. However, while this assumption makes the model computationally efficient, it may not always hold true in real-world datasets. In many real-world situations, features are not truly independent, and this can lead to suboptimal performance. Due to the independence assumption, Naive Bayes may not capture complex relationships between features, which can result in reduced accuracy compared to more sophisticated algorithms.

In this chapter, we explore the concept of Naive Bayes and its application in multiclass classification tasks. The algorithm estimates the probability distribution of the features for each label and employs Bayes' theorem to determine the probability of each label given the features. To implement and evaluate the Naive Bayes classifier, we utilize the Scikit-Learn and PySpark libraries. The analysis will highlight the similarities between the two libraries, enabling a seamless transition from Scikit-Learn to PySpark

A. Testas, *Distributed Machine Learning with PySpark*, https://doi.org/10.1007/978-1-4842-9751-3_11

and harnessing the distributed computing capabilities offered by PySpark. Furthermore, we will contrast the functionalities of Pandas and PySpark, showing that PySpark is a viable choice for handling large-scale data in big data projects.

The Dataset

In this section, we create a simulated dataset for our multiclass classification task. We first generate a Pandas DataFrame for Scikit-Learn and then convert it to a PySpark DataFrame. We use the make_classification built-in function to create the dataset:

```
[In]: from sklearn.datasets import make_classification
[In]: import pandas as pd
[In]: import numpy as np
[In]: X, y = make_classification(
          n_samples=1000,
          n_features=3,
          n_informative=3,
          n_redundant=0,
          n_clusters_per_class=1,
          n_classes=3,
          weights=[0.3, 0.3, 0.4],
          flip_y=0.1,
          random_state=42
          )
[In]: X = np.abs(X)
[In]: pandas_df = pd.DataFrame(data=X, columns=['Feature 1', 'Feature 2',
      'Feature 3'])
[In]: pandas_df['label'] = y
```

The dataset has the following characteristics:

- n_samples=1000: Generates 1000 samples (data points).
- n_features=3: Each data point has three features.
- n_informative=3: All three features are informative (relevant) for classification.

- n_redundant=0: There are no redundant features.
- n_clusters_per_class=1: Each class forms a single cluster.
- n_classes=3: Specifies that there are three classes in the classification problem.
- weights=[0.3, 0.3, 0.4]: Specifies the class weights. Class 1 and Class 2 have a weight of 0.3 each, and Class 3 has a weight of 0.4. This is done to impact the balance of class distribution in the dataset.
- flip_y=0.1: Introduces random noise to the class labels, making them more akin to real-world scenarios.

The `X = np.abs(X)` line ensures that all feature values are non-negative by taking the absolute value of the feature matrix X. This is to ensure that the features are non-negative, as Naive Bayes assumes non-negative inputs. The last two lines create a Pandas DataFrame named pandas_df to organize the dataset.

We can convert the Pandas DataFrame to a PySpark DataFrame using the PySpark `createDataFrame` method. The following code first creates a new Spark Session and then converts pandas_df into spark_df.

```
[In]: from pyspark.sql import SparkSession
[In]: spark = SparkSession.builder.getOrCreate()
[In]: spark_df = spark.createDataFrame(pandas_df)
```

We are now ready to do some exploratory data analysis on the two DataFrames. We begin by inspecting the top five rows of each DataFrame:

```
[In]: print(pandas_df.head())
[Out]:
```

	Feature 1	Feature 2	Feature 3	label
0	0.23	0.39	0.89	2
1	0.75	1.75	3.16	2
2	0.32	0.09	0.96	1
3	4.28	1.59	1.66	2
4	0.32	0.74	1.15	0

```
[In]: spark_df.show(5)
[Out]:
```

Feature 1	Feature 2	Feature 3	label
0.23	0.39	0.89	2
0.75	1.75	3.16	2
0.32	0.09	0.96	1
4.28	1.59	1.66	2
0.32	0.74	1.15	0

We can see that the top five rows generated by the Pandas head() and PySpark show() methods are the same, with the only difference being that Pandas generates an index while PySpark doesn't.

Next, we generate summary statistics using Pandas describe() and PySpark summary() methods:

```
[In]: print(pandas_df.describe())
[Out]:
```

	Feature 1	Feature 2	Feature 3	label
count	1000	1000	1000	1000
mean	1.13	1.12	1.07	1.1
std	0.79	0.72	0.72	0.83
min	0.00	0.01	0.00	0
25%	0.50	0.60	0.50	0
50%	0.98	1.03	0.99	1
75%	1.61	1.48	1.52	2
max	4.28	3.90	3.95	2

```
[In]: spark_df.summary().show()
[Out]:
```

summary	Feature 1	Feature 2	Feature 3	label
count	1000	1000	1000	1000
mean	1.13	1.12	1.07	1.1
stddev	0.79	0.72	0.72	0.83
min	0.00	0.01	0.00	0
25%	0.50	0.60	0.50	0
50%	0.98	1.03	0.99	1
75%	1.61	1.48	1.52	2
max	4.28	3.90	3.95	2

The output from both datasets is a summary statistics table with four columns: Feature 1, Feature 2, Feature 3, and label. Each row in the table provides specific statistics for each column, and the columns represent the following information:

- count: This row displays the count of non-missing values in each column. In this case, the dataset contains 1,000 data points, and all columns have 1,000 non-missing values.
- mean: This shows the average value for each column. It represents the arithmetic mean of all the values in each column. For example, the mean of Feature 1 is approximately 1.13, while that of Feature 2 is roughly 1.12.
- std/stddev: This row represents the standard deviation for each column. The standard deviation measures the spread or dispersion of values around the mean. For example, the standard deviation of Feature 1 is approximately 0.79, while that of Feature 2 is about 0.72.
- min: This row displays the minimum value in each column. It shows the smallest value present in each column. For example, the minimum value in Feature 1 is 0, while that in Feature 2 is 0.01.

- 0.25, 0.50, 0.75: These rows represent the quartiles of the data in each column. The 0.25 row corresponds to the first quartile (25th percentile), the 0.50 row corresponds to the second quartile (median or 50th percentile), and the 0.75 row corresponds to the third quartile (75th percentile). For example, the first quartile for Feature 1 is approximately 0.5, the median is 0.98, and the third quartile is 1.61.
- max: This row displays the maximum value in each column. It shows the largest value present in each column. For example, the maximum value in Feature 1 is 4.28, while that of Feature 2 is 3.9.

Let's now calculate the distribution of the target variable using Pandas `value_counts()` and PySpark `groupBy()` and `count()` methods:

```
[In]: print(pandas_df["label"].value_counts())
[Out]:
```

0	302
1	296
2	402

```
[In]: spark_df.groupBy("label").count().show()
[Out]:
```

label	count
0	302
1	296
2	402

We can see from the output that summing up across the three classes adds up to a total count of 1,000. Labels 0 and 1 each contribute about 30%, and label 2's share is about 40% of the total count. This distribution makes the sample moderately balanced. In other words, while label 2 has a higher proportion compared to labels 0 and 1, it is relatively balanced, especially when compared to scenarios where one class dominates the dataset significantly.

Naive Bayes Classification

In this section, we build, train, and evaluate a Naive Bayes classifier using Scikit-Learn and PySpark libraries. The code utilizes the classification dataset we generated in the previous section, which is split into training and testing sets. A Naive Bayes classifier is trained and then used to make predictions on the test set. Various evaluation metrics including accuracy, precision, recall, and F1 score are calculated to assess the performance of the model.

Naive Bayes with Scikit-Learn

Step 1: Importing necessary libraries

```
[In]: import numpy as np
[In]: from sklearn.model_selection import train_test_split
[In]: from sklearn.naive_bayes import GaussianNB
[In]: from sklearn.metrics import accuracy_score, precision_score, recall_
      score, f1_score
```

In this step, we import the essential libraries required for our classification task. These include numpy to work with array data; `train_test_split` to split the dataset into training and testing sets; `GaussianNB`, which is the Gaussian Naive Bayes classifier we are using for classification; and `accuracy_score`, `precision_score`, `recall_score`, and `f1_score` to assess the model's performance.

Step 2: Generate the data

```
[In]: X, y = make_classification(
    n_samples=1000,
    n_features=3,
    n_informative=3,
    n_redundant=0,
    n_clusters_per_class=1,
    n_classes=3,
    weights=[0.3, 0.3, 0.4],
    flip_y=0.1,
    random_state=42
)
[In]: X = np.abs(X)
```

This step has been explained in detail in the " The Dataset" section. In summary, we generate the classification dataset using the Scikit-Learn built-in `make_classification` function. The dataset has 1,000 samples, 3 features, and the label. The data is non-negative as required by the Naive Bayes algorithm.

Step 3: Prepare data

```
[In]: X_train, X_test, y_train, y_test = train_test_split(X, y, test_
      size=0.2, random_state=42)
```

In this step, we split the generated data into training and testing sets using `train_test_split`. This separation is crucial for training the model on 80% of the data (X_train, y_train) and evaluating its performance on the remaining 20% (X_test, y_test) to ensure it generalizes well. This 80-20 split helps us assess the model's ability to make accurate predictions on unseen data.

Step 4: Train the model

```
[In]: nb = GaussianNB()
[In]: nb.fit(X_train, y_train)
```

Here, we create an instance of the Gaussian Naive Bayes classifier (nb) and fit it to the training data (X_train, y_train) using the `fit()` method.

Step 5: Make predictions on the test data

```
[In]: y_pred = nb.predict(X_test)
```

In this step, we use the trained model (nb) to make predictions on the test data (X_test) by employing the `predict()` method.

Step 6: Calculate and print evaluation metrics

```
[In]: accuracy = accuracy_score(y_test, y_pred)
[In]: precision = precision_score(y_test, y_pred, average='weighted')
[In]: recall = recall_score(y_test, y_pred, average='weighted')
[In]: f1 = f1_score(y_test, y_pred, average='weighted')
[In]: print("Accuracy:", accuracy)
[In]: print("Precision:", precision)
[In]: print("Recall:", recall)
[In]: print("F1-score:", f1)
[Out]:
```

```
Accuracy: 0.44
Precision: 0.45
Recall: 0.44
F1-score: 0.43
```

In this final step, we assess the model's performance using various evaluation metrics. We utilize the following methods to calculate each metric:

- Accuracy: Calculated using the accuracy_score method
- Precision: Calculated using the precision_score method with the weighted averaging strategy
- Recall: Calculated using the recall_score method with the weighted averaging strategy
- F1-score: Calculated using the f1_score method with the weighted averaging strategy

The weighted averaging strategy accounts for class imbalance by considering the proportions of each class in the dataset when calculating the metrics. We print the metrics to evaluate how well the Gaussian Naive Bayes classifier is performing on the test data.

The interpretation of the output is given as follows:

- Accuracy (0.44): Accuracy is a measure of the overall correctness of the model's predictions. It represents the ratio of correctly predicted instances (both true positives and true negatives) to the total number of instances in the test dataset. In this case, the accuracy is 0.44, which means that the model correctly predicted 44% of the instances in the test data.
- Precision (0.45): Precision is a measure of the model's ability to make accurate positive predictions. It is the ratio of true positive predictions to the total number of positive predictions made by the model. The calculated precision is 0.45, indicating that 45% of the positive predictions made by the model were correct.
- Recall (0.44): Recall, also known as sensitivity or true positive rate, measures the model's ability to correctly identify all relevant instances in the dataset. It is the ratio of true positive predictions to

the total number of actual positive instances in the dataset. A recall score of 0.44 suggests that the model successfully identified 44% of all actual positive instances.

- F1-score (0.43): The F1 score is the weighted mean of precision and recall. It is a single metric that balances the trade-off between precision and recall. A higher F1 score indicates a better balance between precision and recall. In this case, the F1 score is 0.43.

Overall, the model's performance is modest, which is unsurprising given that we are working with a simulated dataset.

Naive Bayes with PySpark

The following code builds, trains, and evaluates the Naive Bayes classifier using PySpark:

Step 1: Import necessary libraries

```
[In]: from pyspark.sql import SparkSession
[In]: from pyspark.ml.classification import NaiveBayes
[In]: from pyspark.ml.linalg import Vectors
[In]: from sklearn.datasets import make_classification
[In]: from pyspark.ml.evaluation import MulticlassClassificationEvaluator
[In]: import numpy as np
```

In this first step, the required libraries and modules are imported to set up the environment for the machine learning tasks. These include SparkSession to work with Spark's DataFrame-based API, NaiveBayes to use the Naive Bayes classifier from PySpark's machine learning library, Vectors for handling feature vectors in Spark, make_classification to generate a simulated classification dataset for training and testing, MulticlassClassificationEvaluator to evaluate the model, and numpy to work with arrays.

Step 2: Create a Spark Session

```
[In]: spark = SparkSession.builder.getOrCreate()
```

In this step, a Spark Session is created using SparkSession.builder. This session allows us to interact with Spark's DataFrame and machine learning APIs.

Step 3: Generate the dataset with non-negative features and three classes

```
[In]: X, y = make_classification(
          n_samples=1000,
          n_features=3,
          n_informative=3,
          n_redundant=0,
          n_clusters_per_class=1,
          n_classes=3,
          weights=[0.3, 0.3, 0.4],
          flip_y=0.1,
          random_state=42
)
[In]: X = np.abs(X)
[In]: data = [(Vectors.dense(x.tolist()), float(label)) for x, label in
      zip(X, y)]
[In]: spark_df = spark.createDataFrame(data, ["features", "label"])
```

This step was explained in detail in the "The Dataset" section. In a nutshell, we are creating a simulated dataset using `make_classification` from sklearn.datasets. This dataset contains three classes and non-negative feature values. The features are ensured to be non-negative by applying `np.abs` to the generated feature matrix. The data is then transformed into a Spark DataFrame, spark_df, with features and labels, using the `createDataFrame()` method.

Step 4: Split the data into training and testing sets

```
[In]: train_df, test_df = spark_df.randomSplit([0.8, 0.2], seed=42)
```

The dataset is split into training and testing sets using `randomSplit`. 80% of the data is allocated for training (train_df), and 20% is reserved for testing (test_df). The seed=42 ensures reproducibility.

Step 5: Train the model

```
[In]: nb = NaiveBayes(featuresCol="features", labelCol="label",
      modelType="multinomial")
[In]: nb_model = nb.fit(train_df)
```

Here, we create a Naive Bayes classifier designed for multiclass classification tasks using the `NaiveBayes` class. This classifier is well suited for scenarios where we have multiple classes to predict (in this case, three classes: 0, 1, and 2).

The featuresCol parameter specifies the input features column, the labelCol parameter specifies the target label column, and the modelType parameter is set to "multinomial" to indicate that we are dealing with a multiclass classification problem.

Next, we fit the Naive Bayes model to our training data (represented by train_df) using the `fit()` method. This process involves the model learning from the training data to make predictions on new, unseen data.

Step 6: Make predictions on the test data

```
[In]: predictions = nb_model.transform(test_df)
```

The trained model (nb_model) is used to make predictions on the test data (test_df) using the PySpark `transform()` method. This results in a DataFrame called predictions.

Step 7: Evaluate the model using various metrics

```
[In]: evaluator = MulticlassClassificationEvaluator(metricName="accuracy")
[In]: accuracy = evaluator.evaluate(predictions)
[In]: evaluator = MulticlassClassificationEvaluator(metricName="weighted
      Precision")
[In]: precision = evaluator.evaluate(predictions)
[In]: evaluator = MulticlassClassificationEvaluator(metricName="weighted
      Recall")
[In]: recall = evaluator.evaluate(predictions)
[In]: evaluator = MulticlassClassificationEvaluator(metricName="f1")
[In]: f1 = evaluator.evaluate(predictions)

[In]: print("Accuracy:", accuracy)
[In]: print("Precision:", precision)
[In]: print("Recall:", recall)
[In]: print("F1-score:", f1)

[Out]:
Accuracy: 0.39
Precision: 0.22
Recall: 0.39
F1-score: 0.23
```

In this (final) evaluation step, several metrics are computed to assess the model's performance. We use the `MulticlassClassificationEvaluator` to create an evaluator for each of these metrics, including accuracy, weighted precision, weighted recall, and F1 score.

For each metric, the evaluator is applied to the model's predictions generated on the test data. The results are then stored in their respective variables: accuracy, precision, recall, and f1. These metrics provide insights into different aspects of the model's performance, such as its ability to correctly classify instances, account for class imbalance (weighted metrics), and balance precision and recall (F1 score).

Finally, we print out these metrics to have a comprehensive view of how well the model performs across various evaluation criteria.

Let's understand what these metrics tell us:

- Accuracy (0.39): This tells us that the model correctly predicted 39% of the instances in the test dataset.
- Precision (0.22): A precision of 0.22 indicates that only 22% of the positive predictions made by the model were correct.
- Recall (0.39): A recall of 0.39 suggests that the model successfully captured 39% of all actual positive instances.
- F1-score (0.23): The F1 score provides a balance between precision and recall. In this case, the F1 score is 0.23, which suggests that the model has a trade-off between precision and recall.

Overall, the model's performance is modest, which is somewhat expected given that we are using a simulated dataset. In real-world scenarios, we would delve deeper into the reasons behind this performance level. One potential factor to investigate would be overfitting, which can often lead to reduced performance.

Bringing It All Together

In the previous sections of this chapter, we presented code examples that illustrated each modeling step separately. However, in this section, our goal is to combine all the relevant code from those steps into a single code block. This allows the reader to execute the code as a unified unit.

Scikit-Learn

Step 1: Import necessary libraries

```
[In]: import numpy as np
[In]: from sklearn.model_selection import train_test_split
[In]: from sklearn.naive_bayes import GaussianNB
[In]: from sklearn.metrics import accuracy_score, precision_score, recall_
      score, f1_score, confusion_matrix
```

Step 2: Generate the data

```
[In]: X, y = make_classification(
    n_samples=1000,
    n_features=3,
    n_informative=3,
    n_redundant=0,
    n_clusters_per_class=1,
    n_classes=3,  # Specify three classes
    weights=[0.3, 0.3, 0.4],  # Specify class weights
    flip_y=0.1,
    random_state=42
)
[In]: X = np.abs(X)
```

Step 3: Split the data into training and testing sets

```
[In]: X_train, X_test, y_train, y_test = train_test_split(X, y, test_
      size=0.2, random_state=42)
```

Step 4: Train the model

```
# Create a Gaussian Naive Bayes classifier
[In]: nb = GaussianNB()

# Fit the model to the training data
[In]: nb.fit(X_train, y_train)
```

Step 5: Make predictions on the test data

```
[In]: y_pred = nb.predict(X_test)
```

Step 6: Calculate and print evaluation metrics

```
# Calculate the metrics
[In]: accuracy = accuracy_score(y_test, y_pred)
[In]: precision = precision_score(y_test, y_pred, average='weighted')
[In]: recall = recall_score(y_test, y_pred, average='weighted')
[In]: f1 = f1_score(y_test, y_pred, average='weighted')

[In]: print("Accuracy:", accuracy)
[In]: print("Precision:", precision)
[In]: print("Recall:", recall)
[In]: print("F1-score:", f1)
```

PySpark

Step 1: Import necessary libraries

```
[In]: from pyspark.sql import SparkSession
[In]: from pyspark.ml.classification import NaiveBayes
[In]: from pyspark.ml.linalg import Vectors
[In]: from sklearn.datasets import make_classification
[In]: from pyspark.ml.evaluation import MulticlassClassificationEvaluator,
      MulticlassClassificationEvaluator
[In]: import numpy as np
```

Step 2: Create a Spark Session

```
[In]: spark = SparkSession.builder.appName("NaiveBayesMultiClassExample").
      getOrCreate()
```

Step 3: Generate the dataset with non-negative features and three classes

```
[In]: X, y = make_classification(
          n_samples=1000,
          n_features=3,
          n_informative=3,
          n_redundant=0,
          n_clusters_per_class=1,
          n_classes=3,
```

```
        weights=[0.3, 0.3, 0.4],
        flip_y=0.1,
        random_state=42
        )

# Ensure non-negative feature values
[In]: X = np.abs(X)

# Create a DataFrame from the NumPy array
[In]: data = [(Vectors.dense(x.tolist()), float(label)) for x, label in
      zip(X, y)]
[In]: spark_df = spark.createDataFrame(data, ["features", "label"])
```

Step 4: Split the data into training and testing sets

```
[In]: train_df, test_df = spark_df.randomSplit([0.8, 0.2], seed=42)
```

Step 5: Train the model

```
# Create a Naive Bayes classifier for multi-class classification
[In]: nb = NaiveBayes(featuresCol="features", labelCol="label",
      modelType="multinomial")

# Fit the model to the training data
[In]: nb_model = nb.fit(train_df)
```

Step 6: Make predictions on the test data

```
[In]: predictions = nb_model.transform(test_df)
```

Step 7: Evaluate the model using various metrics

```
# Create a MulticlassClassificationEvaluator to evaluate the model
[In]: evaluator = MulticlassClassificationEvaluator(metricName="accuracy")

# Calculate the metrics
[In]: accuracy = evaluator.evaluate(predictions)
[In]: evaluator = MulticlassClassificationEvaluator(metricName="weighted
      Precision")
[In]: precision = evaluator.evaluate(predictions)
```

```
[In]: evaluator = MulticlassClassificationEvaluator(metricName="weighted
Recall")
[In]: recall = evaluator.evaluate(predictions)
[In]: evaluator = MulticlassClassificationEvaluator(metricName="f1")
[In]: f1 = evaluator.evaluate(predictions)

# Print the metrics
[In]: print("Accuracy:", accuracy)
[In]: print("Precision:", precision)
[In]: print("Recall:", recall)
[In]: print("F1-score:", f1)
```

Summary

In this chapter, we introduced Naive Bayes classification. We discussed the advantages and disadvantages of this algorithm and demonstrated how to build, train, and evaluate it. We utilized the Scikit-Learn and PySpark libraries for implementation. This analysis highlighted the similarities between the two libraries, enabling a seamless transition from Scikit-Learn to PySpark and harnessing the distributed computing capabilities offered by PySpark. Furthermore, we contrasted the functionalities of Pandas and PySpark, showcasing PySpark as a viable choice for handling large-scale data in big data projects.

In the next chapter, we will introduce a new type of supervised learning: the Multilayer Perceptron (MLP). This is a powerful approach to machine learning widely used for classification tasks. MLP classifiers have shown their effectiveness in solving classification problems across different fields because they can capture intricate dependencies and nonlinear relationships within the data.

CHAPTER 12

Neural Network Classification with Pandas, Scikit-Learn, and PySpark

In this chapter, we explore the world of Multilayer Perceptron (MLP) classifiers, a powerful approach to supervised machine learning. MLPs have several advantages, making them versatile for a wide range of tasks, from regression to classification spanning across various domains such as image recognition, natural language processing, and speech recognition, to name a few.

One key advantage is that MLPs can approximate any continuous function given enough hidden neurons and training data. Nonlinearity is another advantage. By using activation functions like ReLU, sigmoid, or tanh, MLPs can model complex, nonlinear relationships between input and output variables and capture intricate patterns in data. Moreover, MLPs automatically learn relevant features from raw data, reducing the need for manual feature engineering. This can be particularly useful when dealing with high-dimensional data. Furthermore, training MLPs can be parallelized across multiple GPUs or CPUs, which can significantly speed up training on large datasets. Additionally, with proper regularization techniques such as dropout and weight decay, MLPs can generalize well to unseen data, reducing overfitting.

However, MLPs aren't without shortcomings. To begin with, MLPs require a substantial amount of data for training, especially when dealing with deep architectures. Insufficient data can lead to overfitting. Even with large datasets, without proper regularization and hyperparameter tuning, MLPs can easily overfit to the training

A. Testas, *Distributed Machine Learning with PySpark*, https://doi.org/10.1007/978-1-4842-9751-3_12

data, resulting in poor generalization. Moreover, MLPs have several hyperparameters, including the number of layers, neurons per layer, and learning rate. Finding the right combination of hyperparameters can be challenging and computationally expensive. There is another disadvantage. MLPs are often considered as black-box models because it can be challenging to interpret the relationships they learn, especially in deep architectures with many layers. This means MLPs aren't the right choice when interpretability is important.

Another important limitation is that deep MLPs suffer from vanishing and exploding gradient problems. In deep MLPs, during the training process, gradients (which indicate how much the model's parameters should be adjusted to minimize the error) can become extremely small (vanishing gradient) or extremely large (exploding gradient) as they are propagated backward through the network, which makes training difficult. Additionally, training deep MLPs may require specialized hardware like GPUs to achieve reasonable training times.

To illustrate the usage of MLP classifiers, we work with the handwritten digits dataset, a well-known benchmark data for digit recognition. This dataset consists of images representing handwritten digits (0–9) and their corresponding labels. By training an MLP classifier on this dataset, we aim to accurately classify handwritten digits based on their pixel representations. We implement the classifiers using Scikit-Learn and PySpark libraries and utilize Pandas and PySpark for data processing and exploration.

The Dataset

The load_digits is a built-in dataset in Scikit-Learn. It is a multiclass classification dataset that contains images of handwritten digits (0–9). Each digit image is represented as an 8×8 matrix of pixel values, and the dataset includes corresponding labels indicating the true digit for each image. Each pixel is treated as a feature, giving 64 pixels in total, which correspond to the 64 features in the dataset. The target variable contains an array of labels corresponding to each image in the dataset. Each label represents the digit value from 0 to 9.

We can load the dataset in several steps:

Step 1: Import pandas and load_digits

```
[In]: import pandas as pd
[In]: from sklearn.datasets import load_digits
```

By importing `load_digits`, we gain access to the function that allows us to load the handwritten digits dataset.

Step 2: Load the dataset

```
[In]: dataset = load_digits()
```

This line of code calls the `load_digits()` function and assigns the returned dataset to the variable named dataset.

Step 3: Create a Pandas DataFrame from the data starting with the features

```
[In]: pandas_df = pd.DataFrame(data=dataset.data, columns=dataset.
      feature_names)
```

Step 4: Add the target column

```
[In]: pandas_df['target'] = dataset.target
```

We can now start learning about the dataset, beginning with Pandas. The `shape` attribute shows that the dataset has 1,797 rows and 65 columns (64 features plus the target variable):

```
[In]: print(pandas_df.shape)
[Out]: (1797, 65)
```

We can print the labels or column names as follows:

```
[In]: print(pandas_df.columns)
[Out]:
Index(['pixel_0_0', 'pixel_0_1', 'pixel_0_2', 'pixel_0_3', 'pixel_0_4',
'pixel_0_5', 'pixel_0_6', 'pixel_0_7', 'pixel_1_0', 'pixel_1_1',
'pixel_1_2', 'pixel_1_3', 'pixel_1_4', 'pixel_1_5', 'pixel_1_6',
'pixel_1_7', 'pixel_2_0', 'pixel_2_1', 'pixel_2_2', 'pixel_2_3',
'pixel_2_4', 'pixel_2_5', 'pixel_2_6', 'pixel_2_7', 'pixel_3_0',
'pixel_3_1', 'pixel_3_2', 'pixel_3_3', 'pixel_3_4', 'pixel_3_5',
'pixel_3_6', 'pixel_3_7', 'pixel_4_0', 'pixel_4_1', 'pixel_4_2',
'pixel_4_3', 'pixel_4_4', 'pixel_4_5', 'pixel_4_6', 'pixel_4_7',
'pixel_5_0', 'pixel_5_1', 'pixel_5_2', 'pixel_5_3', 'pixel_5_4',
'pixel_5_5', 'pixel_5_6', 'pixel_5_7', 'pixel_6_0', 'pixel_6_1',
'pixel_6_2', 'pixel_6_3', 'pixel_6_4', 'pixel_6_5', 'pixel_6_6',
```

```
'pixel_6_7', 'pixel_7_0','pixel_7_1', 'pixel_7_2', 'pixel_7_3',
'pixel_7_4', 'pixel_7_5', 'pixel_7_6', 'pixel_7_7', 'target'],
dtype='object')
```

This output represents the column names or labels for the features (pixels) and the target variable. pixel_0_0, pixel_0_1, ..., pixel_7_7 are the column names for the features (pixels) of the dataset. Each column represents a specific pixel in the images of handwritten digits. The format pixel_x_y indicates the location of the pixel within an image. Here, x represents the row (0 to 7), and y represents the column (0 to 7) of the pixel. In other words, for each row in the dataset, we will have values associated with these columns corresponding to the pixel intensity at the given (x, y) position within the digit image.

The target column is used to store the target labels or the digit labels associated with each image in the dataset. In our handwritten digit recognition task, this column contains the actual digit (0 through 9) that the corresponding image represents. For example, if we have an image of the digit 3, the value in the target column for that row would be 3.

We can print the first five rows and first five columns of pandas_df as follows:

```
[In]: print(pandas_df.iloc[:5, :5])
[Out]:
```

	pixel_0_0	pixel_0_1	pixel_0_2	pixel_0_3	pixel_0_4
0	0	0	5	13	9
1	0	0	0	12	13
2	0	0	0	4	15
3	0	0	7	15	13
4	0	0	0	1	11

In this output, each column value represents the grayscale intensity or darkness of the pixel at the corresponding position within an image. The intensity values range from 0 (completely white) to 16 (completely black). We can confirm the min and max values as follows:

```
[In]: (pandas_df.values.min(), pandas_df.values.max())
[Out]: (0.0, 16.0)
```

We can print distinct values of the target column as follows:

```
[In]: print(pandas_df.iloc[:, -1].unique())
[Out]: [0 1 2 3 4 5 6 7 8 9]
```

This output shows that there are ten digits ranging from 0 to 9.

We can display the first five images from the load_digits dataset using Matplotlib and Scikit-Learn using the following steps:

Step 1: Import necessary libraries

```
[In]: import matplotlib.pyplot as plt
[In]: from sklearn.datasets import load_digits
```

Step 2: Load the dataset

```
dataset = load_digits()
```

Step 3: Display the first five images

```
[In]: fig, axes = plt.subplots(nrows=1, ncols=5, figsize=(10, 3))
[In]: for i, ax in enumerate(axes):
        ax.imshow(dataset.images[i], cmap='gray')
        ax.set_title(f"Label: {dataset.target[i]}")
        ax.axis('off')
[In]: plt.tight_layout()
[In]: plt.show()
[Out]:
```

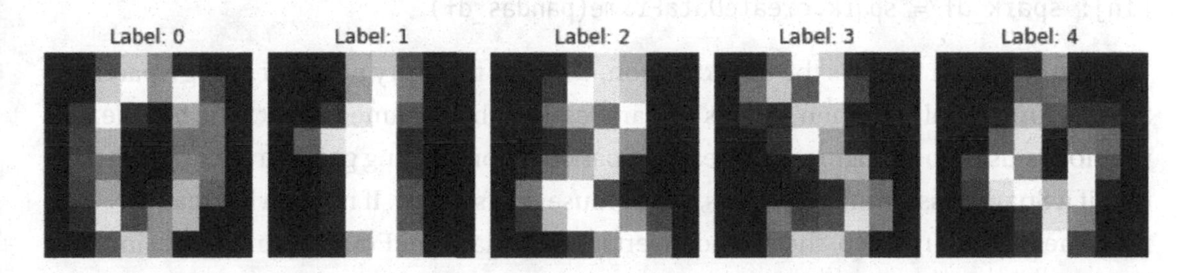

To get the preceding output, we first import the `Matplotlib` library, specifically the submodule `pyplot`, and assign it the alias `plt`. This allows us to use Matplotlib's plotting functions. We then import the `load_digits` function from Scikit-Learn's datasets module. This function is used to load the load_digits dataset, which contains images of handwritten digits along with their labels.

Next, we set dataset = load_digits(), which loads the load_digits dataset and assigns it to the variable dataset. By setting fig, axes = plt.subplots(nrows=1, ncols=5, figsize=(10, 3)), we create a figure with a row of five subplots using the `subplots()` function. The nrows argument specifies the number of rows, ncols specifies the number of columns, and figsize sets the size of the figure.

Finally, the following *for loop* iterates over the five subplots, assigning each subplot to the variable ax:

- ax.imshow(dataset.images[i], cmap='gray') displays the image at index i from the dataset.images array using the `imshow()` function. The cmap='gray' argument sets the colormap to grayscale.
- ax.set_title(f"Label: {dataset.target[i]}") sets the title of the subplot to include the label corresponding to the image at index i from the dataset.target array.
- ax.axis('off') turns off the axis ticks and labels for the subplot.
- plt.tight_layout() adjusts the layout of the subplots to prevent overlapping.
- plt.show() displays the figure with the subplots.

To explore the handwritten digits dataset using PySpark, we need to first convert the pandas_df to a PySpark DataFrame using the following lines of code:

```
[In]: from pyspark.sql import SparkSession
[In]: spark = SparkSession.builder.getOrCreate()
[In]: spark_df = spark.createDataFrame(pandas_df)
```

The code first imports the SparkSession, which is the entry point for using Apache Spark's functionality. It then creates a SparkSession object named spark. The `builder` method is used to configure and create a Spark Session. Calling `getOrCreate()` ensures that if a Spark Session already exists, it will reuse that session. If not, it will create a new one. In the final step, the code converts the Pandas DataFrame (pandas_df) into a PySpark DataFrame (spark_df) using the `createDataFrame()` method provided by the SparkSession.

Having converted the Pandas DataFrame to a PySpark DataFrame, we can now perform the same operations as with Pandas on the spark_df.

Let's start by displaying the shape of the DataFrame:

```
[In]: print((spark_df.count(), len(spark_df.columns)))
[Out]: (1797, 65)
```

This indicates that the dataset has 1,797 rows and 65 columns (64 features or 8×8 pixels and a target variable, which is the handwritten digit ranging from 0 to 9).

Next, we will print the column names:

```
[In]: print(spark_df.columns)
[Out]:
['pixel_0_0', 'pixel_0_1', 'pixel_0_2', 'pixel_0_3', 'pixel_0_4',
'pixel_0_5', 'pixel_0_6', 'pixel_0_7', 'pixel_1_0', 'pixel_1_1',
'pixel_1_2', 'pixel_1_3', 'pixel_1_4', 'pixel_1_5', 'pixel_1_6',
'pixel_1_7', 'pixel_2_0', 'pixel_2_1', 'pixel_2_2', 'pixel_2_3',
'pixel_2_4', 'pixel_2_5', 'pixel_2_6', 'pixel_2_7', 'pixel_3_0',
'pixel_3_1', 'pixel_3_2', 'pixel_3_3', 'pixel_3_4', 'pixel_3_5',
'pixel_3_6', 'pixel_3_7', 'pixel_4_0', 'pixel_4_1', 'pixel_4_2',
'pixel_4_3', 'pixel_4_4', 'pixel_4_5', 'pixel_4_6', 'pixel_4_7',
'pixel_5_0', 'pixel_5_1', 'pixel_5_2', 'pixel_5_3', 'pixel_5_4',
'pixel_5_5', 'pixel_5_6', 'pixel_5_7', 'pixel_6_0', 'pixel_6_1',
'pixel_6_2', 'pixel_6_3', 'pixel_6_4', 'pixel_6_5', 'pixel_6_6',
'pixel_6_7', 'pixel_7_0', 'pixel_7_1', 'pixel_7_2', 'pixel_7_3',
'pixel_7_4', 'pixel_7_5', 'pixel_7_6', 'pixel_7_7', 'label']
```

Now, let's print the first five rows and first five columns of the PySpark DataFrame just like we did with Pandas:

```
[In]: spark_df.select(spark_df.columns[:5]).show(5)
[Out]:
+---------+---------+---------+---------+---------+
|pixel_0_0|pixel_0_1|pixel_0_2|pixel_0_3|pixel_0_4|
+---------+---------+---------+---------+---------+
|      0.0|      0.0|      5.0|     13.0|      9.0|
|      0.0|      0.0|      0.0|     12.0|     13.0|
|      0.0|      0.0|      0.0|      4.0|     15.0|
```

```
|       0.0|       0.0|       7.0|     15.0|     13.0|
|       0.0|       0.0|       0.0|      1.0|     11.0|
+---------+---------+---------+---------+---------+
only showing top 5 rows
```

In the following steps, we show the distinct values of the target variable:

```
[In]: distinct_values = spark_df.select(spark_df.columns[-1]).distinct().
      orderBy(spark_df.columns[-1])
[In]: distinct_values.show()
[Out]:
+-----+
|label|
+-----+
|    0|
|    1|
|    2|
|    3|
|    4|
|    5|
|    6|
|    7|
|    8|
|    9|
+-----+
```

To retrieve the distinct values from the target column as shown previously, we first selected the last column of the DataFrame using spark_df.columns[-1]. Then, we applied the `distinct()` method to retrieve a new DataFrame containing only the unique values from the selected column. We used the `orderBy()` method to sort the distinct values in ascending order based on the last column. The resulting DataFrame, containing the distinct values in ascending order, is stored in the variable `distinct_values`. Finally, the `show()` method is called on the distinct_values DataFrame.

MLP Classification

In this section, we build, train, and evaluate a Multilayer Perceptron (MLP) model and use it to predict the ten handwritten digits (0–9). The MLP model makes predictions by utilizing the architecture of a neural network. The MLP consists of multiple layers of interconnected artificial neurons, also known as nodes or units. Each neuron applies a nonlinear activation function to the weighted sum of its inputs to introduce nonlinearity into the model's decision-making process.

When making predictions using the MLP model for the handwritten digit classification task, the following steps occur:

- Input layer: The 8×8 pixel representation of a digit image is flattened into a 1D vector and serves as the input to the MLP model. Each pixel value becomes a feature that contributes to the prediction.
- Forward propagation: The input features are propagated through the network in a forward direction, layer by layer. Each layer consists of multiple neurons, and the outputs of the previous layer serve as inputs to the subsequent layer.
- Weighted sum and activation: At each neuron, the inputs are multiplied by associated weights, and these weighted inputs are summed up. The sum is then passed through an activation function to introduce nonlinearity. The activation function determines whether the neuron fires or activates based on the weighted sum.
- Hidden layers: The MLP typically consists of one or more hidden layers sandwiched between the input and output layers. The hidden layers allow the model to learn complex relationships and representations by extracting higher-level features from the input data.
- Output layer: The final hidden layer is connected to the output layer, which produces the predictions for each digit class. The number of neurons in the output layer corresponds to the number of possible classes (in our case, digits 0–9). Each output neuron represents the probability or confidence score of the input belonging to a particular class.

- Activation function in the output layer: For multiclass classification, the softmax activation function is commonly used to transform the output values into probabilities that sum up to 1.
- Prediction: The digit class with the highest probability or confidence score in the output layer is selected as the predicted class for the input digit.

During the training process, the MLP model adjusts the weights of the connections between neurons using backpropagation and optimization algorithms such as gradient descent. This iterative process aims to minimize the difference between the predicted outputs and the true labels in the training data, improving the model's ability to make accurate predictions. By learning from a large number of labeled examples, the MLP model develops an understanding of the underlying patterns and features that distinguish different handwritten digits. This enables it to generalize and make predictions on unseen digit images with reasonable accuracy.

We utilize both Scikit-Learn and PySpark libraries to build, train, and evaluate the MLP model. The steps to predict the handwritten digits using the load_digits dataset in both libraries are the same:

- Load the dataset: We start by loading the load_digits dataset from Scikit-Learn. This dataset contains grayscale images of handwritten digits, along with their corresponding labels.
- Split the data: Split the dataset into training and testing sets. This division allows us to train the MLP classifier on a portion of the data and evaluate its performance on unseen data. Typically, around 80% of the data is used for training, while the remaining 20% is used for testing.
- Create and train the MLP classifier: Initialize an MLP classifier, and configure the desired number of hidden layers and neurons in each layer. Train the MLP classifier using the training data, where the model learns to recognize patterns and associations between the input features and their corresponding target labels.
- Predict the digits: Apply the trained MLP classifier to the testing set to make predictions. The model will take the input features (pixel values of handwritten digits) and assign a predicted label to each sample.

- Evaluate the performance: Compare the predicted labels with the true labels of the testing set to assess the performance of the MLP classifier. Common evaluation metrics include accuracy, precision, recall, F1 score, and confusion matrix.

The fact that both Scikit-Learn and PySpark follow similar guidelines for deep learning implementation will assist data scientists who aim to migrate from Scikit-Learn to PySpark. This migration will enable the deep learning practitioners to leverage PySpark's powerful distributed computing API in significant ways. While deep learning frameworks like TensorFlow, Keras, and PyTorch are commonly used for training deep neural networks, Spark can complement these frameworks in several key areas:

- Data preprocessing: Spark can efficiently handle large-scale data preprocessing tasks. We can use Spark to clean, transform, and preprocess data before feeding it into a deep learning model. This can be particularly useful when dealing with massive datasets that don't fit into memory.
- Data loading: Spark can be used to load and distribute data to multiple worker nodes, allowing for efficient data distribution and parallelism. This is crucial when working with large datasets that need to be processed by deep learning models.
- Distributed training: While deep learning frameworks like TensorFlow and PyTorch have their own distributed training capabilities, Spark can be integrated to distribute deep learning training across multiple nodes in a Spark cluster. This can be beneficial when training large models on distributed computing resources.
- Model serving: After training deep learning models, we can deploy the deep learning models using Spark's streaming capabilities. This allows us to serve predictions or perform real-time inference on incoming data, making it suitable for applications like recommendation systems or fraud detection.
- Integration with big data ecosystem: Spark can seamlessly integrate with other components of the big data ecosystem, such as Hadoop Distributed File System (HDFS), Hive, and HBase. This integration allows data scientists to leverage existing data pipelines and infrastructure when working on deep learning projects.

MLP Classification with Scikit-Learn

We first demonstrate the use of the Scikit-Learn library for training and evaluating a neural network classifier on the digits dataset. We will follow these steps:

Step 1: Import the necessary libraries

```
[In]: from sklearn.datasets import load_digits
[In]: from sklearn.model_selection import train_test_split
[In]: from sklearn.neural_network import MLPClassifier
[In]: from sklearn.metrics import accuracy_score, precision_score, recall_
      score, f1_score, confusion_matrix
```

In this step, we imported the necessary libraries. These include `load_digits` for loading the digits dataset, `train_test_split` for splitting the data into training and testing sets, `MLPClassifier` for creating and training the neural network classifier, and `accuracy_score`, `precision_score`, `recall_score`, `f1_score`, `confusion_matrix` for evaluating the classifier's performance.

Step 2: Load the dataset

```
[In]: dataset = load_digits()
```

In this step, the digits dataset is loaded using `load_digits()`, and the data is assigned to the variable dataset:

Step 3: Split data into training and testing sets

```
[In]: X_train, X_test, y_train, y_test = train_test_split(dataset.data,
      dataset.target, test_size=0.2, random_state=42)
```

The data is split into training and testing sets using `train_test_split()`. The input features are stored in X_train and X_test, while the corresponding target values are stored in y_train and y_test. The test_size parameter specifies the proportion of the data to be allocated for testing (20%), and random_state 42 ensures reproducibility of the split.

Step 4: Create a neural network

```
[In]: classifier = MLPClassifier(hidden_layer_sizes=(64, 64))
[In]: classifier.fit(X_train, y_train)
```

In this step, a neural network classifier is created using `MLPClassifier()`, with the hidden_layer_sizes parameter specifying the architecture of the neural network. There are two hidden layers, each with 64 neurons. The classifier is then trained on the training data using the `fit()` method, with X_train as the input features and y_train as the target values.

Step 5: Make predictions

```
[In]: y_pred = classifier.predict(X_test)
```

The predictions are made on the test set using `predict()`, and the predicted values are stored in y_pred.

Step 6: Evaluate the model's performance

```
[In]: accuracy = accuracy_score(y_test, y_pred)
[In]: print("Accuracy:", accuracy)
[In]: precision = precision_score(y_test, y_pred, average='weighted')
[In]: recall = recall_score(y_test, y_pred, average='weighted')
[In]: f1 = f1_score(y_test, y_pred, average='weighted')
[In]: print("Precision:", precision)
[In]: print("Recall:", recall)
[In]: print("F1 Score:", f1)
[In]: confusion_mat = confusion_matrix(y_test, y_pred)
[In]: print("Confusion Matrix:")
[In]: print(confusion_mat)
```

To evaluate the performance of the Scikit-Learn MLP classifier, the accuracy metric is first obtained using `accuracy_score()` by comparing the predicted values y_pred with the actual target values y_test. The accuracy value is printed using the `print()` method.

More evaluation metrics (precision, recall, and F1 score) are then computed using `precision_score()`, `recall_score()`, and `f1_score()`, respectively. These metrics are calculated by comparing the predicted values with the true values. The precision, recall, and F1 score values are printed using the `print()` method.

Finally, the confusion matrix is calculated using `confusion_matrix()` by comparing the predicted values with the true values. The confusion matrix provides a tabular representation of the classifier's performance across different classes. The confusion matrix is printed using `print()` for visualizing the results.

Once the code is executed, we get the following results:

```
[Out]:
Accuracy: 0.98
Precision: 0.98
Recall: 0.98
F1 Score: 0.98
Confusion Matrix:
[[32  0  0  0  1  0  0  0  0  0]
 [ 0 27  1  0  0  0  0  0  0  0]
 [ 0  0 32  0  0  0  0  1  0  0]
 [ 0  0  0 33  0  1  0  0  0  0]
 [ 0  0  0  0 46  0  0  0  0  0]
 [ 0  0  0  0  0 46  1  0  0  0]
 [ 1  0  0  0  0  0 34  0  0  0]
 [ 0  0  0  0  0  0  0 33  0  1]
 [ 0  1  0  0  0  0  0  0 29  0]
 [ 0  0  0  0  0  0  0  1  0 39]]
```

The explanation of this output is as follows:

- Accuracy (0.98): This is a measure of how often the classifier correctly predicts the target values. The accuracy is 0.98, which means that the classifier achieved an accuracy of 98%. It indicates that the classifier made correct predictions for approximately 98% of the instances in the test set.
- Precision (0.98): This is a measure of the classifier's ability to correctly identify positive instances (true positives) out of all instances it labeled as positive (true positives + false positives). Precision is calculated for each class and then averaged. The precision score is 0.98, indicating that, on average, the classifier achieved a precision of approximately 98% across all classes.
- Recall (0.98): This metric, also known as sensitivity or true positive rate, measures the classifier's ability to identify positive instances correctly out of all actual positive instances (true positives + false

negatives). Like precision, recall is calculated for each class and then averaged. The recall score here is 0.98, indicating that, on average, the classifier achieved a recall of approximately 98% across all classes.

- F1 Score (0.98): This is the weighted mean of precision and recall. It provides a balanced measure that considers both precision and recall. The F1 score is 0.98, indicating a balanced performance between precision and recall.
- Confusion Matrix: This matrix evaluates the performance of the classifier by showing the counts of instances that were correctly or incorrectly classified for each class:
    - Row 1 corresponds to class 0. The value 32 in the first column indicates that there were 32 instances of class 0 that were correctly predicted as class 0. The value 1 in the fifth column suggests that one instance of class 0 was misclassified as another class.
    - Row 2 corresponds to class 1. The value 27 in the second column indicates that there were 27 instances of class 1 that were correctly predicted as class 1. The value 1 in the third column suggests that one instance of class 1 was misclassified as another class.
    - Row 3 corresponds to class 2. The value 32 in the third column indicates that there were 32 instances of class 2 that were correctly predicted as class 2. The value 1 in the eighth column suggests that one instance of class 2 was misclassified as another class.
    - Row 4 corresponds to class 3. The value 33 in the fourth column indicates that there were 33 instances of class 3 that were correctly predicted as class 3. The value 1 in the sixth column suggests that one instance of class 3 was misclassified as another class.
    - Row 5 corresponds to class 4. The value 46 in the fifth column indicates that there were 46 instances of class 4 that were correctly predicted as class 4. There were no misclassifications into other classes, as indicated by the zeros in the other columns.

- Row 6 corresponds to class 5. The value 46 in the sixth column indicates that there were 46 instances of class 5 that were correctly predicted as class 5. The value 1 in the seventh column suggests that one instance of class 5 was misclassified as another class.
- Row 7 corresponds to class 6. The value 34 in the seventh column indicates that there were 34 instances of class 6 that were correctly predicted as class 6. The value 1 in the first column indicates that one instance of class 6 was misclassified as another class.
- Row 8 corresponds to class 7. The value 33 in the eighth column indicates that there were 33 instances of class 7 that were correctly predicted as class 7. The value 1 in the tenth column indicates that one instance of class 7 was misclassified as another class.
- Row 9 corresponds to class 8. The value 29 in the ninth column indicates that there were 29 instances of class 8 that were correctly predicted as class 8. The value 1 in the second column indicates that one instance of class 8 was misclassified as another class.
- Row 10 corresponds to class 9. The value 39 in the tenth column indicates that there were 39 instances of class 9 that were correctly predicted as class 9. The value 1 in the eighth column indicates that one instance of class 9 was misclassified as another class.

MLP Classification with PySpark

In this subsection, we build, train, and evaluate an MLP classifier with PySpark using the same handwritten digits dataset. This approach serves as a parallel to the earlier Scikit-Learn code, allowing readers to see the similarities in how machine learning tasks can be performed using both frameworks. As with Scikit-Learn, we start by importing the necessary libraries:

Step 1: Import the necessary libraries

```
[In]: from pyspark.sql import SparkSession
[In]: from pyspark.ml.feature import VectorAssembler
[In]: from pyspark.ml.classification import MultilayerPerceptronClassifier
[In]: from pyspark.ml.evaluation import MulticlassClassificationEvaluator
import pandas as pd
from sklearn.datasets import load_digits
```

In this first step, each of the imported libraries has its own functionality: `SparkSession` is used to create a Spark Session, `VectorAssembler` is used to assemble the input features into a vector column, `MultilayerPerceptronClassifier` is the MLP classifier used for classification, `MulticlassClassificationEvaluator` is used to evaluate the performance of the classifier, pd is used to create a Pandas DataFrame, and `load_digits` is used to load the handwritten digits dataset.

Step 2: Create a Spark Session

In this step, we create a Spark Session and give it a name DigitClassification:

```
[In]: spark = SparkSession.builder.appName("DigitClassification").
      getOrCreate()
```

Step 3: Load the digits dataset

```
[In]: dataset = load_digits()
```

The `load_digits()` function is part of Scikit-Learn's built-in datasets module and is specifically designed to load the handwritten digits dataset. Once we load the dataset, we assign the result to the variable dataset.

Step 4: Create a DataFrame

```
[In]: df = pd.DataFrame(data=dataset.data, columns=dataset.feature_names)
[In]: df['label'] = dataset.target
```

In this code, the first line creates a DataFrame called df using the `DataFrame()` function from the Pandas library. The argument `data-dataset.data` specifies the data for the DataFrame, which corresponds to the pixel values of the digit images. The argument `columns=dataset.feature_names` sets the column names of the DataFrame as the feature names provided in the dataset object. These feature names represent the pixel positions in the image.

The second line `df['label'] = dataset.target` adds a new column called label to the DataFrame df. `dataset.target` represents the target variable or the true labels associated with each digit image. Each label corresponds to the numerical value of the digit represented in the image.

Step 5: Convert the Pandas DataFrame to PySpark

In this step, we convert the Pandas df to a PySpark DataFrame:

```
[In]: spark_df = spark.createDataFrame(df)
```

This line of code creates a PySpark DataFrame called spark_df from the existing Pandas DataFrame df using the `createDataFrame()` method. By converting the Pandas DataFrame to a PySpark DataFrame, we can leverage the distributed computing capabilities of Spark and apply scalable operations on the data using Spark's built-in functions and machine learning algorithms. This is particularly useful when working with large datasets that may not fit into memory on a single machine.

Step 6: Split data into training and testing sets

In this step, we split the data into training and testing sets:

```
[In]: train, test = spark_df.randomSplit([0.8, 0.2], seed=42)
```

This line of code splits the PySpark DataFrame spark_df into two separate DataFrames, train and test, for training and testing the MLP classifier, respectively. The `randomSplit()` method allows us to randomly partition a DataFrame into multiple parts based on the given weights. The weights [0.8, 0.2] indicate that approximately 80% of the data will be assigned to the train DataFrame, and the remaining 20% will be assigned to the test DataFrame. The seed=42 parameter sets the random seed for reproducibility. By providing the same seed value, we ensure that the data split remains consistent across multiple runs of the code, which is useful for comparison and debugging purposes.

Step 7: Create a Vector Assembler

In this step, we create a Vector Assembler:

```
[In]: input_features = spark_df.columns[:-1]
[In]: assembler = VectorAssembler(inputCols=input_features,
      outputCol="features")
[In]: train_transformed = assembler.transform(train)
[In]: test_transformed = assembler.transform(test)
```

This code creates a Vector Assembler in three steps:

1. The input features column name is defined.
2. A VectorAssembler is used to combine the features into a single vector column.
3. The training and testing data are transformed using the VectorAssembler.

Let's break this down a bit further:

1. Defining the input features column name:

   `input_features = spark_df.columns[:-1]` assigns the column names of the spark_df DataFrame, except for the last column (which is the target column), to the input_features variable. This step extracts the names of the columns that represent the input features of the MLP model.
2. Creating a Vector Assembler:

   assembler = VectorAssembler(inputCols=input_features, outputCol="features") creates a `VectorAssembler` object called assembler. inputCols=input_features specifies the input columns (input features) to be assembled into a vector, while outputCol="features" sets the name of the output vector column that will contain the assembled features.
3. Transforming the training and testing data:

   train_transformed = assembler.transform(train) applies the `VectorAssembler` transformation to the train DataFrame. It takes the input features from the specified columns and creates a new column named features that contains a vector representation of the assembled features. Similarly, test_transformed = assembler.transform(test) applies the `VectorAssembler` transformation to the test DataFrame.

Step 8: Create an MLP classifier

In this step, we create a neural network:

```
[In]: layers = [len(input_features), 64, 64, len(dataset.target_names)]
[In]: classifier = MultilayerPerceptronClassifier(layers=layers, seed=42,
    labelCol="label")
```

This code creates an MLP classifier in two steps:

1. The layers for the neural network model are defined.
2. A `MultilayerPerceptronClassifier` is created based on those layers.

Let's explain these two steps further:

1. Defining the layers for the neural network:

   `layers=[len(input_features), 64, 64, len(dataset.target_names)]` defines a list called layers that specifies the number of nodes or units in each layer of the neural network. The first element of the list corresponds to the number of input features, which is equal to the length of the input_features list. The subsequent elements represent the number of nodes in each hidden layer of the neural network. The last element corresponds to the number of output classes, which is equal to the length of dataset.target_names.

2. Creating the Multilayer Perceptron classifier:

   classifier = MultilayerPerceptronClassifier(layers=laye rs, seed=42, labelCol="label") creates an instance of the `MultilayerPerceptronClassifier`, which is a Multilayer Perceptron (MLP) neural network model provided by PySpark's ML library. layers=layers specifies the layers configuration for the MLP model, which was defined earlier. seed=42 sets the random seed for reproducibility of the model, while labelCol="label" indicates the name of the label column in the DataFrame, which is the column representing the true labels of the digits.

Step 9: Train the MLP classifier

In this step, we train the neural network:

```
[In]: model = classifier.fit(train_transformed)
```

In this line of code, the `fit()` method is called on the classifier object to train the neural network model using the training data. The `train_transformed` is the training data obtained after applying the `VectorAssembler` transformation to the original training data. The `fit()` method uses this preprocessed training data to optimize the weights and biases of the MLP through an iterative process called backpropagation. During training, the model learns to map the input features to the target labels by adjusting its internal parameters.

After the training process is completed, the `fit()` method returns a trained model object, which is assigned to the model variable. This trained model can then be used to make predictions on new, unseen data.

Step 10: Make predictions

In this step, we make predictions on the test data:

```
[In]: predictions = model.transform(test_transformed)
```

In this line of code, the `transform()` method is called on the trained model object, passing the test_transformed DataFrame as the input. The code generates predictions for the test data using the trained neural network model.

Step 11: Evaluate the model

In this step, we evaluate the model using various metrics:

```
[In]: evaluator = MulticlassClassificationEvaluator(metricName="accuracy")
[In]: accuracy = evaluator.evaluate(predictions)
[In]: evaluator_precision = MulticlassClassificationEvaluator(metricName=
      "weightedPrecision")
[In]: precision = evaluator_precision.evaluate(predictions)
[In]: evaluator_recall = MulticlassClassificationEvaluator(metricName=
      "weightedRecall")
[In]: recall = evaluator_recall.evaluate(predictions)
[In]: evaluator_f1 = MulticlassClassificationEvaluator(metricName="f1")
[In]: f1 = evaluator_f1.evaluate(predictions)
[In]: print("Accuracy:", accuracy)
[In]: print("Precision:", precision)
```

```
[In]: print("Recall:", recall)
[In]: print("F1 Score:", f1)
[In]: confusion_matrix = predictions.groupBy("label").pivot("prediction").
      count().fillna(0).orderBy("label")
[In]: confusion_matrix.show(truncate=False)
[Out]:
Accuracy: 0.95
Precision: 0.96
Recall: 0.95
F1 Score: 0.95
+-----+---+---+---+---+---+---+---+---+---+
|label|0.0|1.0|2.0|3.0|4.0|5.0|6.0|7.0|8.0|9.0|
+-----+---+---+---+---+---+---+---+---+---+
|0    |32 |0  |0  |0  |1  |0  |0  |0  |0  |0  |
|1    |0  |30 |0  |0  |0  |0  |0  |0  |1  |0  |
|2    |0  |1  |40 |0  |0  |0  |0  |0  |0  |0  |
|3    |0  |0  |0  |36 |0  |0  |0  |0  |1  |1  |
|4    |0  |1  |0  |0  |36 |0  |0  |0  |0  |0  |
|5    |0  |0  |0  |0  |0  |35 |0  |0  |0  |0  |
|6    |1  |0  |0  |0  |2  |0  |37 |0  |1  |0  |
|7    |0  |0  |0  |0  |0  |0  |0  |33 |2  |1  |
|8    |0  |0  |0  |1  |0  |0  |0  |0  |28 |0  |
|9    |0  |0  |0  |1  |0  |0  |0  |1  |0  |30 |
+-----+---+---+---+---+---+---+---+---+---+
```

In this final step, several evaluation metrics are calculated, and a confusion matrix is computed based on the predictions generated by the trained MLP model. The following is the three-step process of creating the evaluators, calculating the evaluation metrics, and printing the evaluation metrics:

Creating the evaluators:

- evaluator = MulticlassClassificationEvaluator(metricName="accuracy") creates an instance of the `MulticlassClassificationEvaluator` with the accuracy metric. The evaluator will be used to evaluate the accuracy of the model's predictions.

- evaluator_precision = MulticlassClassificationEvaluator (metricName="weightedPrecision") creates an evaluator with the weightedPrecision metric. It will be used to calculate the precision of the model's predictions.
- evaluator_recall = MulticlassClassificationEvaluator(metricName= "weightedRecall") creates an evaluator with the weightedRecall metric. It will be used to calculate the recall of the model's predictions.
- evaluator_f1 = MulticlassClassificationEvaluator(metricName="f1") creates an evaluator with the f1 metric. It will be used to calculate the F1 score of the model's predictions.

Calculating the evaluation metrics:

- accuracy = evaluator.evaluate(predictions) calculates the accuracy of the model's predictions by applying the evaluator to the predictions DataFrame. The evaluator compares the predicted labels with the true labels and computes the accuracy score.
- precision = evaluator_precision.evaluate(predictions) calculates the precision of the model's predictions using the evaluator_precision.
- recall = evaluator_recall.evaluate(predictions) calculates the recall of the model's predictions using the evaluator_recall.
- f1 = evaluator_f1.evaluate(predictions) calculates the F1 score of the model's predictions using the evaluator_f1.
- confusion_matrix = predictions.groupBy("label").pivot("prediction"). count().fillna(0).orderBy("label") calculates the confusion matrix based on the predictions DataFrame. The DataFrame is grouped by the true labels ("label") and pivoted against the predicted labels ("prediction") to count the occurrences. The resulting DataFrame represents the confusion matrix.

Printing the evaluation metrics:

- print("Precision:", precision) displays the precision metric.
- print("Recall:", recall) displays the recall metric.
- print("F1 Score:", f1) displays the f1 metric.

- confusion_matrix.show(truncate=False) displays the confusion matrix with each true label ("label") as rows and the predicted labels ("prediction") as columns. Each cell of the matrix contains the count of samples with that specific combination of true and predicted labels.

Once the code in this step is executed, the evaluation metrics are displayed. The preceding output can be interpreted as follows:

- *Accuracy* is a metric that measures the overall correctness of the model's predictions. It calculates the ratio of correctly predicted samples to the total number of samples. The model achieved an accuracy of approximately 0.95, indicating that it correctly predicted the label of around 95% of the samples.
- *Precision* is a metric that measures the proportion of true positive predictions out of all positive predictions made by the model. It evaluates the ability of the model to minimize false positives. The precision score of approximately 0.96 suggests that the model has a high percentage of correct positive predictions compared to the total positive predictions made.
- *Recall,* also known as sensitivity or true positive rate, measures the proportion of true positive predictions out of all actual positive samples in the dataset. It assesses the model's ability to capture positive instances. With a recall score of approximately 0.95, the model is effective at identifying the majority of positive samples in the dataset.
- The *F1 score* is the weighted mean of precision and recall. It provides a single metric that balances the trade-off between precision and recall. The F1 score of approximately 0.95 indicates that the model achieves a good balance between precision and recall, with a relatively high overall performance.
- The *confusion matrix* shows the counts of true positive, true negative, false positive, and false negative for each class. The MLP's confusion matrix is a 10×10 matrix representing the classification results for digits 0 to 9. The row headers represent the true labels,

while the column headers represent the predicted labels. Each cell in the matrix represents the count of instances where the true label and predicted label coincide. The following is a breakdown of the confusion matrix:

- Row 0: Out of the instances with true label 0, 32 were correctly predicted as 0, while there was 1 instance incorrectly predicted as 4.
- Row 1: Out of the instances with true label 1, 30 were correctly predicted as 1, while there was 1 instance incorrectly predicted as 8.
- Row 2: Out of the instances with true label 2, 40 were correctly predicted as 2, while 1 was misclassified as 1.
- Row 3: Out of the instances with true label 3, 36 were correctly predicted as 3, while there were 2 instances incorrectly predicted as 8 and 9, respectively.
- Row 4: Out of the instances with true label 4, 36 were correctly predicted as 4, while 1 was incorrectly predicted as 1.
- Row 5: Out of the instances with true label 5, 35 were correctly predicted as 5, and none was misclassified.
- Row 6: Out of the instances with true label 6, 37 were correctly predicted as 6, and 1 instance was incorrectly predicted as 0, 2 as 4, and 1 as 8.
- Row 7: Out of the instances with true label 7, 33 were correctly predicted as 7, 2 incorrectly predicted as 8, and 1 as 9.
- Row 8: Out of the instances with true label 8, 28 were correctly predicted as 8, while 1 instance was incorrectly predicted as 3.
- Row 9: Out of the instances with true label 9, 30 were correctly predicted as 9, 1 was incorrectly predicted as 3, and 1 was misclassified as 7.

Bringing It All Together

In this section, we consolidate all the relevant code from the previous steps into a single code block. This enables the reader to execute the code as a cohesive unit.

Scikit-Learn

Step 1: Import required libraries

```
[In]: from sklearn.datasets import load_digits
[In]: from sklearn.model_selection import train_test_split
[In]: from sklearn.neural_network import MLPClassifier
[In]: from sklearn.metrics import accuracy_score, precision_score, recall_
      score, f1_score, confusion_matrix
```

Step 2: Load the dataset

```
[In]: dataset = load_digits()
```

Step 3: Split the data into training and testing sets

```
[In]: X_train, X_test, y_train, y_test = train_test_split(dataset.data,
      dataset.target, test_size=0.2, random_state=42)
```

Step 4: Train a neural network classifier

```
[In]: classifier = MLPClassifier(hidden_layer_sizes=(64, 64))  # Specify
      the hidden layer sizes
[In]: classifier.fit(X_train, y_train)
```

Step 5: Make predictions on the test set

```
[In]: y_pred = classifier.predict(X_test)
```

Step 6: Calculate and print evaluation metrics

```
# Calculate and print accuracy
[In]: accuracy = accuracy_score(y_test, y_pred)
[In]: print("Accuracy:", accuracy)
```

```
# Calculate and print precision, recall and F1 scores:
[In]: precision = precision_score(y_test, y_pred, average='weighted')
[In]: recall = recall_score(y_test, y_pred, average='weighted')
[In]: f1 = f1_score(y_test, y_pred, average='weighted')
[In]: print("Precision:", precision)
[In]: print("Recall:", recall)
[In]: print("F1 Score:", f1)
# Calculate and print the confusion matrix
[In]: confusion_mat = confusion_matrix(y_test, y_pred)
[In]: print("Confusion Matrix:")
[In]: print(confusion_mat)
```

PySpark

Step 1: Import required libraries

```
[In]: from pyspark.sql import SparkSession
[In]: from pyspark.ml.feature import VectorAssembler
[In]: from pyspark.ml.classification import MultilayerPerceptronClassifier
[In]: from pyspark.ml.evaluation import MulticlassClassificationEvaluator
[In]: import pandas as pd
[In]: from sklearn.datasets import load_digits
```

Step 2: Create a SparkSession

```
[In]: spark = SparkSession.builder.appName("DigitClassification").
      getOrCreate()
```

Step 3: Load the dataset using Scikit-Learn

```
[In]: dataset = load_digits()
```

Step 4: Create a DataFrame from the data using Pandas

```
# Create features
[In]: df = pd.DataFrame(data=dataset.data, columns=dataset.feature_names)
# Add the target column
[In]: df['label'] = dataset.target
```

Step 5: Convert the Pandas DataFrame to a PySpark DataFrame

```
[In]: spark_df = spark.createDataFrame(df)
```

Step 6: Split the data into training and testing sets

```
[In]: train, test = spark_df.randomSplit([0.8, 0.2], seed=42)
```

Step 7: Combine features into a vector column

```
# Define the input features column name
[In]: input_features = spark_df.columns[:-1]
# Create a VectorAssembler to combine features into a single vector column
[In]: assembler = VectorAssembler(inputCols=input_features,
      outputCol="features")
# Transform the training and testing data using the VectorAssembler
[In]: train_transformed = assembler.transform(train)
[In]: test_transformed = assembler.transform(test)
```

Step 8: Create a neural network

```
# Define the layers for the neural network
[In]: layers = [len(input_features), 64, 64, len(dataset.target_names)]
# Create the MultilayerPerceptronClassifier
[In]: classifier = MultilayerPerceptronClassifier(layers=layers, seed=42,
      labelCol="label")
```

Step 9: Train the classifier

```
[In]: model = classifier.fit(train_transformed)
```

Step 10: Make predictions on the test data

```
[In]: predictions = model.transform(test_transformed)
```

Step 11: Evaluate the accuracy of the model

```
# Calculate accuracy
[In]: evaluator = MulticlassClassificationEvaluator(metricName="accuracy")
[In]: accuracy = evaluator.evaluate(predictions)
[In]: print("Accuracy:", accuracy)
```

```
# Calculate and print precision, recall, and F1 score
[In]: evaluator_precision = MulticlassClassificationEvaluator(metricName=
      "weightedPrecision")
[In]: precision = evaluator_precision.evaluate(predictions)
[In]: evaluator_recall = MulticlassClassificationEvaluator(metricName=
  "weightedRecall")
[In]: recall = evaluator_recall.evaluate(predictions)
[In]: evaluator_f1 = MulticlassClassificationEvaluator(metricName="f1")
[In]: f1 = evaluator_f1.evaluate(predictions)
[In]: print("Precision:", precision)
[In]: print("Recall:", recall)
[In]: print("F1 Score:", f1)
# Calculate and print the confusion matrix
[In]: confusion_matrix = predictions.groupBy("label").pivot("prediction").
      count().fillna(0).orderBy("label")
[In]: confusion_matrix.show(truncate=False)
```

Summary

In this chapter, we explored the world of Multilayer Perceptron (MLP) classifiers, a powerful approach to supervised machine learning. We examined the advantages and disadvantages of MLPs and concluded that deep learning remains a versatile technique for both regression and classification.

The chapter illustrated the usage of MLP classifiers, working with the handwritten digits dataset. We implemented the classifiers using Scikit-Learn and PySpark libraries and utilized Pandas and PySpark for data processing and exploration. The trained model achieved very high accuracy, leading to accurate predictions of handwritten digits. We highlighted that, since Scikit-Learn and PySpark follow the same guidelines for deep learning, transitioning to PySpark to harness its distributed capabilities was worthwhile. We also explained the important ways in which PySpark can complement deep learning frameworks like TensorFlow, Keras, and PyTorch.

This chapter concludes our journey with classification. In the next chapter, we will switch to another machine learning technique: recommender systems. These algorithms are crucial in today's digital landscape as they enhance user experiences and drive business success. By analyzing user preferences and behavior, recommender systems can provide personalized recommendations, improving content discovery, increasing user engagement, and boosting sales and customer satisfaction. They play a pivotal role in e-commerce, content streaming, and various online platforms, helping users find relevant products, services, or content in a crowded and information-rich digital world.

CHAPTER 13

Recommender Systems with Pandas, Surprise, and PySpark

In this chapter, we explore a new area of supervised learning, that of recommender systems. Even though recommender systems fall under supervised learning, they do not typically fall under either regression (Chapters 3–6) or classification (Chapters 7–12). They are considered a distinct area within machine learning called collaborative filtering.

While recommender systems involve predicting ratings or preferences, they are different from traditional regression tasks because the output is not a continuous numerical value. They also differ from classification tasks because they don't involve categorizing or assigning discrete labels to instances. Rather, they focus on generating personalized recommendations based on user-item interactions. In other words, the goal is to suggest items or content that the user is likely to be interested in but may not have discovered on their own.

There are several types of recommender systems, including the following:

- Content-based filtering: This approach recommends items similar to the ones a user has liked or interacted with in the past. It analyzes the characteristics or features of items and recommends items with similar attributes. For example, if a user has previously liked action movies, a content-based recommender system might suggest other action movies to that user.

A. Testas, *Distributed Machine Learning with PySpark*, https://doi.org/10.1007/978-1-4842-9751-3_13

- Collaborative filtering: This approach recommends items based on the preferences of similar users. It looks for patterns and similarities among users' behaviors and makes recommendations based on those patterns. Collaborative filtering can be further classified into two types:
    - User-based collaborative filtering: This method finds users who have similar preferences and recommends items that those similar users have liked or interacted with.
    - Item-based collaborative filtering: This method identifies items that are similar to the ones a user has liked and recommends those similar items.
- Hybrid recommender systems: These systems combine multiple approaches to leverage the strengths of different methods. For example, a hybrid system might use content-based filtering to make initial recommendations and then refine them using collaborative filtering techniques.

In this chapter, we use collaborative filtering to develop a recommender system for Amazon electronic products. To implement the algorithm, we use a matrix factorization technique, which is commonly used in collaborative filtering recommender systems. In matrix factorization, the user-item interactions are modeled as a matrix.

We implement the recommender system using the Surprise and PySpark libraries. Since Scikit-Learn does not have built-in functionality for building recommender systems, we have chosen to use the Python Surprise library. Surprise, like Scikit-Learn, is designed for small to medium-sized datasets, and unlike PySpark, it is not typically used for large-scale, big data scenarios. Both Scikit-Learn and Surprise are well suited for tasks where the user can comfortably load the entire dataset into memory on a single machine.

In Surprise, we use SVD (Singular Value Decomposition), while in PySpark, we use ALS (Alternating Least Squares), both of which are based on the matrix factorization technique. In SVD, the goal is to factorize this matrix into two lower-rank matrices, typically referred to as the user matrix and the item matrix, while in ALS, the aim is to iteratively alternate between updating the user matrix and the item matrix in order to solve for the latent factors of users and items by minimizing the reconstruction error between observed and predicted ratings. Both are powerful libraries, but ALS is more commonly used in collaborative filtering recommender systems due to its ability to handle large-scale datasets and its parallelizable nature.

In addition, we utilize both Pandas and PySpark for data processing and exploration. By comparing the implementations in the Pandas and PySpark libraries, we can gain insights into their similarities. This assists data scientists who aim to migrate from Pandas to PySpark, enabling them to leverage the latter's powerful distributed computing API.

The Dataset

The dataset for this project is an open source dataset containing ratings of various Amazon electronic products provided by UC San Diego's Computer Science Department, available at the following link:

URL: `http://jmcauley.ucsd.edu/data/amazon/`

Contributor: Julian McAuley

We begin by reading the CSV file into a Pandas DataFrame from a GitHub location where we have stored a copy of the dataset:

```
[In]: import pandas as pd
[In]: column_names = ['user_id', 'product_id', 'rating', 'timestamp']
[In]: pandas_df = pd.read_csv(
'https://raw.githubusercontent.com/abdelaziztestas/'
'spark_book/main/amazon_electronics.csv', names=column_names)
```

We first import the Pandas library as pd and then define a list called column_names, which contains the names of the columns for the DataFrame. Next, we read the CSV file using the `read_csv()` method and assign the resulting DataFrame to the variable pandas_df.

We should now be able to explore the dataset by first looking at the shape of the DataFrame:

```
[In]: print(pandas_df.shape)
[Out]: (1048576, 4)
```

The output indicates a large dataset with over 1 million rows (1,048,576) and 4 columns. We can print the names of these columns by using the Pandas `columns` attribute:

```
[In]: print(pandas_df.columns)
[Out]: Index(['user_id', 'product_id', 'rating', 'timestamp'],
       dtype='object')
```

The names of the four columns are user_id, product_id, rating, and timestamp. The following is a description of each column:

- user_id: This represents a unique user id.
- product_id: This represents a product distinct id.
- rating: This represents the rating of the product by a user.
- timestamp: This is the time the user rated the product. It is irrelevant for this project.

We can display the first five rows of each of these columns by using the Pandas `head()` method:

```
[In]: print(pandas_df.head())
[Out]:
          user_id product_id  rating   timestamp
0   AKM1MP6P0OYPR  132793040       5  1365811200
1  A2CX7LUOHB2NDG  321732944       5  1341100800
2  A2NWSAGRHCP8N5  439886341       1  1367193600
3  A2WNBOD3WNDNKT  439886341       3  1374451200
4  A1GI0U4ZRJA8WN  439886341       1  1334707200
```

The rating column in the sample data ranges from 1 to 5 stars, where 1 star is Very Poor, suggesting that the customer had a highly negative experience with the product, and 5 is Excellent, indicating that the customer had a highly positive experience with the product. The rating of 3 means Average, suggesting that the product met the customer's basic expectations but may not have exceeded them.

The other two ratings (not shown in the preceding output) are as follows:

- 2 stars: Poor. This indicates dissatisfaction with the product but may not be as severe as a 1-star rating.
- 4 stars: Good. This indicates that the customer had a positive experience with the product.

We can confirm that the dataset has ratings that range from 1 to 5 as follows:

```
[In]: import numpy as np
[In]: sorted_unique_ratings = np.sort(pandas_df["rating"].unique())
[In]: print(sorted_unique_ratings)
[Out]: [1 2 3 4 5]
```

In this code, we import the NumPy library as np and then create a variable called sorted_unique_ratings to store the unique values of the rating column from a Pandas DataFrame (pandas_df). We use NumPy's `sort()` function to sort these unique values in ascending order. Finally, we print the sorted and unique ratings.

We can check the data types of these columns using the Pandas `dtypes` attribute. This helps us understand if we need to modify these data types before building the machine learning algorithm:

```
[In]: print(pandas_df.dtypes)
[Out]:
user_id       object
product_id    object
rating         int64
timestamp      int64
dtype: object
```

The output indicates that the rating column is numeric while the user_id and product_id are categorical. For Surprise's Singular Value Decomposition (SVD) algorithm, there is no need for manual label encoding or one-hot encoding of these categorical variables as Surprise is designed to handle categorical variables like user IDs and item IDs directly. Surprise internally manages the mapping of these categorical values to numerical indices during the model training and prediction process.

However, the situation is different when working with the PySpark recommendation algorithm as it requires explicit indexing of categorical variables, including user IDs and item IDs. We would need to use PySpark's StringIndexer to convert these categorical values into numerical indices before training the model.

Let's perform a count of occurrences for each unique value in the user_id column of the pandas_df DataFrame.

We first sort by descending order:

```
[In]: pandas_result_desc =
pandas_df.groupby('user_id').size().reset_index(name='count').sort_
values(by='count', ascending=False).head(5)
[In]: print(pandas_result_desc)
[Out]:
               user_id  count
611113   A5JLAU2ARJOBO    412
226172  A231WM2Z2JLOU3    249
240295  A25HBO5V8S8SEA    164
616190     A6FIAB28IS79    146
747120    AT6CZDCP4TRGA    128
```

Then sort by ascending order:

```
[In]: pandas_result_asc =
pandas_df.groupby('user_id').size().reset_index(name='count').sort_
values(by='count', ascending=True).head(5)
[In]: print(pandas_result_asc)
[Out]:
                        user_id  count
0         A00037441I8XOQJSUWCAG      1
499070            A3EEUTOF9898GS      1
499071            A3EEUZUBZQ4N4D      1
499072            A3EEVD827ZC4JY      1
499073            A3EEW1G63825UL      1
```

The output indicates that the top user has rated 412 Amazon electronic products and each user has rated at least 1 product.

We can group by products instead of users.

First, we sort by descending order:

```
[In]: pandas_result_desc = pandas_df.groupby('product_id').size().reset_
      index(name='count').sort_values(by='count', ascending=False).head(5)
[In]: print(pandas_result_desc)
[Out]:
       product_id  count
30276  B0002L5R78   9487
24439  B0001FTVEK   5345
61285  B000I68BD4   4903
46504  B000BQ7GW8   4275
14183  B00007E7JU   3523
```

and then sort by ascending order:

```
[In]: pandas_result_asc = pandas_df.groupby('product_id').size().reset_
      index(name='count').sort_values(by='count', ascending=True).head(5)
[In]: print(pandas_result_asc)
[Out]:
       product_id  count
46105  B000BMQOMO      1
17738  B00009R8AJ      1
43278  B000AAJ27W      1
17736  B00009R8AH      1
43279  B000AAJ2CW      1
```

We can see from the output that the top product has been rated 9,487 times and each product has been rated at least once.

Now let's translate the previous Pandas steps to PySpark:

```
[In]: from pyspark.sql import SparkSession
[In]: spark = SparkSession.builder.getOrCreate()
[In]: spark_df = spark.createDataFrame(pandas_df)
```

We begin by importing the `SparkSession` class, which serves as the entry point for Spark operations. Then, we create a `SparkSession` instance named spark with default configuration settings, ensuring that we either retrieve an existing session or create a new one if none exists. Finally, we convert a Pandas DataFrame (pandas_df) into a PySpark DataFrame (spark_df) using the `createDataFrame()` method.

We should now be able to print the shape of spark_df:

```
[In]: print((spark_df.count(), len(spark_df.columns)))
[Out]: (1048576, 4)
```

This confirms the counts produced by Pandas: 1,048,576 rows and 4 columns.

In the next step, we display the column names using the PySpark `columns` attribute:

```
[In]: print(spark_df.columns)
[Out]: ['user_id', 'product_id', 'rating', 'timestamp']
```

We can show the first five rows of spark_df as follows:

```
[In]: spark_df.show(5)
[Out]:
+--------------+----------+------+----------+
|       user_id|product_id|rating| timestamp|
+--------------+----------+------+----------+
| AKM1MP6P0OYPR| 132793040|     5|1365811200|
|A2CX7LUOHB2NDG| 321732944|     5|1341100800|
|A2NWSAGRHCP8N5| 439886341|     1|1367193600|
|A2WNBOD3WNDNKT| 439886341|     3|1374451200|
|A1GI0U4ZRJA8WN| 439886341|     1|1334707200|
+--------------+----------+------+----------+
only showing top 5 rows
```

We can check that there are indeed five unique ratings ranging from 1 to 5 stars as follows:

```
[In]: from pyspark.sql.functions import col
[In]: unique_sorted_ratings = spark_df.select("rating").distinct().
      orderBy(col("rating"))
[In]: unique_sorted_ratings.show()
[Out]:
+------+
|rating|
+------+
|     1|
|     2|
```

```
|     3|
|     4|
|     5|
+------+
```

To get these unique ratings, we first import the `col` function, which is used for column operations. Next, we select the rating column from the DataFrame, ensuring that we obtain unique values using the `distinct()` method. Afterward, we use the `orderBy()` function with the col("rating") expression to sort the unique rating values in ascending order. Finally, we display the result using the `show()` method.

We can print the schema of the PySpark DataFrame as follows:

```
[In]: spark_df.printSchema()
[Out]:
root
 |-- user_id: string (nullable = true)
 |-- product_id: string (nullable = true)
 |-- rating: long (nullable = true)
 |-- timestamp: long (nullable = true)
```

This output confirms the data types indicated by Pandas: two numerical columns (rating and timestamp) and two categorical columns (user_id and product_id). The PySpark model requires the user_id and product_id to be in numerical format, so we will need to make this conversion before training the model.

We can calculate the occurrence count for each user in spark_df.

Let's first do the counts in descending order:

```
[In]: spark_df.groupBy("user_id").count().orderBy("count",
      ascending=False).show(5)
[Out]:
+--------------+-----+
|       user_id|count|
+--------------+-----+
| A5JLAU2ARJOBO|  412|
|A231WM2Z2JLOU3|  249|
|A25HBO5V8S8SEA|  164|
|  A6FIAB28IS79|  146|
```

```
| AT6CZDCP4TRGA|  128|
+--------------+-----+
only showing top 5 rows
```

Then do the same in ascending order:

```
[In]: spark_df.groupBy("user_id").count().orderBy("count",
      ascending=True).show(5)
[Out]:
+--------------+-----+
|       user_id|count|
+--------------+-----+
| AFS5ZGZ2M3ZGV|    1|
|A19TDRIKW64Z6I|    1|
|A1K775TKUNZL43|    1|
|A2VS1WG9VKINXI|    1|
| A76QA8ID3NTCC|    1|
+--------------+-----+
only showing top 5 rows
```

These results are similar to the output produced by Pandas; that is, the most active user had 412 occurrence counts while the least active user had 1.

Instead of grouping by users, we can count the number of ratings by product. We first do this in descending order by setting the ascending parameter to False:

```
[In]: spark_df.groupBy("product_id").count().orderBy("count",
      ascending=False).show(5)
[Out]:
+----------+-----+
|product_id|count|
+----------+-----+
|B0002L5R78| 9487|
|B0001FTVEK| 5345|
|B000I68BD4| 4903|
|B000BQ7GW8| 4275|
|B00007E7JU| 3523|
+----------+-----+
only showing top 5 rows
```

Then do the same in ascending order:

```
[In]: spark_df.groupBy("product_id").count().orderBy("count",
    ascending=False).show(5)
[Out]:
+----------+-----+
|product_id|count|
+----------+-----+
|B00004SC3R|    1|
|B00005KHSJ|    1|
|B00004YK3Q|    1|
|B00000J025|    1|
|B000050GDM|    1|
+----------+-----+
only showing top 5 rows
```

The results confirm the Pandas output: the top electronic product has been rated 9,487, and each product has been rated at least once.

Building a Recommender System

In this section, we build, train, and evaluate a collaborative filtering recommender system using Amazon's electronic products dataset. The model can then be used to recommend some of these products to users who may not have seen them based on their previous ratings for other products. We demonstrate how this is done in both Surprise and PySpark.

Recommender System with Surprise

We start by building a recommender system using the Python Surprise library. Surprise provides a simple and convenient API for implementing collaborative filtering algorithms, matrix factorization techniques, and various recommendation algorithms. It is commonly used for tasks like movie recommendations, book recommendations, and other personalized recommendation systems.

Surprise can be installed with the following command:

```
[In]: pip install scikit-surprise
```

To start the modeling process of creating a recommender system, we need to import the necessary libraries:

Step 1: Import necessary libraries

```
[In]: import pandas as pd
[In]: from surprise import Dataset, Reader, SVD
[In]: from surprise.model_selection import train_test_split
[In]: from surprise.accuracy import rmse
```

This code sets the stage for building and evaluating a recommendation system using the Surprise library. The typical workflow is to proceed by loading the dataset, defining the data reader, splitting the data into training and testing sets using `train_test_split`, and then training and evaluating the SVD recommendation model using the imported functions and classes.

In the code provided previously, we first import the Pandas library for data loading. We then import `Dataset`, `Reader`, and `SVD` from the `surprise` library.

The following are descriptions of the imported Surprise modules:

- Dataset: This provides a convenient way to load and manage datasets for recommender systems.
- Reader: This defines the format of the input data to be used with Surprise.
- SVD: This is an algorithm in Surprise that stands for Singular Value Decomposition. SVD is a matrix factorization technique used in recommender systems to predict user ratings for items.

Next, we import the `train_test_split` function to split the dataset into training and testing sets for evaluating the performance of the recommender system. We then import the `rmse` function to compute the root mean squared error (RMSE), a commonly used metric to evaluate the accuracy of predictions made by a recommender system. It measures the average difference between the predicted and actual ratings.

Step 2: Read data from the CSV file

```
[In]: column_names = ['user_id', 'product_id', 'rating', 'timestamp']
[In]: pandas_df =   pd.read_csv('https://raw.githubusercontent.com/
      abdelaziztestas/'
      'spark_book/main/amazon_electronics.csv', names=column_names)
```

This code involves creating a DataFrame using the Pandas library. The first line creates a list called column_names that contains the names of the columns in the DataFrame. The columns are user_id, product_id, rating, and timestamp. This list will be used to assign column names to the DataFrame.

The second line reads the CSV file from a URL that points to the location of the CSV file on GitHub and creates a Pandas DataFrame named pandas_df. The `read_csv` function is used to read the CSV file. The argument names=column_names specifies that the names in the column_names list should be used as the column names for the Pandas DataFrame.

Step 3: Define the rating scale and load data into Surprise Dataset

```
[In]: reader = Reader(rating_scale=(1, 5))
[In]: data = Dataset.load_from_df(pandas_df[['user_id', 'product_id',
      'rating']], reader)
```

The purpose of this step is to create a `Dataset` object, which will later be used to train and evaluate the recommender system model in Surprise. We can break this into a two-step process:

1. Create an instance of the `Reader` class from the Surprise library using the first line of code, that is, reader = Reader(rating_scale=(1, 5)). The `Reader` class is used to specify the format of the input data. We have set the rating_scale parameter to (1, 5) because the rating variable in the Amazon electronic product data ranges from 1 to 5. This step is necessary because Surprise expects data in a specific format, and using the `Reader` class helps define that format.

2. Create a Surprise `Dataset` object by loading data from pandas_df using the second line of code, that is, data = Dataset.load_from_df(pandas_df[['user_id', 'product_id', 'rating']], reader). The `load_from_df` method is provided by the `Dataset` class in

Surprise and is used to load the data into the required format. In the same line of code, pandas_df[['user_id', 'product_id', 'rating']] selects the columns user_id, product_id, and rating from the pandas_df DataFrame. This subset contains the data needed for the recommendation system. The `reader` object is passed as an argument to `load_from_df` to specify the format of the data.

Step 4: Split data into train and test sets

```
[In]: trainset, testset = train_test_split(data, test_size=0.2)
```

This line of code uses the `train_test_split` function from the Surprise library to split the data into training and testing sets. It uses data (the `Dataset` object) that contains the data to be split as an argument. The test_size=0.2 specifies the proportion of the data that should be allocated for testing. It is set to 0.2, which means 20% of the data will be used for testing, and the remaining 80% will be used for training. The resulting train and test datasets are trainset and testset, respectively. Each contains three elements: user_id, product_id, and rating.

Step 5: Build the SVD model

```
[In]: model = SVD()
```

This line creates an object of the SVD class from the Surprise library and assigns it to the variable model. SVD is a matrix factorization technique commonly used in recommender systems.

Step 6: Train the model

```
[In]: model.fit(trainset)
```

This line of code is used to train the SVD model on the training set using the `fit()` method of the model object, which is an instance of the SVD class in the Surprise library. The trainset contains user-item ratings required by the SVD model to learn and make predictions.

During the training process, the SVD model analyzes the user-item interactions in the training set and learns the underlying patterns and relationships. It uses the matrix factorization technique to decompose the rating matrix and estimate latent factors associated with users and items. The model adjusts its internal parameters to minimize the difference between predicted ratings and actual ratings in the training set.

Step 7: Evaluate the model

```
[In]: predictions = model.test(testset)
[In]: rmse_score = rmse(predictions)
[In]: print('RMSE:', rmse_score)
[Out]: RMSE: 1.31
```

This code is used to generate predictions using the trained SVD model and calculate the root mean squared error (RMSE) as a measure of prediction accuracy.

Let's start with the predictions.

We can print the predictions for the first five test cases and compare them with the actual predictions as follows:

```
[In]: for prediction in test_predictions[:5]:
          rounded_prediction = round(prediction.est, 1)
          print(f"{prediction.uid}, {prediction.iid}, " +
                f"Actual Rating: {prediction.r_ui}, " +
                f"Predicted Rating: {rounded_prediction}")
[Out]:
AL7WWD4J3NSOY, BOOOAYSGHA, Actual Rating: 4.0, Predicted Rating: 4.1
A2OTHXZ7NONID, BOOODZB8BW, Actual Rating: 3.0, Predicted Rating: 3.6
A3GY647TN2ECRO, BOOOO3006E, Actual Rating: 4.0, Predicted Rating: 4.2
AK6PVIKDN83MW, BOOO7POE6O, Actual Rating: 1.0, Predicted Rating: 3.6
A28GX1L3BBZHY3, BOOOO5O7HO, Actual Rating: 5.0, Predicted Rating: 4.4
```

The code prints predictions for the first five test cases generated by the SVD recommendation system. It begins with a for loop iterating over the first five predictions in the test_predictions collection. For each prediction, it calculates the predicted rating, rounding it to one decimal place. The script then prints key information: the user ID, item ID, actual rating (ground truth), and the rounded predicted rating. The code uses formatted strings (f-strings) for clean and readable output. It includes the following details:

- prediction.uid: The user ID associated with the prediction
- prediction.iid: The item (or product) ID associated with the prediction
- prediction.r_ui: The actual rating (ground truth) for this user-item pair

We can see from the output that there are discrepancies between the actual and predicted ratings provided by the user. Some of these discrepancies are significant. For example, the user AK6PVIKDN83MW gave an actual rating of 1.0 for the item B0007POE6O, while the model predicted a much higher rating of 3.6. Similarly, the user A28GX1L3BBZHY3 assigned a high actual rating of 5.0 to the item B00005O7H0, and the model predicted a lower rating of 4.4.

Even though this is just a sample, it gives us an idea of how the model is performing. Large discrepancies lead to larger RMSE values. In our case, an RMSE of approximately 1.31 indicates that, on average, the predictions made by the SVD model have an error (or residual) of approximately 1.31 units when compared to the actual ratings in the test set. This suggests that the model's predictions are, on average, deviating by a significant margin from the actual ratings in the test set.

In a real-world scenario, we would typically investigate why the model overfits and strive to enhance its performance by minimizing its RMSE. However, let's assume for the moment that we are satisfied with the model's performance. The following code demonstrates how this trained model can be utilized for making predictions on unseen data:

```
[In]: existing_user_id = 'AK6PVIKDN83MW'
[In]: new_product_id = 'B660005667RX'
[In]: predicted_rating = model.predict(existing_user_id, new_
      product_id).est
[In]: print(f'Predicted Rating for User {existing_user_id} and Product
      {new_product_id}: {predicted_rating:.2f}')
[Out]:
Predicted Rating for User AK6PVIKDN83MW and Product B660005667RX: 3.79
```

We begin by defining an existing_user_id and a new_product_id to represent a specific user and a product for which we want to make a prediction. Using the trained recommendation model (model), we call the `predict` method, passing in the user and product IDs as arguments. The prediction stored in predicted_rating is then printed using `print()`.

In this example, the model predicts a rating of 3.79 for the user AK6PVIKDN83MW and new product B660005667RX. This predicted value represents an estimated rating or score that the model believes the user would assign to the new product. Since the predicted rating is high (indicating a strong likelihood that the user will like the product), we might recommend B660005667RX to the user as a personalized recommendation.

Recommender System with PySpark

In this subsection, we build, train, and evaluate a recommender system using PySpark. The same Amazon electronic products dataset we used for Scikit-Learn will be used to train and test the PySpark model.

We use ALS (Alternating Least Squares)—a popular collaborative filtering algorithm designed for building recommendation systems at scale. Collaborative filtering methods make recommendations by leveraging the historical behavior and preferences of users, such as their interactions with items (e.g., ratings). ALS specifically focuses on matrix factorization to learn latent factors that represent users and items. ALS decomposes the user-item interaction matrix into two lower-dimensional matrices: one representing users and the other representing items. These matrices capture latent factors that explain the observed user-item interactions. ALS tries to find these latent factors such that their product approximates the original user-item matrix.

To implement this algorithm, we start by importing the necessary libraries.

Step 1: Import necessary libraries

```
[In]: from pyspark.ml.feature import StringIndexer
[In]: from pyspark.ml.recommendation import ALS
[In]: from pyspark.ml.evaluation import RegressionEvaluator
[In]: from pyspark.sql import SparkSession
```

This code is used to import necessary modules from the PySpark library to work with the recommendation system. The `StringIndexer` class is used to convert categorical variables into numerical indices. This is necessary as PySpark, unlike Surprise, needs the data in an indexed format. The `StringIndexer` will be used to index both user and item IDs. The `ALS` is the Alternating Least Squares class used to build the collaborative filtering algorithm based on matrix factorization. The `RegressionEvaluator` class is used to evaluate the performance of the model by calculating the RMSE (root mean squared error). The `SparkSession` class is the entry point for working with structured data in PySpark. It provides the functionality to create DataFrames and perform various data processing operations.

Step 2: Create a SparkSession

```
[In]: spark = SparkSession.builder.getOrCreate()
```

This line of code is used to create a SparkSession object in PySpark. It provides a way to interact with various Spark functionality, such as creating DataFrames and performing distributed data processing operations. The `builder` is a method of the `SparkSession` class that allows the user to configure and set various options for the SparkSession. The `getOrCreate()` is a method of the `builder` that checks if a SparkSession already exists. If it does, it returns the existing SparkSession. If not, it creates a new SparkSession.

Step 3: Load data from DataFrame

```
[In]: spark_df = spark.createDataFrame(pandas_df)
```

This line of code is used to create a Spark DataFrame (spark_df) from an existing Pandas DataFrame (pandas_df) using the `createDataFrame()`method in PySpark.

Step 4: Convert user_id and product_id columns to numeric values

```
[In]: user_indexer = StringIndexer(inputCol="user_id",
      outputCol="user_index")
[In]: product_indexer = StringIndexer(inputCol="product_id",
      outputCol="product_index")
[In]: indexed_data = user_indexer.fit(spark_df).transform(spark_df)
[In]: indexed_data = product_indexer.fit(indexed_data).
      transform(indexed_data)
```

This code uses the `StringIndexer` class to convert categorical columns in spark_df into numerical indices. The following is the process:

- The first line creates an instance of `StringIndexer` and assigns it to the variable user_indexer. It specifies that the input column to be indexed is user_id from the DataFrame, and the resulting indexed values will be stored in a new column named user_index.
- The second line creates another instance of `StringIndexer` and assigns it to the variable product_indexer. Similar to the previous line, it specifies that the input column to be indexed is product_id, and the indexed values will be stored in a new column named product_index.
- The third line applies the `fit()` and `transform()` methods of the user_indexer object to the PySpark DataFrame spark_df. The `fit()` method is used to learn and fit the transformation on the DataFrame, while the `transform()` method applies the transformation and

returns a new DataFrame indexed_data. The user_id column is indexed, and the resulting indexed values are stored in the user_index column.

- The last line applies the `fit()` and `transform()` methods of the `product_indexer` object to the indexed_data DataFrame obtained in the previous step. Similar to the previous line, it fits the transformation on the DataFrame and applies the transformation to create a new DataFrame indexed_data. The product_id column is indexed, and the resulting indexed values are stored in the product_index column.

Step 5: Split data into train and test sets

```
[In]: train, test = indexed_data.randomSplit([0.8, 0.2], seed=42)
```

This line of code is used to split the indexed_data DataFrame into training and testing datasets. The indexed_data DataFrame represents the data where categorical columns user_id and product_id have been converted into numerical indices using `StringIndexer`. To create separate sets for training and testing, the `randomSplit()` function is used. This splits the indexed_data DataFrame into two datasets: train and test. The first argument [0.8, 0.2] specifies the proportions for the split. In this case, 80% of the data is allocated to the training set (train), while the remaining 20% is allocated to the testing set (test). The second argument seed=42 sets a specific random seed to ensure reproducibility, meaning the split will be the same every time the code is executed with the same seed value.

Step 6: Build an ALS model

```
[In]: als = ALS(userCol='user_index', itemCol='product_index',
      ratingCol='rating', coldStartStrategy='drop', nonnegative=True)
```

This code is used to create an instance of the `ALS` (Alternating Least Squares) algorithm for collaborative filtering in PySpark. The `ALS` class from PySpark's MLlib is being instantiated with the following arguments:

- userCol='user_index' specifies the name of the column in the DataFrame that represents the user indices. It is set to user_index, indicating that the user_index column will be used as the user identifier in the ALS model.

- itemCol='product_index' specifies the name of the column in the DataFrame that represents the item (product) indices. It is set to product_index, indicating that the product_index column will be used as the item identifier in the ALS model.
- ratingCol='rating' specifies the name of the column in the DataFrame that contains the ratings. It is set to rating, indicating that the rating column will be used as the target variable (ratings) in the ALS model.
- coldStartStrategy='drop' specifies the strategy to handle cold start scenarios, where the model encounters new users or items that were not present in the training data. The drop strategy is used here, which means that any rows with missing user or item indices in the DataFrame will be dropped during model fitting or prediction.
- nonnegative=True ensures that predictions are non-negative since our actual ratings range from 1 to 5 stars.

Step 7: Train the model

```
[In]: model = als.fit(train)
```

This line of code is used to train the ALS (Alternating Least Squares) model on the training dataset (train). It calls the `fit()` method on the `als` object, which is an instance of the ALS algorithm in PySpark. The `fit()` method trains the ALS model using the training dataset.

During the training process, the ALS algorithm will optimize the model's parameters to minimize the difference between the predicted ratings and the actual ratings in the training dataset. The algorithm uses an iterative approach, alternating between updating the user factors and item factors to find the optimal values that capture the underlying patterns in the data. Once the training process is completed, the `fit()` method returns a trained ALS model, which is assigned to the variable model. This trained model can then be used to make predictions on new, unseen data or to generate recommendations based on user-item interactions.

Step 8: Evaluate the model

```
[In]: predictions = model.transform(test)
[In]: evaluator = RegressionEvaluator(labelCol='rating', metricName='rmse')
[In]: rmse_score = evaluator.evaluate(predictions)
```

```
[In]: print('RMSE:', rmse_score)
[Out]: RMSE: 1.96
```

This code is used to generate predictions using the trained ALS model and evaluate its performance using the RMSE (root mean squared error) metric.

Here is the process in more detail:

- The first line applies the trained ALS model (model) to the test dataset (test) using the `transform()` method. This method applies the model to the test data and generates predictions for the user-item combinations in the test dataset. The resulting predictions are stored in the predictions DataFrame.
- The second line creates an instance of the `RegressionEvaluator` class to evaluate the performance of regression models. It is instantiated with the following arguments:
  - labelCol='rating' specifies the name of the column in the predictions DataFrame that contains the actual ratings. It is set to rating, indicating that the actual ratings are stored in the rating column.
  - metricName='rmse' specifies the evaluation metric to be used, which is RMSE (root mean squared error) in this case.
- The third line applies the `evaluate()` method of the evaluator object to the predictions DataFrame. This method calculates the RMSE between the predicted ratings and the actual ratings in the predictions DataFrame. The resulting RMSE score is assigned to the variable rmse_score.
- The last line prints the RMSE score. It provides an indication of how well the ALS model performed in predicting the ratings on the test dataset. Lower RMSE values indicate better model performance, as they represent smaller differences between the predicted and actual ratings.

Let's start with the predictions.

We can compare the predicted and actual ratings to have some idea about the model's performance. The following code prints a sample of the first five records:

```
[In]: predictions.select("user_id", "product_id", "rating",
      "prediction").show(5)
[Out]:
+--------------+----------+------+----------+
|       user_id|product_id|rating|prediction|
+--------------+----------+------+----------+
|A2AY4YUOX2N1BQ|B000050ZS3|     4|  4.310914|
|A2AY4YUOX2N1BQ|B0000665V7|     5|   4.56616|
|A2AY4YUOX2N1BQ|B00008R9ML|     4|  5.730056|
|A2AY4YUOX2N1BQ|B00009R76N|     5| 4.4490294|
|A2AY4YUOX2N1BQ|B00009R7BD|     5|  6.400287|
+--------------+----------+------+----------+
only showing top 5 rows
```

The output indicates discrepancies between the actual and predicted ratings, although some are relatively small after rounding. For example, user A2AY4YUOX2N1BQ rated product B000050ZS3 as 4, while the model predicted a rating of 4.3, which rounds to 4. Similarly, user A2AY4YUOX2N1BQ rated product B0000665V7 as 5, whereas the model rated it as 4.6, rounding to 5. However, there are somewhat significant discrepancies for other ratings. For example, A2AY4YUOX2N1BQ rated product B00009R7BD as 5, while the model predicted it as 6.4, which remains significant even after rounding.

As for the RMSE, the reported value is approximately 1.96, which means that, on average, the predicted ratings by the ALS model deviate from the actual ratings by approximately 1.96 units. This value provides an indication of the model's accuracy in predicting user-item ratings. The closer the RMSE is to zero, the better the model's predictions align with the true ratings.

Bringing It All Together

Let's now consolidate all the code snippets from the previous steps into a single code block. This way, the reader can execute the code as a single block in both Surprise and PySpark.

Surprise

Step 1: Import necessary libraries

```
[In]: import pandas as pd
[In]: from surprise import Dataset, Reader, SVD
[In]: from surprise.model_selection import train_test_split
[In]: from surprise.accuracy import rmse
```

Step 2: Load data

```
[In]: column_names = ['user_id', 'product_id', 'rating', 'timestamp']
[In]: pandas_df = pd.read_csv('https://raw.githubusercontent.com/
      abdelaziztestas/'
'spark_book/main/amazon_electronics.csv', names=column_names)
```

Step 3: Define the rating scale

```
[In]: reader = Reader(rating_scale=(1, 5))
```

Step 4: Load data into Surprise Dataset

```
[In]: data = Dataset.load_from_df(pandas_df[['user_id', 'product_id',
      'rating']], reader)
```

Step 5: Split data into train and test sets

```
[In]: trainset, testset = train_test_split(data, test_size=0.2)
```

Step 6: Build SVD model

```
[In]: model = SVD()
```

Step 7: Train the model

```
[In]: model.fit(trainset)
```

Step 8: Evaluate the model on the testset

```
[In]: predictions = model.test(testset)
[In]: rmse_score = rmse(predictions)
[In]: print('RMSE:', rmse_score)
```

Step 9: Print actual and predicted ratings for first five test cases

```
[In]: for prediction in predictions[:5]:
          rounded_prediction = round(prediction.est, 1)
          print(f"{prediction.uid}, {prediction.iid}, " +
                f"Actual Rating: {prediction.r_ui}, " +
                f"Predicted Rating: {rounded_prediction}")
```

Step 10: Make new prediction

```
[In]: existing_user_id = 'AK6PVIKDN83MW'
[In]: new_product_id = 'B660005667RX'
[In]: predicted_rating = model.predict(existing_user_id, new_
      product_id).est
[In]: print(f'Predicted Rating for User {existing_user_id} and Product
      {new_product_id}: {predicted_rating:.2f}')
```

PySpark

Step 1: Import necessary libraries

```
[In]: from pyspark.ml.feature import StringIndexer
[In]: from pyspark.ml.recommendation import ALS
[In]: from pyspark.ml.evaluation import RegressionEvaluator
[In]: from pyspark.sql import SparkSession
```

Step 2: Create a SparkSession

```
[In]: spark = SparkSession.builder.getOrCreate()
```

Step 3: Load data from DataFrame

```
[In]: spark_df = spark.createDataFrame(pandas_df)
```

Step 4: Convert user_id and product_id columns to numeric values

```
[In]: user_indexer = StringIndexer(inputCol="user_id",
      outputCol="user_index")
[In]: product_indexer = StringIndexer(inputCol="product_id",
      outputCol="product_index")
```

```
[In]: indexed_data = user_indexer.fit(spark_df).transform(spark_df)
[In]: indexed_data = product_indexer.fit(indexed_data).
      transform(indexed_data)
```

Step 5: Split data into train and test sets

```
[In]: train, test = indexed_data.randomSplit([0.8, 0.2], seed=42)
```

Step 6: Build the ALS model

```
[In]: als = ALS(userCol='user_index', itemCol='product_index',
      ratingCol='rating', coldStartStrategy='drop', nonnegative=True)
```

Step 7: Train the model

```
[In]: model = als.fit(train)
```

Step 8: Evaluate the model

```
[In]: predictions = model.transform(test)
[In]: evaluator = RegressionEvaluator(labelCol='rating', metricName='rmse')
[In]: rmse_score = evaluator.evaluate(predictions)
[In]: print('RMSE:', rmse_score)
```

Step 9: Print the first five predictions

```
[In]: predictions.select("user_id", "product_id", "rating",
      "prediction").show(5)
```

Summary

In this chapter, we explored a new area of supervised learning: recommender systems. We utilized collaborative filtering to develop a recommender system for Amazon electronic products. To implement the algorithm, we employed a matrix factorization technique, commonly used in collaborative filtering recommender systems.

We implemented the recommender system using the Surprise and PySpark libraries. Surprise, much like Scikit-Learn, is designed for small to medium-sized datasets, and unlike PySpark, it is not typically used for large-scale, big data scenarios. Additionally, we utilized both Pandas and PySpark for data processing and exploration. By comparing the implementations in these libraries, we gained insights into their similarities. The goal is to assist data scientists who intend to transition from Pandas to PySpark, enabling them to leverage the latter's powerful distributed computing API.

In the next chapter, we will delve into yet another area of supervised learning: natural language processing (NLP), exploring how to analyze and extract valuable insights from text data.

CHAPTER 14

Natural Language Processing with Pandas, Scikit-Learn, and PySpark

In this chapter, we move to a new area of machine learning, namely, that of processing text data and applying an algorithm to it. This area of machine learning is known as natural language processing (NLP), which finds uses in many business applications including speech recognition, chatbots, language translation, and email spam detection (ham or spam).

The project of this chapter is to examine the key steps involved in processing text data using an open source dataset known as the 20 Newsgroups. This is a collection of newsgroup documents partitioned into 20 different topics. We build, train, and evaluate a Multinomial Naive Bayes algorithm and use it to predict the topic categories in this dataset. Even though any other supervised learning classification model covered in the previous chapters can be used, Naive Bayes is often the model of choice for NLP tasks such as topic modeling (the task at hand).

Before training the model, we need to ensure that the text data that feeds into it is in the correct format. This requires cleaning, tokenization, and vectorization. Cleaning in the context of NLP is the act of preprocessing text data by removing irrelevant or noisy information that does not add much to the meaning, such as punctuation marks and stop words. Tokenization involves breaking down the text into smaller units called tokens, which can be words, phrases, or even characters, which helps to organize the text and enables further analysis. Vectorization encompasses converting the text data into numerical representations that the machine learning algorithm can understand. We explore two methods commonly used for this purpose: Bag of Words (BoW) and Term Frequency-Inverse Document Frequency (TF-IDF).

A. Testas, *Distributed Machine Learning with PySpark*, https://doi.org/10.1007/978-1-4842-9751-3_14

To demonstrate the use of Naive Bayes with NLP, we use three libraries: Pandas, Scikit-Learn, and PySpark. The goal is to help data scientists familiar with the small data library tools (Pandas and Scikit-Learn) to transition to PySpark, enabling them to leverage the advantages of distributed programming and parallel computing for NLP tasks.

The Dataset

The 20 Newsgroups dataset comes installed in Scikit-Learn. It consists of a collection of close to 20,000 newsgroup documents, each representing a single post from one of the newsgroups. These documents are spread across 20 different categories or topics, covering a diverse range of subjects such as sports, politics, religion, and science.

To be able to explore this dataset, we need to first create a Pandas DataFrame through the following steps:

Step 1: Import necessary libraries

```
[In]: import pandas as pd
[In]: from sklearn.datasets import fetch_20newsgroups
```

The code imports the Pandas library as well as the `fetch_20newsgroups` function, which allows us to download and load the 20 Newsgroups dataset.

Step 2: Fetch the 20 Newsgroups dataset using Scikit-Learn

```
[In]: categories = ['alt.atheism', 'comp.graphics', 'comp.os.ms-
                    windows.misc', 'comp.sys.ibm.pc.hardware',
                    'comp.sys.mac.hardware', 'comp.windows.x',
                    'misc.forsale', 'rec.autos', 'rec.motorcycles',
                    'rec.sport.baseball', 'rec.sport.hockey',
                    'sci.crypt', 'sci.electronics', 'sci.med',
                    'sci.space', 'soc.religion.christian',
                    'talk.politics.guns', 'talk.politics.mideast',
                    'talk.politics.misc', 'talk.religion.misc']
[In]: newsgroups_all = fetch_20newsgroups(subset='all',
                    categories=categories, shuffle=True, random_state=42)
```

The first line defines a list called categories that contains the names of the 20 newsgroup categories. Each category represents a different topic or subject discussed in the newsgroups. For example, alt.atheism represents discussions related to atheism, while comp.graphics represents discussions related to computer graphics.

The second line uses the `fetch_20newsgroups` function to load the 20 Newsgroups dataset. By specifying subset='all', we are indicating that we want to load the entire dataset (specifying subset='train' would load the training set, while specifying subset='test' would load the test set). The categories=categories argument tells the function to fetch the documents that belong to the categories listed in the categories list we defined earlier. The shuffle=True parameter ensures that the documents are randomly shuffled, while random_state=42 sets a specific random seed for reproducibility.

Step 3: Create a Pandas DataFrame from the Scikit-Learn data and target arrays

```
[In]: newsgroups_df = pd.DataFrame({'text': newsgroups_all.data, 'label':
    newsgroups_all.target})
```

This line creates a Pandas DataFrame named newsgroups_df using the `DataFrame()` constructor. It combines the text data and label information from the `newsgroups_all` object.

Now that we have successfully created a Pandas DataFrame that holds the 20 newsgroups data, we begin our exploration by printing the shape of the DataFrame:

```
[In]: newsgroups_df.shape
[Out]: (18846, 2)
```

This output indicates that the DataFrame has 18,846 rows and 2 columns (text and label).

Next, we show the top five rows of the DataFrame using the Pandas `head()` method:

```
[In]: print(newsgroups_df.head())
[Out]:
                                                text  label
0  From: Mamatha Devineni Ratnam <mr47+@andrew.cm...     10
1  From: mblawson@midway.ecn.uoknor.edu (Matthew ...      3
2  From: hilmi-er@dsv.su.se (Hilmi Eren)\nSubject...     17
3  From: guyd@austin.ibm.com (Guy Dawson)\nSubjec...      3
4  From: Alexander Samuel McDiarmid <am2o+@andrew...      4
```

We can confirm from the preceding output that the DataFrame has two columns: text and label. The labels correspond to the topic categories, which should be 20. We can verify this number by using the Pandas nunique() method:

```
[In]: newsgroups_df.label.nunique()
[Out]: 20
```

The text column is truncated and doesn't tell us much about its content. To print the entire first line, we can run the following code, which extracts the value of the text column for the first row (index 0) in the Pandas DataFrame:

```
[In]: print(newsgroups_df.iloc[0]['text'])
[Out]:
I am sure some bashers of Pens fans are pretty confused about the lack of
any kind of posts about the recent Pens massacre of the Devils. Actually,
I am bit puzzled too and a bit relieved. However, I am going to put an end
to non-PIttsburghers' relief with a bit of praise for the Pens. Man, they
are killing those Devils worse than I thought. Jagr just showed you why he
is much better than his regular season stats. He is also a lot fo fun to
watch in the playoffs. Bowman should let JAgr have a lot of fun in the next
couple of games since the Pens are going to beat the pulp out of Jersey
anyway. I was very disappointed not to see the Islanders lose the final
regular season game. PENS RULE!!!
```

The output contains text related to the Pittsburgh Penguins (referred to as "Pens") and their victory over the New Jersey Devils in a hockey game. The content of the sample document includes opinions and reactions to the Pens' performance, expressing surprise, relief, and praise for the team. It mentions specific players like Jagr and talks about the anticipation of upcoming games. Additionally, there is a mention of disappointment regarding the outcome of the final regular season game involving the Islanders.

To replicate the Pandas results using PySpark, we need to first convert the Pandas newsgroups_df to a PySpark DataFrame.

Let's go through the steps:

Step 1: Import necessary libraries

```
[In]: from pyspark.sql import SparkSession
```

We import the SparkSession class to manage the connection to Spark and provide the interface for working with structured data.

Step 2: Create a Spark Session

```
[In]: spark = SparkSession.builder.getOrCreate()
```

This line creates an instance of the SparkSession and assigns it to the variable spark. The builder method returns a SparkSession.Builder object, and the getOrCreate() method is called on this object to either create a new SparkSession or retrieve an existing one. By assigning the result to the variable spark, we can use the spark object throughout our code to access the functionality provided by SparkSession, such as reading data and applying transformations on it. One such transformation is converting the newsgroups_df from Pandas to PySpark using the createDataFrame() method, as shown in the following.

Step 3: Convert the Pandas DataFrame to PySpark:

```
[In]: spark_df = spark.createDataFrame(newsgroups_df)
```

Now that we have successfully created a Spark DataFrame, we can perform similar explorations as done earlier with Pandas.

Let's begin by displaying the shape of the spark_df using the count() and len() methods:

```
[In]: row_count = spark_df.count()
[In]: column_count = len(spark_df.columns)
[In]: shape = (row_count, column_count)
[In]: print(shape)
[Out]: (18846, 2)
```

This output confirms Pandas output (18,846 rows and 2 columns) using the shape attribute. Since PySpark doesn't have a built-in shape attribute, we combined the count() and len() methods to calculate the number of rows and columns, respectively.

Next, we display the top five rows of the PySpark DataFrame using the PySpark show() method:

```
[In]: spark_df.show(5)
[Out]:
+--------------------+-----+
|                text|label|
+--------------------+-----+
|From: Mamatha Dev...|   10|
|From: mblawson@mi...|    3|
|From: hilmi-er@ds...|   17|
|From: guyd@austin...|    3|
|From: Alexander S...|    4|
+--------------------+-----+
only showing top 5 rows
```

The output is truncated. We can specify a larger value for the truncate parameter in the show() method to increase the truncation limit. For example, the following code will set truncation to the maximum length, effectively displaying the full content of each column:

```
[In]: spark_df.show(5, truncate=False)
[Out]:
From: Mamatha Devineni Ratnam <mr47+@andrew.cmu.edu>\nSubject: Pens fans
reactions\nOrganization: Post Office, Carnegie Mellon, Pittsburgh, PA\
nLines: 12\nNNTP-Posting-Host: po4.andrew.cmu.edu\n\n\n\n
I am sure some bashers of Pens fans are pretty confused about the lack\nof
any kind of posts about the recent Pens massacre of the Devils. Actually,\
nI am  bit puzzled too and a bit relieved. However, I am going to put an
end\nto non-PIttsburghers' relief with a bit of praise for the Pens. Man,
they\nare killing those Devils worse than I thought. Jagr just showed you
why\nhe is much better than his regular season stats. He is also a lot\nfo
fun to watch in the playoffs. Bowman should let JAgr have a lot of\nfun in
the next couple of games since the Pens are going to beat the pulp out of
Jersey anyway. I was very disappointed not to see the Islanders lose the
final\nregular season game.          PENS RULE!!!\n\n|10
```

Alternatively, we can print sample content using the following code:

```
[In]: text_value = spark_df.select('text').first()['text']
[In]: print(text_value)
[Out]:
From: Mamatha Devineni Ratnam <mr47+@andrew.cmu.edu>
Subject: Pens fans reactions
Organization: Post Office, Carnegie Mellon, Pittsburgh, PA
Lines: 12
NNTP-Posting-Host: po4.andrew.cmu.edu

I am sure some bashers of Pens fans are pretty confused about the lack of
any kind of posts about the recent Pens massacre of the Devils. Actually, I
am  bit puzzled too and a bit relieved. However, I am going to put an end
to non-PIttsburghers' relief with a bit of praise for the Pens. Man, they
are killing those Devils worse than I thought. Jagr just showed you why he
is much better than his regular season stats. He is also a lot fo fun to
watch in the playoffs. Bowman should let JAgr have a lot of fun in the next
couple of games since the Pens are going to beat the pulp out of Jersey
anyway. I was very disappointed not to see the Islanders lose the final
regular season game.          PENS RULE!!!
```

The first line of code selects the text column from spark_df and retrieves the first row as a Row object. The `select()` method is used to select only the text column from the DataFrame, and `first()` retrieves the first row as a Row object. The second line prints the value stored in text_value, which represents the text content from the text column of the first row in the DataFrame.

Finally, we can use the `distinct()` and `count()` methods in PySpark to calculate the number of unique categories in the newsgroups dataset:

```
[In]: spark_df.select("label").distinct().count()
[Out]: 20
```

This confirms the number of categories computed by Pandas (i.e., 20 categories).

Cleaning, Tokenization, and Vectorization

As stated in the introduction to this chapter, prior to constructing, training, and assessing a Multinomial Naive Bayes model for classifying topics into one of the 20 newsgroup categories, it is essential to ensure that the text data is formatted in a way compatible with the algorithm. This entails three essential steps:

1. Cleaning: The initial step involves preprocessing the text data to eliminate irrelevant or noisy elements that do not significantly contribute to its meaning. These elements encompass punctuation marks (such as periods, commas, and question marks) and stop words (like "and," "the," and "is").
2. Tokenization: Following cleaning, the text is broken down into smaller units referred to as tokens. These tokens can encompass words, phrases, or even individual characters. Tokenization serves the purpose of structuring the text and facilitating subsequent analysis.
3. Vectorization: Lastly, the text data must be transformed into numerical representations that can be understood by the machine learning algorithm. This transformation enables the algorithm to process and learn from the textual information effectively.

There are two main approaches to the last step (vectorization): Bag of Words (BoW) and Term Frequency-Inverse Document Frequency (TF-IDF).

Let's explain these methods:

In the BoW approach, each document is represented as a vector where each dimension or feature corresponds to a unique word in the entire corpus. The value in each dimension represents the frequency of that word in the document.

Let's use a few sentences about visiting the Eiffel Tower in Paris as an example for creating a BoW representation. In this example, we'll tokenize the text, remove punctuation and common stop words, and create a simple BoW vector:

Sentences:

"I visited the Eiffel Tower in Paris last summer."

"The Eiffel Tower is an iconic landmark in France."

"Paris is a beautiful city with many attractions."

Steps to create BoW:

1. Tokenization: Break down the sentences into individual words.

   Sentence 1: ["I", "visited", "the", "Eiffel", "Tower", "in", "Paris", "last", "summer"]

   Sentence 2: ["The", "Eiffel", "Tower", "is", "an", "iconic", "landmark", "in", "France"]

   Sentence 3: ["Paris", "is", "a", "beautiful", "city", "with", "many", "attractions"]

2. Removing stop words: Remove common stop words that do not carry much meaning.

   Sentence 1: ["visited", "Eiffel", "Tower", "Paris", "summer"]

   Sentence 2: ["Eiffel", "Tower", "iconic", "landmark", "France"]

   Sentence 3: ["Paris", "beautiful", "city", "attractions"]

3. Create BoW vector: Here, we represent the text data into numerical form so that it can be used by a machine learning algorithm. We first create a list of unique words (vocabulary) that appear in all three sentences:

   [visited Eiffel Tower Paris summer iconic landmark France beautiful city attractions]

The vocabulary consists of 11 distinct words. Next, we represent each word with a Boolean value 0 or 1:

Sentence 1: ["visited", "Eiffel", "Tower", "Paris", "summer"]

[visited 1, Eiffel 1, Tower 1, Paris 1, summer 1, iconic 0, landmark 0, France 0, beautiful 0, city 0, attractions 0]

Binary Vector for Sentence 1: [1, 1, 1, 1, 1, 0, 0, 0, 0, 0, 0]

Sentence 2: ["Eiffel", "Tower", "iconic", "landmark", "France"]

[visited 0, Eiffel 1, Tower 1, Paris 0, summer 0, iconic 1, landmark 1, France 1, beautiful 0, city 0, attractions 0]

Binary Vector for Sentence 2: [0, 1, 1, 0, 0, 1, 1, 1, 0, 0, 0]

Sentence 3: ["Paris", "beautiful", "city", "attractions"]

[visited 0, Eiffel 0, Tower 0, Paris 1, summer 0, iconic 0, landmark 0, France 0, beautiful 1, city 1, attractions 1]

Binary Vector for Sentence 3: [0, 0, 0, 1, 0, 0, 0, 0, 1, 1, 1]

Combining the three sentences, we get the following BoW matrix:

```
[
  [1, 1, 1, 1, 1, 0, 0, 0, 0, 0, 0],
  [0, 1, 1, 0, 0, 1, 1, 1, 0, 0, 0],
  [0, 0, 0, 1, 0, 0, 0, 0, 1, 1, 1]
]
```

This matrix represents the frequency of words in each sentence, where 1 indicates the presence of a word and 0 indicates its absence. The matrix can then be used as numerical input for machine learning algorithms.

However, while the BoW approach is a straightforward method for text vectorization, it has its limitations, especially when dealing with larger and more complex text data. BoW represents text using binary values, indicating word presence or absence, which can be simplistic in capturing the nuances of language. Put it differently, BoW treats text as a collection of individual words, disregarding word order and context. This method is simple and efficient but lacks information about word importance.

A more sophisticated vectorization technique is the Term Frequency-Inverse Document Frequency (TF-IDF). This is a more advanced vectorization technique that goes beyond simple word frequency. It assigns weights to words based on their frequency within a document (Term Frequency) and their rarity across the entire corpus (Inverse Document Frequency). Words that are frequent within a specific document but rare in the corpus as a whole receive higher TF-IDF scores, making them more informative and significant in the context of that document. TF-IDF effectively addresses the limitations of BoW by downweighting common words (such as stop words) and emphasizing rare, important terms.

The formulas for Term Frequency (TF) and Inverse Document Frequency (IDF) are as follows:

Term Frequency (TF):

TF(t, d) = (Number of times term t appears in document d) / (Total number of terms in document d)

Inverse Document Frequency (IDF):

IDF(t, D) = log_e (Total number of documents in the corpus D / Number of documents containing term t)

TF-IDF is then calculated as the product of TF and IDF:

$$\text{TF-IDF}(t, d, D) = \text{TF}(t, d) * \text{IDF}(t, D)$$

TF-IDF is a widely used technique in text analysis with several advantages. First and foremost, TF-IDF is effective in capturing the importance of words within a document by assigning higher weights to terms that are both frequent within the document and rare in the overall corpus. This helps in highlighting the most relevant and distinctive words for a specific document or context, making it suitable for tasks such as information retrieval, search engines, and text classification. Additionally, TF-IDF is relatively simple to implement and computationally efficient, making it a practical choice for a wide range of natural language processing applications.

Despite its strengths, TF-IDF has certain limitations. One drawback is that it treats each term independently, failing to capture semantic relationships between words or phrases in a document. Additionally, TF-IDF relies on word frequency, which may not always accurately reflect the importance of terms in more complex language structures. Furthermore, TF-IDF can be sensitive to document length, potentially biasing longer documents that naturally contain more terms. Lastly, TF-IDF does not consider word order or context, which can be crucial in tasks like sentiment analysis or understanding the meaning of phrases. As such, while TF-IDF is a valuable tool, it may not be the best choice for every text analysis scenario, especially those requiring a deeper understanding of language semantics.

We can now write some Python code to apply the preprocessing steps to the 20 Newsgroups dataset. We use the Natural Language Toolkit (NLTK) library for this purpose. NLTK provides easy-to-use interfaces to over 50 corpora and lexical resources, such as WordNet. It also includes a wide range of text processing libraries for tasks like tokenization, stemming, tagging, parsing, and semantic reasoning. NLTK is widely used in natural language processing (NLP) and text analysis tasks, including sentiment analysis, text classification, and language modeling.

NLTK can be installed with the following command:

```
[In]: pip install nltk
```

We also need to download punkt and stopwords resources from the NLTK library:

```
[In]: nltk.download('punkt')
[In]: nltk.download('stopwords')
```

The first command downloads the `punkt` resource, which is a data package that contains pre-trained models for tokenization (the process of splitting text into individual tokens, such as words and punctuation). The second command downloads the `stopwords` resource, which includes a predefined list of stop words commonly used in the English language, such as "the", "is", and "and". These words are commonly removed in natural language processing tasks as they have little or no significant meaning in the context of language analysis.

In the next step, we import the necessary libraries:

Step 1: Import necessary libraries

```
[In]: import nltk
[In]: from sklearn.datasets import fetch_20newsgroups
[In]: from sklearn.feature_extraction.text import TfidfVectorizer
[In]: import string
```

The first line of code imports NLTK. The second line imports the `fetch_20newsgroups` function, which is used to download and retrieve the 20 Newsgroups dataset. The third line imports the `TfidfVectorizer`, which is a feature extraction class commonly used in text analysis. It converts a collection of raw documents into a matrix of TF-IDF features, which can be used as input for the machine learning algorithm. The last line imports the string module, which provides a collection of common string operations in Python. It is used to access constants such as `string.punctuation`, which represents a string of all ASCII punctuation characters, during the punctuation removal step.

Step 2: Load the 20 Newsgroups dataset

```
[In]: newsgroups_data = fetch_20newsgroups(subset='all',
                remove=('headers', 'footers', 'quotes'))
```

This line of code is used to fetch the 20 Newsgroups dataset and store it in the variable newsgroups_data using the `fetch_20newsgroups` function. The subset='all' parameter specifies which subset of the dataset (train, test, or all) to fetch. In this case, 'all' indicates that all the data from the 20 Newsgroups dataset will be fetched.

The remove=('headers', 'footers', 'quotes') parameter specifies which parts of each document should be removed, in this case, headers, signature blocks, and quotation blocks, respectively. This metadata is removed as part of the data cleaning process because it has little to do with topic classification.

It is important to note that the remove parameter is specific to the 20 Newsgroups dataset. The re (regular expressions) in Python is a general-purpose module for working with regular expressions to remove specific patterns including quotes, headers, and footers. This is a more manual approach that gives us more control over the cleaning process, but it requires defining the patterns and applying them to the document text.

Here is an example of how the re module can be used to clean data:

```
[In]: import re
[In]: text = "Check out this website: https://www.example.com. It has
          great content."
[In]: pattern = r"http[s]?://\S+"
[In]: clean_text = re.sub(pattern, "", text)
[In]: print(clean_text)
[Out]: Check out this website: It has great content.
```

In the preceding example, we want to remove any URLs from the given text because it doesn't add to the quality of the input for the algorithm (on the contrary, unnecessary patterns can create noise in the data and increase processing time). We define a regular expression pattern http[s]?://\S+ that matches URLs starting with either "http://" or "https://". The \S+ part matches one or more non-whitespace characters, ensuring that the URL is captured entirely. We then use the re.sub() function to substitute all matches of the pattern with an empty string, effectively removing them from the text. The resulting cleaned text, without any URLs, is then displayed.

Step 3: Select a single document from the dataset

```
[In]: document_index =0
[In]: document = newsgroups_data.data[document_index]
```

In the first line of code, document_index is assigned the value 0. This variable is used as an index to indicate the position of a specific single document within the newsgroups_data dataset. We have chosen for the sake of illustration to work with just one document. In reality, however, the entire corpus will be used.

The second line retrieves this document from the newsgroups_data dataset based on the value of document_index.

In the next steps, we move to the tokenization process. We start by defining a function called tokenize:

Step 4: Define the tokenize function

```
[In]: def tokenize(text):
        tokens = nltk.word_tokenize(text)
        stop_words = set(nltk.corpus.stopwords.words('english'))
        punctuations = set(string.punctuation)
        filtered_tokens = [token for token in tokens if
        token.lower() not in stop_words and token not in
        punctuations]
        return filtered_tokens
```

The function takes a text parameter as input. This function is responsible for tokenizing the given text into individual words and filtering out stop words and punctuation.

Inside the function, we have the following:

- nltk.word_tokenize(text) tokenizes the input text into a list of words using the NLTK library's word tokenizer.
- nltk.corpus.stopwords.words('english') retrieves a set of common English stop words from the NLTK corpus.
- string.punctuation contains a string of all ASCII punctuation characters.

The tokens are then filtered using a list comprehension. Each token is checked if it is not in the set of stop words (after converting it to lowercase) and not in the set of punctuation characters. The filtered tokens are stored in the filtered_tokens list, which is returned by the tokenize function.

Step 5: Tokenize the document

```
[In]: tokens = tokenize(document)
```

This line calls the tokenize function, passing the document as the input text. The document variable contains the text of the document. The resulting tokens are stored in the tokens variable.

Step 6: Print the tokens

```
[In]: print("Tokens:")
[In]: print(tokens)
```

The first line prints the string "Tokens:", serving as a header, while the second prints the contents of the tokens variable.

Let's take a look at the cleaned and tokenized output before moving on to the vectorization process:

```
[Out]:
Tokens:
['sure', 'bashers', 'Pens', 'fans', 'pretty', 'confused', 'lack', 'kind',
'posts', 'recent', 'Pens', 'massacre', 'Devils', 'Actually', 'bit',
'puzzled', 'bit', 'relieved', 'However', 'going', 'put', 'end', 'non-
PIttsburghers', 'relief', 'bit', 'praise', 'Pens', 'Man', 'killing',
'Devils', 'worse', 'thought', 'Jagr', 'showed', 'much', 'better',
'regular', 'season', 'stats', 'also', 'lot', 'fo', 'fun', 'watch',
'playoffs', 'Bowman', 'let', 'JAgr', 'lot', 'fun', 'next', 'couple',
'games', 'since', 'Pens', 'going', 'beat', 'pulp', 'Jersey', 'anyway',
'disappointed', 'see', 'Islanders', 'lose', 'final', 'regular', 'season',
'game', 'PENS', 'RULE']
```

Here is the original text:

I am sure some bashers of Pens fans are pretty confused about the lack of any kind of posts about the recent Pens massacre of the Devils. Actually, I am bit puzzled too and a bit relieved. However, I am going to put an end to non-PIttsburghers' relief with a bit of praise for the Pens. Man, they are killing those Devils worse than I thought. Jagr just showed you why he is much better than his regular season stats. He is also a lot fo fun to watch in the playoffs. Bowman should let JAgr have a lot of fun in the next couple of games since the Pens are going to beat the pulp out of Jersey anyway. I was very disappointed not to see the Islanders lose the final regular season game. PENS RULE!!!

Upon comparing the original text and the generated tokens, we can see that the tokenization process has split the text into individual words and removed both punctuation marks and stop words.

Moving to vectorization, we start by creating a vectorizer to convert the tokenized text into a matrix of TF-IDF features:

Step 7: Create a vectorizer

```
[In]: vectorizer = TfidfVectorizer()
```

This line creates an instance of the `TfidfVectorizer` class from the Scikit-Learn library. This is used for converting a collection of raw documents into a matrix of TF-IDF (Term Frequency-Inverse Document Frequency) features. TF-IDF is a numerical statistic that reflects the importance of a term in a document within a larger collection of documents.

Step 8: Apply vectorization on the document

```
[In]: tfidf_matrix = vectorizer.fit_transform([document])
```

This line performs the transformation of a single document into a TF-IDF matrix using the previously created vectorizer instance. First, a document is passed as an argument to the `fit_transform()` method of the vectorizer object. This method combines two steps: fitting and transforming.

- Fitting: The `fit_transform()` method fits the vectorizer to the given document or corpus. This means it learns the vocabulary from the document and computes the IDF values based on the term frequencies observed in the document. In this case, since we have a single document, the vocabulary and IDF values will be learned from that document.
- Transforming: Once the vectorizer is fitted, the `transform()` step converts the document into a TF-IDF matrix representation. The resulting tfidf_matrix is a sparse matrix where each row corresponds to a document and each column represents a unique term from the vocabulary. The matrix contains the TF-IDF values for each term in the document.

Step 9: Print the TF-IDF matrix

```
[In]: print("\nTF-IDF Matrix:")
[In]: print(tfidf_matrix.toarray())
[Out]:
TF-IDF Matrix:
[[0.10585122 0.05292561 0.05292561 0.15877684 0.05292561 0.05292561
  0.05292561 0.05292561 0.15877684 0.05292561 0.05292561 0.05292561
  0.15877684 0.05292561 0.05292561 0.05292561 0.10585122 0.05292561
  0.05292561 0.05292561 0.05292561 0.05292561 0.05292561 0.10585122
  0.05292561 0.05292561 0.10585122 0.05292561 0.10585122 0.05292561
```

```
0.05292561 0.10585122 0.10585122 0.05292561 0.10585122 0.05292561
0.05292561 0.05292561 0.05292561 0.05292561 0.05292561 0.05292561
0.10585122 0.05292561 0.05292561 0.05292561 0.05292561 0.05292561
0.05292561 0.4234049  0.05292561 0.26462806 0.05292561 0.05292561
0.05292561 0.05292561 0.05292561 0.05292561 0.05292561 0.05292561
0.05292561 0.10585122 0.05292561 0.05292561 0.05292561 0.10585122
0.05292561 0.05292561 0.05292561 0.05292561 0.05292561 0.05292561
0.05292561 0.10585122 0.52925612 0.05292561 0.05292561 0.05292561
0.26462806 0.05292561 0.05292561 0.05292561 0.05292561 0.05292561
0.05292561 0.05292561 0.05292561]]
```

This TF-IDF matrix represents the TF-IDF values for the terms in the document. Each value in the matrix represents the TF-IDF score for the corresponding term in the document. TF-IDF combines two measures: Term Frequency (TF) and Inverse Document Frequency (IDF). TF measures the occurrence of a term in the document, while IDF measures the rarity of the term across all documents in the dataset.

In this specific matrix, the values range from 0.05292561 to 0.52925612. Higher values indicate that a term is more important or significant within the document relative to the rest of the terms. The matrix is sparse, meaning that most of the values are close to zero because a document typically contains a limited subset of the entire vocabulary. By representing the document in a TF-IDF matrix, it becomes possible to analyze or compare it with other documents based on the importance of terms in the context of the document collection.

Naive Bayes Classification

Having demonstrated how data cleaning, tokenization, and vectorization work in Scikit-Learn using one single document for illustration, we can move on to building, training, and evaluating a Multinomial Naive Bayes algorithm in both Scikit-Learn and PySpark using the entire 20 Newsgroups dataset.

The two libraries are similar in their overall structure as they both perform text classification using Naive Bayes and handle preprocessing steps like data cleaning, tokenization, and vectorization. One key difference is that Scikit-Learn directly uses `TfidfVectorizer` to perform tokenization, stop word removal, and TF-IDF feature extraction in a single step, while PySpark utilizes separate transformers like `Tokenizer`, `StopWordsRemover`, `HashingTF`, and `IDF` to achieve similar functionality.

The following is a summary of the steps involved in Scikit-Learn and PySpark for topic classification using the 20 Newsgroups dataset:

In Scikit-Learn:

1. Load the 20 Newsgroups dataset, specifying the desired categories.
2. Split the dataset into training and testing sets.
3. Create TF-IDF vectors from the text data using the `TfidfVectorizer`. Remove stop words during this process.
4. Train a Naive Bayes classifier using the `MultinomialNB` algorithm.
5. Predict the labels for the test set.
6. Evaluate the model's accuracy by comparing the predicted labels with the true labels using `accuracy_score`.

In PySpark:

1. Create a SparkSession.
2. Fetch the 20 Newsgroups dataset using Scikit-Learn and convert it into a Pandas DataFrame.
3. Convert the Pandas DataFrame into a PySpark DataFrame.
4. Perform NLP tasks on the PySpark DataFrame:
    - Tokenize the text column using the `Tokenizer` and remove punctuation.
    - Remove stop words using the `StopWordsRemover`.
    - Perform TF-IDF feature extraction using `HashingTF` and `IDF`.
5. Split the dataset into training and testing sets.
6. Train a Naive Bayes classifier using the `NaiveBayes` algorithm, specifying the features and label columns.
7. Make predictions on the test set.
8. Evaluate the model's accuracy using the `MulticlassClassificationEvaluator` with the accuracy metric.

Naive Bayes with Scikit-Learn

To build a Multinomial Naive Bayes model in Scikit-Learn, we begin by importing the required libraries.

Step 1: Import necessary libraries

```
[In]: import nltk
[In]: from sklearn.datasets import fetch_20newsgroups
[In]: from sklearn.feature_extraction.text import TfidfVectorizer
[In]: from sklearn.naive_bayes import MultinomialNB
[In]: from sklearn.metrics import accuracy_score
[In]: from sklearn.model_selection import train_test_split
[In]: import string
```

This code sets up the necessary imports to work with the NLTK (Natural Language Toolkit) and Scikit-Learn packages for text classification using the 20 Newsgroups dataset. We first import nltk—a widely used library for working with human language data, such as text processing and tokenization. We then import the `fetch_20newsgroups` function to download and load the 20 Newsgroups dataset. Next, we import the `TfidfVectorizer` class to convert text documents into numerical feature vectors using the TF-IDF (Term Frequency-Inverse Document Frequency) representation. In the next step, we import the `MultinomialNB` class, which is an algorithm commonly used for text classification based on the Naive Bayes principle. Moving on, we import the `accuracy_score` function to compute the accuracy of the model by comparing the predicted labels with the true labels. We also import the `train_test_split` function to split the dataset into training and testing subsets. Finally, we import the `string` module to access a string containing all ASCII punctuation characters, which will help us in removing punctuation marks.

Step 2: Define a tokenization function

```
def tokenize(text):
    tokens = nltk.word_tokenize(text)
    tokens = [token.lower() for token in tokens if token not in
              string.punctuation]
    return tokens
```

This code defines a function named tokenize that takes a text string as input. Inside this function, the `word_tokenize` function from the `nltk` library is used to split the input text into individual words or tokens. The second line creates a new list comprehension that iterates over each token in the tokens list. It checks if each token is not present in the `string.punctuation` string, which contains all ASCII punctuation characters. If a token is not a punctuation mark, it converts it to lowercase using the `lower()` method and adds it to the new list. Finally, the tokenize function returns the list of tokens obtained after filtering out punctuation marks and converting them to lowercase.

Step 3: Load the 20 Newsgroups dataset

```
[In]: categories = ['alt.atheism', 'comp.graphics',
'comp.os.ms-windows.misc', 'comp.sys.ibm.pc.hardware', 'comp.sys.
mac.hardware', 'comp.windows.x', 'misc.forsale', 'rec.autos', 'rec.
motorcycles', 'rec.sport.baseball', 'rec.sport.hockey', 'sci.crypt', 'sci.
electronics', 'sci.med', 'sci.space', 'soc.religion.christian', 'talk.
politics.guns', 'talk.politics.mideast', 'talk.politics.misc', 'talk.
religion.misc']
[In]: newsgroups_all = fetch_20newsgroups(subset='all',
      categories=categories, remove=('headers', 'footers', 'quotes'),
      shuffle=True, random_state=42)
```

In this code, the categories list contains the names of different topics or categories that are part of the 20 Newsgroups dataset. Each category represents a specific topic. The `fetch_20newsgroups` function is used to fetch the 20 Newsgroups dataset. It takes the categories list as an argument to specify which categories to include in the dataset. The subset='all' parameter indicates that all documents from the specified categories should be included. The remove=('headers', 'footers', 'quotes') parameter specifies to remove certain parts of the documents (headers, footers, and quotes) that are not relevant for the task at hand. The shuffle=True parameter shuffles the dataset randomly, and random_state=42 sets a seed for reproducibility.

Step 4: Split the dataset into train and test sets

```
[In]: X_train, X_test, y_train, y_test = train_test_split(newsgroups_all.
      data, newsgroups_all.target, test_size=0.2, random_state=42)
```

This line uses the `train_test_split` function from Scikit-Learn to split the 20 Newsgroups dataset into training and testing sets. The newsgroups_all.data refers to the input data, which contains the features or input variables for each sample in the dataset. The newsgroups_all.target represents the target or output variable that we want to predict or classify. It contains the corresponding labels or classes for each sample in the dataset. The test_size=0.2 parameter specifies the proportion of the dataset that should be allocated for the test set. It is set to 0.2, indicating that 20% of the data will be used for testing, while the remaining 80% will be used for training. Finally, the random_state=42 parameter sets the random seed to ensure reproducibility. The `train_test_split` function then returns four sets of data: X_train, X_test, y_train, and y_test.

Step 5: Create TF-IDF vectors from the tokenized text data

```
[In]: vectorizer = TfidfVectorizer(stop_words='english',
      tokenizer=tokenize)
[In]: X_train_vectors = vectorizer.fit_transform(X_train)
[In]: X_test_vectors = vectorizer.transform(X_test)
```

In this code, an instance of the `TfidfVectorizer` class is created and assigned to the variable vectorizer. This class is used for converting a collection of text documents into a matrix of TF-IDF (Term Frequency-Inverse Document Frequency) features. It takes two parameters: (1) the stop_words='english' parameter specifies that common English words, such as articles, pronouns, and prepositions, should be excluded from the analysis. These words are considered noise and do not carry significant information for NLP tasks, and (2) the tokenizer=tokenize parameter refers to the tokenize function (created in step 2) that will be used to split the input text into individual tokens.

The `fit_transform` method is used to fit the vectorizer on the training data and transform it, while the `transform` method is used to apply the same transformation to the test data. This allows the text data to be ready for use in our machine learning model.

Step 6: Train a Naive Bayes classifier

```
[In]: clf = MultinomialNB()
[In]: clf.fit(X_train_vectors, y_train)
```

Only two lines of code are required to train the algorithm. In the first line, an instance of the `MultinomialNB` class is created and assigned to the variable clf. The `MultinomialNB` class implements the Multinomial Naive Bayes algorithm. Naive Bayes is a probabilistic machine learning algorithm commonly used for text classification tasks.

The second line fits the Multinomial Naive Bayes classifier to the training data. The `fit()` method of the `MultinomialNB` class takes two arguments: (1) X_train_vectors represents the input features, which are the TF-IDF vectors obtained from the training data. Each row of X_train_vectors corresponds to a document, and each column represents a TF-IDF feature, and (2) y_train represents the target variable or labels associated with each document in the training data. It contains the class labels or categories to which the documents belong.

The `fit()` method trains the Multinomial Naive Bayes classifier by estimating the probabilities of each class label based on the input features and the corresponding target labels. After the `fit()` method is executed, the clf object is trained and ready to make predictions on new, unseen data.

Step 7: Predict the labels for the test set

```
[In]: y_pred = clf.predict(X_test_vectors)
```

In this line of code, the trained classifier clf is used to predict the class labels for the test data X_test_vectors. The `predict()` method of the classifier is called, and it takes the test data as its argument. The X_test_vectors variable represents the TF-IDF vectors of the test data obtained from the `TfidfVectorizer`. Each row of X_test_vectors corresponds to a document, and each column represents a TF-IDF feature.

The `predict()` method applies the learned model on the test data and assigns a predicted class label to each instance. It uses the trained classifier's learned parameters to calculate the posterior probability of each class given the input features. Based on these probabilities, it predicts the most probable class label for each instance. The resulting predicted class labels are stored in the y_pred variable, which now holds the predicted labels for the test instances based on the trained classifier.

Step 8: Print a sample of actual vs. predicted labels

```
[In]: for i in range(5):
          actual_label = newsgroups_all.target_names[y_test[i]]
          predicted_label = newsgroups_all.target_names[y_pred[i]]
          print("Actual:", actual_label)
          print("Predicted:", predicted_label)
          print("---")
[Out]:
Actual   : rec.sport.baseball
Predicted: rec.sport.baseball
```

```
---
Actual   : sci.electronics
Predicted: comp.os.ms-windows.misc
---
Actual   : sci.space
Predicted: sci.space
---
Actual   : talk.politics.misc
Predicted: talk.politics.misc
---
Actual   : alt.atheism
Predicted: sci.space
---
Accuracy: 0.70
```

In this step, we wanted to compare a sample of actual labels with the predicted labels to see how the model is doing in terms of performance. A loop is set up to iterate over the first five instances of the test data. For each iteration, the following actions are performed:

- actual_label = newsgroups_all.target_names[y_test[i]] retrieves the actual label of the i-th test instance. The y_test array contains the true labels of the test instances. The y_test[i] index is used to access the label index for the i-th test instance, and newsgroups_all.target_names is a list or array that maps label indices to their corresponding label names. By indexing newsgroups_all.target_names with y_test[i], the actual label name is obtained and assigned to the variable actual_label.

- predicted_label = newsgroups_all.target_names[y_pred[i]] retrieves the predicted label of the i-th test instance. The y_pred array contains the predicted labels obtained from the classifier. Similar to the previous step, the y_pred[i] index is used to access the predicted label index for the i-th test instance, and newsgroups_all.target_names is indexed with y_pred[i] to obtain the corresponding label name. The predicted label name is assigned to the variable predicted_label.

- print("Actual:", actual_label) prints the actual label name for the current test instance.
- print("Predicted:", predicted_label) prints the predicted label name for the current test instance.
- print("---") prints a separator to visually distinguish between different instances.

Comparing the predicted labels with the true labels for the top five instances in the preceding output, we can see that the model has made two mistakes. The sci.electronics category is predicted as comp.os.ms-windows.misc, and the alt.atheism category is predicted as sci.space.

Step 9: Calculate accuracy score

```
[In]: accuracy = accuracy_score(y_test, y_pred)
[In]: print("Accuracy:", accuracy)
[Out]: 0.70
```

In the first line of code, the `accuracy_score` function is used to calculate the accuracy of the classifier's predictions. The function takes two arguments: (1) y_test, which represents the true labels or target values of the test instances, and (2) y_pred, which represents the predicted labels obtained from the classifier for the corresponding test instances. The function compares the predicted labels (y_pred) with the true labels (y_test) and computes the accuracy of the predictions. It calculates the fraction of correctly predicted labels over the total number of instances. The resulting accuracy value is assigned to the variable accuracy.

The second line prints the accuracy value calculated in the previous step. The accuracy is displayed as a decimal value between 0 and 1, where 1 represents a perfect classification accuracy. The resulting accuracy value is approximately 0.70, indicating that the Naive Bayes classifier's predictions match the true labels for about 70% of the test instances.

Naive Bayes with PySpark

We now implement the same Multinomial Naive Bayes algorithm in PySpark using the same dataset. Just like with Scikit-Learn, we begin by importing the necessary libraries.

Step 1: Import necessary libraries

```
[In]: from pyspark.sql import SparkSession
[In]: from pyspark.ml.feature import HashingTF, IDF, RegexTokenizer,
      StopWordsRemover, VectorAssembler
[In]: from pyspark.ml.classification import NaiveBayes
[In]: from pyspark.ml.evaluation import MulticlassClassificationEvaluator
[In]: import pandas as pd
[In]: from pyspark.sql.functions import rand
[In]: from sklearn.datasets import fetch_20newsgroups
```

This code imports the libraries and modules used in a PySpark implementation of the Multinomial Naive Bayes algorithm. We start by importing the `SparkSession` class so that we are able to interact with Spark, create DataFrames, and execute Spark operations. We then import several feature extraction and transformation classes, including `HashingTF` for term frequency calculation, `IDF` for Inverse Document Frequency calculation, `RegexTokenizer` for tokenization, `StopWordsRemover` for removing stop words, and `VectorAssembler` for assembling feature vectors.

Next, we import the `NaiveBayes` class for predicting class labels based on input features. In the next step, we import the `MulticlassClassificationEvaluator` class, which provides evaluation metrics for the multiclass classification model. We also import the Pandas library for data manipulation as well as the `rand` function (this will be used to generate a random sample of actual and predicted class labels as part of the evaluation step). Finally, we import the `fetch_20newsgroups` function to fetch the 20 Newsgroups dataset.

Step 2: Create a SparkSession

```
[In]: spark = SparkSession.builder.appName("TextClassification").
      getOrCreate()
```

In this line of code, a `SparkSession` object named spark is created using the `builder()` method, which allows configuration and customization of the SparkSession. The `appName()` method is used to set the name of the Spark application as "TextClassification". If a SparkSession with the specified name already exists, it will be retrieved; otherwise, a new SparkSession will be created. The SparkSession serves as the entry point for interacting with Spark and provides functionality for working with distributed data structures, executing operations, and accessing various Spark functionalities.

Step 3: Fetch the 20 Newsgroups dataset using Scikit-Learn

```
[In]: categories = ['alt.atheism', 'comp.graphics', 'comp.os.ms-
                    windows.misc', 'comp.sys.ibm.pc.hardware', 'comp.sys.
                    mac.hardware', 'comp.windows.x', 'misc.forsale', 'rec.
                    autos', 'rec.motorcycles', 'rec.sport.baseball', 'rec.
                    sport.hockey', 'sci.crypt', 'sci.electronics', 'sci.
                    med', 'sci.space', 'soc.religion.christian', 'talk.
                    politics.guns', 'talk.politics.mideast', 'talk.
                    politics.misc', 'talk.religion.misc']
[In]: newsgroups_all = fetch_20newsgroups(subset='all',
                    categories=categories, remove=('headers', 'footers',
                    'quotes'), shuffle=True, random_state=42)
```

This is the same as in step 3 of the Scikit-Learn code. Here, the categories list contains the names of different topics or categories that are part of the 20 Newsgroups dataset. Each category represents a specific topic. The `fetch_20newsgroups` function is used to fetch the 20 Newsgroups dataset. It takes the categories list as an argument to specify which categories to include in the dataset. The subset='all' parameter indicates that all documents from the specified categories should be included. The remove=('headers', 'footers', 'quotes') parameter specifies to remove certain parts of the documents (headers, footers, and quotes) that are not relevant for the task at hand. The shuffle=True parameter shuffles the dataset randomly, and random_state=42 sets a seed for reproducibility.

Step 4: Create a Pandas DataFrame from the Scikit-Learn data and target arrays

```
[In]: newsgroups_df = pd.DataFrame({'text': newsgroups_all.data, 'label':
      newsgroups_all.target})
```

This line of code creates a new Pandas DataFrame named newsgroups_df using the `DataFrame()` function. This combines the text and label data from the fetched newsgroup documents. The text column is populated with the data from newsgroups_all.data, while the label column is populated with the data from newsgroups_all.target.

Step 5: Convert the Pandas DataFrame to a PySpark DataFrame

```
[In]: pyspark_df = spark.createDataFrame(newsgroups_df)
```

In this step, we convert the Pandas DataFrame to a PySpark DataFrame so we can process it using PySpark NLP code and feed it to the PySpark Naive Bayes algorithm as input. The code creates a new DataFrame named pyspark_df using the `createDataFrame()` method. The Pandas DataFrame newsgroups_df is passed as an argument to this method.

Step 6: Tokenize the text and remove punctuation from the text column

```
[In]: tokenizer = RegexTokenizer(inputCol='text', outputCol='tokens',
      pattern='\\W')
[In]: tokenized_df = tokenizer.transform(pyspark_df)
```

These lines of code demonstrate the tokenization process using the `RegexTokenizer` in PySpark. In the first line, a tokenizer object named tokenizer is first created. The inputCol parameter is set to text, indicating that the tokenizer will take the text data from the text column of the DataFrame. The outputCol parameter is set to tokens, specifying that the tokenized output will be stored in a new column named tokens. The pattern parameter is set to '\\W,' which is a regular expression pattern matching non-word characters. This pattern will be used to split the text into tokens.

The second line applies the tokenizer to the pyspark_df DataFrame using the `transform()` method. It takes the input DataFrame, pyspark_df, and performs the tokenization process using the tokenizer object created earlier. The resulting tokenized DataFrame is assigned to a new DataFrame called tokenized_df, which contains the original columns from pyspark_df along with a new tokens column that holds the tokenized version of the text.

We can take a look at the top five rows of the tokenized_df DataFrame:

```
[In]: tokenized_df.show(5)
[Out]:
+--------------------+-----+--------------------+
|                text|label|              tokens|
+--------------------+-----+--------------------+
|\n\nI am sure som...|   10|[i, am, sure, som...|
|My brother is in ...|    3|[my, brother, is,...|
|\n\n\n\n\tFinally...|   17|[finally, you, sa...|
|\nThink!\n\nIt's ...|    3|[think, it, s, th...|
|1)    I have an o...|    4|[1, i, have, an, ...|
+--------------------+-----+--------------------+
only showing top 5 rows
```

We can observe from the preceding output, although it is truncated, that the text column has been divided into tokens and converted to lowercase, as indicated in the tokens column. We can also observe that non-word characters, such as \n\n, have been removed. The pattern '\W' in the RegexTokenizer specifies that any non-word character should be considered as a delimiter for tokenization. Non-word characters typically include punctuation marks, symbols, and whitespace. By applying this pattern, the tokenizer will split the text into tokens based on word boundaries and exclude any non-word characters.

Step 7: Apply stop words removal

```
[In]: stopwords_remover = StopWordsRemover(inputCol='tokens',
      outputCol='filtered_tokens')
[In]: filtered_df = stopwords_remover.transform(tokenized_df)
```

In this step, we remove stop words using the StopWordsRemover in PySpark. The first line creates an object named stopwords_remover. The inputCol parameter is set to tokens, indicating that the stop words removal will be applied to the tokens column of the DataFrame. The outputCol parameter is set to filtered_tokens, specifying that the resulting tokens after stop words removal will be stored in a new column named filtered_tokens.

The second line applies the stop words remover to the tokenized_df DataFrame using the transform() method. It takes the input DataFrame, tokenized_df, and performs the stop words removal process using the stopwords_remover object created earlier. The resulting DataFrame named filtered_df contains the original columns from tokenized_df along with a new filtered_tokens column that holds the tokens after stop words removal.

Let's take a look at the top five rows of the filtered_df DataFrame:

```
[In]: filtered_df.show(5)
[Out]:
+--------------------+-----+--------------------+--------------------+
|                text|label|              tokens|     filtered_tokens|
+--------------------+-----+--------------------+--------------------+
|\n\nI am sure som...|   10|[i, am, sure, som...|[sure, bashers, p...|
|My brother is in ...|    3|[my, brother, is,...|[brother, market,...|
|\n\n\n\n\tFinally...|   17|[finally, you, sa...|[finally, said, d...|
|\nThink!\n\nIt's ...|    3|[think, it, s, th...|[think, scsi, car...|
|1)    I have an o...|    4|[1, i, have, an, ...|[1, old, jasmine,...|
+--------------------+-----+--------------------+--------------------+
only showing top 5 rows
```

The filtered_tokens column contains the tokens after the removal of stop words. By comparing this column with the tokens column, we can observe that stop words, such as "I", "am", and "some", have been removed in the first line, albeit truncated. Similarly, in the second line, we can see that stop words like "my" and "is" have also been eliminated.

Step 8: Apply TF-IDF vectorization

```
[In]: hashing_tf = HashingTF(inputCol='filtered_tokens',
      outputCol='rawFeatures')
[In]: tf_data = hashing_tf.transform(filtered_df)
[In]: idf = IDF(inputCol='rawFeatures', outputCol='features')
[In]: idf_model = idf.fit(tf_data)
[In]: tfidf_data = idf_model.transform(tf_data)
```

In this step, we transform the tokenized and filtered DataFrame into a TF-IDF (Term Frequency-Inverse Document Frequency) representation. The first line of code creates a `HashingTF` object named hashing_tf that takes the filtered_tokens column as input and generates a sparse vector representation called rawFeatures as output.

In the second line, the HashingTF transformation is applied to the filtered_df DataFrame using the `transform()` method. It generates a new DataFrame called tf_data that includes the original columns from filtered_df and a new column rawFeatures containing the sparse vector representation.

In the third line, an IDF (Inverse Document Frequency) object named idf is created, specifying the rawFeatures column as input and the features column as output. The IDF measures the importance of each term in a document collection.

In the fourth line, the IDF transformation is fitted to the tf_data DataFrame using the `fit()` method, creating an IDF model named idf_model. This step calculates the IDF weights for each term in the document collection.

Finally, the IDF model is applied to the tf_data DataFrame using the `transform()` method. It generates a new DataFrame named tfidf_data that includes the original columns from tf_data along with a new column called features, which contains the TF-IDF vector representation for each document.

Let's take a look at the vectorized columns, rawFeatures and features, from the tfidf_data DataFrame. Starting with the rawFeatures column, the following code retrieves the first row of the DataFrame using the `first()` method, selects the rawFeatures column, and then accesses the first element [0] in that column:

```
[In]: tfidf_data.select('rawFeatures').first()[0]
[Out]:
SparseVector(262144, {2710: 1.0, 8538: 1.0, 11919: 1.0, 12454: 1.0, 23071:
1.0, 23087: 2.0, 31739: 2.0, 45916: 1.0, 51178: 2.0, 54961: 1.0, 55270:
2.0, 57448: 1.0, 58394: 2.0, 76764: 1.0, 77751: 1.0, 79132: 1.0, 82237:
1.0, 84685: 1.0, 91431: 1.0, 92726: 1.0, 96760: 1.0, 102382: 2.0, 105901:
1.0, 107855: 1.0, 109753: 1.0, 109906: 1.0, 116581: 1.0, 127337: 1.0,
132975: 1.0, 134125: 1.0, 136496: 1.0, 137765: 3.0, 138895: 1.0, 142239:
1.0, 142343: 1.0, 143478: 5.0, 147136: 1.0, 153969: 1.0, 156917: 1.0,
158102: 1.0, 169300: 1.0, 173339: 1.0, 174802: 1.0, 181001: 1.0, 194361:
1.0, 194710: 1.0, 196689: 1.0, 196839: 1.0, 208344: 1.0, 216372: 1.0,
217435: 1.0, 223985: 1.0, 229604: 1.0, 233967: 1.0, 235375: 1.0, 245599:
2.0, 245973: 1.0, 256965: 1.0})
```

The values in the SparseVector represent the term frequencies (TF) of each term in the document, before applying the IDF transformation. Each value in the vector represents the TF of a specific term at the corresponding index in the vector. The TF is a count of how many times a term appears in the document. For example, if we have an entry {2710: 1.0, 8538: 1.0}, it means that the term at index 2710 appears once in the document, and the term at index 8538 also appears once.

The rawFeatures column provides the raw term frequencies, which can be useful for certain tasks or analysis. However, TF alone does not consider the importance of a term across the entire corpus. This is why the TF-IDF transformation is applied to obtain the features column, which incorporates both the term frequencies and inverse document frequencies to better represent the importance of terms in the document within the context of the entire corpus.

Let's take a look at the data in the features column:

```
[In]: tfidf_data.select('features').first()[0]
[Out]:
SparseVector(262144, {2710: 6.1064, 8538: 2.1065, 11919: 8.7455, 12454:
7.5415, 23071: 3.2227, 23087: 9.0717, 31739: 8.7021, 45916: 5.0401, 51178:
10.8993, 54961: 3.0953, 55270: 8.1641, 57448: 7.2051, 58394: 12.8202,
76764: 2.0493, 77751: 2.8383, 79132: 3.1372, 82237: 4.4781, 84685: 3.4,
91431: 4.3803, 92726: 3.5055, 96760: 4.613, 102382: 5.1199, 105901: 5.5003,
107855: 7.2792, 109753: 3.306, 109906: 6.586, 116581: 4.6401, 127337:
```

```
5.4621, 132975: 2.7923, 134125: 2.5627, 136496: 6.4429, 137765: 8.6343,
138895: 3.2621, 142239: 2.6709, 142343: 3.0585, 143478: 27.3733, 147136:
1.7614, 153969: 4.2606, 156917: 2.972, 158102: 6.4429, 169300: 3.6416,
173339: 2.6717, 174802: 7.5415, 181001: 4.7879, 194361: 4.9766, 194710:
4.5458, 196689: 5.0483, 196839: 4.0696, 208344: 5.5536, 216372: 4.2834,
217435: 4.428, 223985: 3.0953, 229604: 4.1951, 233967: 2.3233, 235375:
2.5808, 245599: 5.63, 245973: 4.1881, 256965: 5.3443})
```

The values in the SparseVector represent the Term Frequencies-Inverse Document Frequencies (TF-IDF) weights for each term in the document. TF-IDF is a numerical statistic that reflects the importance of a term in a document within a collection or corpus.

In the context of the features column in tfidf_data, each value represents the TF-IDF weight for a specific term at the corresponding index in the vector. A higher TF-IDF weight indicates that the term is more important or distinctive in the document compared to the rest of the corpus. For example, let's consider the entry {2710: 6.1064, 8538: 2.1065}. This means that the term at index 2710 has a TF-IDF weight of 6.1064, and the term at index 8538 has a TF-IDF weight of 2.1065. These weights indicate the relative importance of these terms within the document, with a higher weight suggesting greater significance.

Step 9: Assemble features into a vector column

```
[In]: assembler = VectorAssembler(inputCols=['features'],
      outputCol='vectorized_features')
[In]: assembled_data = assembler.transform(tfidf_data)
```

In this code, the `VectorAssembler` class is used to assemble the features column into a single vector column called vectorized_features. The class takes the input columns specified in inputCols, which in this case is ['features'], and combines their values into a vector. The resulting vector column is appended to the DataFrame as assembled_data.

Step 10: Split the dataset into train and test sets

```
[In]: train_data, test_data = assembled_data.randomSplit([0.8, 0.2],
      seed=42)
```

In this line of code, the dataset assembled_data is split into two separate sets, a training set and a test set, using the `randomSplit` function. This takes two arguments: the first argument is a list of proportions that determines the relative sizes of the resulting splits, and the second argument is an optional seed value for reproducibility. [0.8, 0.2] is passed as the list of proportions, indicating that approximately 80% of the data will be allocated to the training set and 20% to the test set. The seed parameter is set to 42 for reproducible random splits.

The resulting splits are assigned to the variables train_data and test_data, representing the training dataset and the test dataset, respectively. These sets can be used for training and evaluating the machine learning model.

Step 11: Train a Naive Bayes classifier

```
[In]: nb = NaiveBayes(featuresCol='vectorized_features', labelCol='label')
[In]: model = nb.fit(train_data)
```

In this code, a Naive Bayes classifier is trained using the training data. First, an instance of the `NaiveBayes` class is created with the features column specified as vectorized_features and the label column specified as label. This configuration determines which columns of the dataset will be used as features (input variables) and labels (output variables) for training the classifier.

Next, the `fit` method is called on the NaiveBayes instance, passing the training data (train_data) as an argument. This step trains the Naive Bayes classifier using the labeled training examples. The model learns the statistical properties of the input features and their corresponding labels. The resulting trained model is assigned to the variable model and can be used to make predictions on new, unseen data.

Step 12: Make predictions on the test set

```
[In]: predictions = model.transform(test_data)
```

In this line of code, the trained Naive Bayes model is used to make predictions on the test dataset. The `transform` method is called on the trained model, passing the test dataset (test_data) as an argument. This step applies the trained model to the test data and generates predictions for each example in the test dataset.

The resulting predictions are assigned to the variable predictions. This DataFrame contains the original features, labels, and the predicted labels for each example in the test dataset. It provides a comparison between the actual labels and the predicted labels based on the learned model.

Step 13: Print a sample of actual and predicted values

In this step, we print a random sample of the actual and predicted values to have an idea about how the model is doing in terms of performance. We start by mapping label and prediction values to label names:

```
[In]: label_map = {i: category for i, category in enumerate(newsgroups_all.
      target_names)}
[In]: map_label = spark.udf.register("map_label", lambda label: label_
      map[label])
[In]: predictions = predictions.withColumn("label_name", map_
      label(predictions.label))
[In]: predictions = predictions.withColumn("prediction_name", map_
      label(predictions.prediction))
```

We then select five random rows of actual vs. predicted labels.

```
[In]: selected_predictions = predictions.select("label_name", "prediction_
      name").orderBy(rand()).limit(5)
[In]: selected_predictions.show(truncate=False)
[Out]:
+---------------------+---------------------+
|label_name           |prediction_name      |
+---------------------+---------------------+
|soc.religion.christian|soc.religion.christian|
|comp.graphics        |comp.graphics        |
|sci.space            |sci.med              |
|alt.atheism          |alt.atheism          |
|sci.med              |sci.med              |
+---------------------+---------------------+
```

In this two-step process, a mapping is first created to associate the numerical label values with their corresponding label names. The label_map dictionary is created, where the keys are the numerical label values and the values are the corresponding label names (e.g., alt.atheism, comp.graphics, etc.). Then, a user-defined function (UDF) named map_label is registered using the `spark.udf.register` method. This UDF takes a label value as input and returns the corresponding label name using the label_map dictionary. Next, the UDF is applied to the label column and prediction column of the predictions DataFrame using the withColumn method. Two new columns, label_name and prediction_name, are added to the DataFrame, which contain the label names corresponding to the numerical label values and predicted label values, respectively.

In the second step, the predictions DataFrame is being used to select five random rows containing the actual and predicted label names. The select method is used to specify the columns to be included in the resulting DataFrame, which are label_name and prediction_name columns. The orderBy method with the rand() function is used to randomize the order of rows in the DataFrame. The limit method is used to limit the number of rows to be selected, which in this case is set to 5. Finally, the show method is called to display the selected rows from the DataFrame. The truncate=False parameter ensures that the full contents of the columns are displayed without truncation.

Comparing the actual and predicted categories in the preceding output, we can see that out of the five categories, the model had one prediction wrong: it predicted sci.space as sci.med.

Step 14: Calculate and print accuracy

We first calculate the accuracy metric with this code:

```
[In]: evaluator = MulticlassClassificationEvaluator(labelCol='label',
      predictionCol='prediction', metricName='accuracy')
[In]: accuracy = evaluator.evaluate(predictions)
```

The MulticlassClassificationEvaluator class is used to evaluate the accuracy of the predictions made by the model. The evaluator object is created with the following parameters:

- labelCol='label' specifies the name of the column in the predictions DataFrame that contains the actual labels.
- predictionCol='prediction' specifies the name of the column in the predictions DataFrame that contains the predicted labels.
- metricName='accuracy' specifies the evaluation metric to be used, which in this case is accuracy.

The evaluate method is called on the evaluator object, passing the predictions DataFrame as the argument. This computes the accuracy of the predictions by comparing the predicted labels with the actual labels in the DataFrame. The computed accuracy value is assigned to the variable accuracy.

Now, let's do the printing:

```
[In]: print("Accuracy:", accuracy)
[Out]: 0.71
```

The preceding line is used to display the value of the accuracy variable. It outputs the string "Accuracy:" followed by the value stored in the accuracy variable. This value is approximately 0.71, which means that the model's predictions were correct for approximately 71% of the test data instances. This value represents the proportion of correctly classified instances out of the total number of instances in the test dataset.

Finally, it's worth noting that there is a negligible difference between PySpark accuracy and that of Scikit-Learn (71% vs. 70%).

Bringing It All Together

In this section, we combine all the important code snippets from the previous steps into a single code block. By doing so, the reader can conveniently execute the code as a single block in both Scikit-Learn and PySpark, ensuring a streamlined and cohesive workflow.

Scikit-Learn

Step 1: Import necessary libraries

```
[In]: import nltk
[In]: from sklearn.datasets import fetch_20newsgroups
[In]: from sklearn.feature_extraction.text import TfidfVectorizer
[In]: from sklearn.naive_bayes import MultinomialNB
[In]: from sklearn.metrics import accuracy_score
[In]: from sklearn.model_selection import train_test_split
[In]: import string
```

Step 2: Tokenization

```
[In]: def tokenize(text):
        tokens = nltk.word_tokenize(text)
        tokens = [token.lower() for token in tokens if token not in
           string.punctuation]
        return tokens
```

Step 3: Load the 20 Newsgroups dataset

```
[In]: categories = ['alt.atheism', 'comp.graphics', 'comp.os.ms-
                    windows.misc', 'comp.sys.ibm.pc.hardware', 'comp.sys.
                    mac.hardware', 'comp.windows.x', 'misc.forsale', 'rec.
                    autos', 'rec.motorcycles', 'rec.sport.baseball', 'rec.
                    sport.hockey', 'sci.crypt', 'sci.electronics', 'sci.
                    med', 'sci.space', 'soc.religion.christian', 'talk.
                    politics.guns', 'talk.politics.mideast', 'talk.
                    politics.misc', 'talk.religion.misc']
[In]: newsgroups_all = fetch_20newsgroups(subset='all',
                    categories=categories, remove=('headers', 'footers',
                    'quotes'), shuffle=True, random_state=42)
```

Step 4: Split the dataset into train and test sets

```
[In]: X_train, X_test, y_train, y_test =
      train_test_split(newsgroups_all.data, newsgroups_all.target, test_
      size=0.2, random_state=42)
```

Step 5: Create TF-IDF vectors from the text data with tokenization

```
[In]: vectorizer = TfidfVectorizer(stop_words='english',
      tokenizer=tokenize)
[In]: X_train_vectors = vectorizer.fit_transform(X_train)
[In]: X_test_vectors = vectorizer.transform(X_test)
```

Step 6: Train a Naive Bayes classifier

```
[In]: clf = MultinomialNB()
[In]: clf.fit(X_train_vectors, y_train)
```

Step 7: Predict the labels for the test set

```
[In]: y_pred = clf.predict(X_test_vectors)
```

Step 8: Print the top five rows of actual vs. predicted labels

```
[In]: for i in range(5):
          actual_label = newsgroups_all.target_names[y_test[i]]
          predicted_label = newsgroups_all.target_names[y_pred[i]]
```

```
    print("Actual:", actual_label)
    print("Predicted:", predicted_label)
    print("---")
```

Step 9: Calculate accuracy score

```
[In]: accuracy = accuracy_score(y_test, y_pred)
[In]: print("Accuracy:", accuracy)
```

PySpark

Step 1: Import necessary libraries

```
[In]: from pyspark.sql import SparkSession
[In]: from pyspark.ml.feature import HashingTF, IDF, RegexTokenizer,
      StopWordsRemover, VectorAssembler
[In]: from pyspark.ml.classification import NaiveBayes
[In]: from pyspark.ml.evaluation import MulticlassClassificationEvaluator
[In]: import pandas as pd
[In]: from pyspark.sql.functions import rand
[In]: from sklearn.datasets import fetch_20newsgroups
```

Step 2: Create a SparkSession

```
[In]: spark = SparkSession.builder.appName("TextClassification").
      getOrCreate()
```

Step 3: Fetch the 20 Newsgroups dataset using Scikit-Learn

```
[In]: categories = ['alt.atheism', 'comp.graphics', 'comp.os.ms-windows.
      misc', 'comp.sys.ibm.pc.hardware', 'comp.sys.mac.hardware', 'comp.
      windows.x', 'misc.forsale', 'rec.autos', 'rec.motorcycles', 'rec.
      sport.baseball', 'rec.sport.hockey', 'sci.crypt', 'sci.electronics',
      'sci.med', 'sci.space', 'soc.religion.christian', 'talk.politics.
      guns', 'talk.politics.mideast', 'talk.politics.misc', 'talk.
      religion.misc']
[In]: newsgroups_all = fetch_20newsgroups(subset='all',
      categories=categories, remove=('headers', 'footers', 'quotes'),
      shuffle=True, random_state=42)
```

Step 4: Create a Pandas DataFrame from the Scikit-Learn data and target arrays

```
[In]: newsgroups_df = pd.DataFrame({'text': newsgroups_all.data, 'label':
      newsgroups_all.target})
```

Step 5: Convert the Pandas DataFrame to a PySpark DataFrame

```
[In]: pyspark_df = spark.createDataFrame(newsgroups_df)
```

Step 6: Tokenize the text plus remove punctuation from the text column

```
[In]: tokenizer = RegexTokenizer(inputCol='text', outputCol='tokens',
      pattern='\\W')
[In]: tokenized_df = tokenizer.transform(pyspark_df)
```

Step 7: Apply stop words removal

```
[In]: stopwords_remover = StopWordsRemover(inputCol='tokens',
      outputCol='filtered_tokens')
[In]: filtered_df = stopwords_remover.transform(tokenized_df)
```

Step 8: Apply TF-IDF vectorization

```
[In]: hashing_tf = HashingTF(inputCol='filtered_tokens',
      outputCol='rawFeatures')
[In]: tf_data = hashing_tf.transform(filtered_df)
[In]: idf = IDF(inputCol='rawFeatures', outputCol='features')
[In]: idf_model = idf.fit(tf_data)
[In]: tfidf_data = idf_model.transform(tf_data)
```

Step 9: Assemble features into a vector column

```
[In]: assembler = VectorAssembler(inputCols=['features'],
      outputCol='vectorized_features')
[In]: assembled_data = assembler.transform(tfidf_data)
```

Step 10: Split the dataset into train and test sets

```
[In]: train_data, test_data = assembled_data.randomSplit([0.8, 0.2],
      seed=42)
```

Step 11: Train a Naive Bayes classifier

```
[In]: nb = NaiveBayes(featuresCol='vectorized_features', labelCol='label')
[In]: model = nb.fit(train_data)
```

Step 12: Make predictions on the test set

```
[In]: predictions = model.transform(test_data)
```

Step 13: Map label and prediction values to label names

```
[In]: label_map = {i: category for i, category in enumerate(newsgroups_all.
      target_names)}
[In]: map_label = spark.udf.register("map_label", lambda label: label_
      map[label])
[In]: predictions = predictions.withColumn("label_name", map_
      label(predictions.label))
[In]: predictions = predictions.withColumn("prediction_name", map_
      label(predictions.prediction))
```

Step 14: Select five random rows of actual vs. predicted labels

```
[In]: selected_predictions = predictions.select("label_name", "prediction_
      name").orderBy(rand()).limit(10)
[In]: selected_predictions.show(truncate=False)
```

Step 15: Calculate accuracy

```
[In]: evaluator = MulticlassClassificationEvaluator(labelCol='label',
      predictionCol='prediction', metricName='accuracy')
[In]: accuracy = evaluator.evaluate(predictions)
```

Step 16: Print accuracy

```
[In]: print("Accuracy:", accuracy)
```

Summary

In this chapter, we delved into text data processing, an area of machine learning known as natural language processing (NLP). We examined the key steps involved in NLP using an open source dataset known as the 20 Newsgroups. We built, trained, and evaluated a Multinomial Naive Bayes algorithm and used it to predict the topic categories in this dataset. Prior to training the model, it was essential to ensure that the text data fed into the model was in the correct format. This involved cleaning, tokenization, and vectorization. We explored two commonly used methods for vectorization: Bag of Words (BoW) and Term Frequency-Inverse Document Frequency (TF-IDF), highlighting their advantages and disadvantages.

To demonstrate the use of Naive Bayes with NLP, we employed three libraries: Pandas, Scikit-Learn, and PySpark. Our goal was to assist data scientists who are familiar with small data library tools (Pandas and Scikit-Learn) in transitioning to PySpark, enabling them to leverage the advantages of distributed programming and parallel computing for NLP tasks.

In the next chapter, we will delve into the process of building, training, and evaluating a k-means clustering algorithm for effective data segmentation. Clustering is a commonly used technique in segmentation analysis, grouping similar observations together based on their characteristics or proximity in the feature space. The result is a set of clusters, with each observation assigned to a specific cluster.

CHAPTER 15

k-Means Clustering with Pandas, Scikit-Learn, and PySpark

In this chapter, we delve into the process of building, training, and evaluating a k-means clustering algorithm for effective data segmentation. Clustering is a commonly used technique in segmentation analysis to group similar observations together based on their characteristics or their proximity in the feature space. The result is a set of clusters, with each observation assigned to a specific cluster. By organizing data into clusters, we can gain a deeper understanding of complex datasets and identify key subgroups.

Building a k-means algorithm involves defining the number of clusters (k) and iteratively optimizing cluster centers to minimize the within-cluster sum of squares. Training the algorithm involves fitting the data to the k-means model and assigning each data point to its corresponding cluster. Evaluating the algorithm is a critical step to assess the quality of the clustering results.

Clustering differs from supervised learning in at least two key areas. First, the input into a clustering algorithm is unlabeled. In supervised learning, such as regression and classification, we predict the label. However, in cluster analysis, there is no label to predict. Instead, we group similar observations together based on their inherent characteristics or patterns.

Second, typical supervised learning algorithms' evaluation metrics cannot be used to evaluate the performance of clustering algorithms. This is because these metrics rely on comparing actual labels with predicted labels, which are not always available in unsupervised learning. Instead, clustering algorithms employ specific evaluation metrics designed for unsupervised learning. One commonly used evaluation metric, which will be used in this chapter, is the Silhouette Score. This ranges between -1 and 1, where a

A. Testas, *Distributed Machine Learning with PySpark*, https://doi.org/10.1007/978-1-4842-9751-3_15

score close to 1 indicates well-separated and compact clusters, while a score close to -1 suggests poor clustering. A score of 0 suggests overlapping or ambiguous clusters.

In the upcoming sections of this chapter, we build, train, and evaluate a k-means algorithm using the well-known open-source Iris dataset. The purpose of this exercise is to group the species of Iris flowers into clusters based on their characteristics.

We leverage the power of three different tools: Pandas, Scikit-Learn, and PySpark. The main objective of this approach is to cater to data scientists who are versed in Pandas and Scikit-Learn and wish to transition to PySpark. By exploring the similarities in steps between Pandas/Scikit-Learn and PySpark, we aim to provide these data scientists with a seamless transition to PySpark and empower them to harness its distributed computing capabilities for enhanced scalability and performance.

The Dataset

For this chapter, we use the open source Iris dataset to implement a k-means clustering algorithm. We have already explored this dataset in Chapter 8. The dataset contains measurements of four features (sepal length, sepal width, petal length, and petal width) from three different species of Iris flowers (setosa, versicolor, and virginica).

We can access the Iris dataset by following these steps:

Step 1: Import the necessary libraries

```
[In]: from sklearn.datasets import load_iris
[In]: import pandas as pd
```

Step 2: Load the Iris dataset

```
[In]: iris = load_iris()
```

Step 3: Access the feature data (X) and target labels (y)

```
[In]: X = iris.data
[In]: y = iris.target
```

Step 4: Define column names

```
[In]: feature_names = iris.feature_names
[In]: target_names = iris.target_names
```

Step 5: Create a DataFrame with column names

```
[In]: pandas_df = pd.DataFrame(X, columns=feature_names)
[In]: pandas_df['target'] = y
[In]: pandas_df['target_names'] = pandas_df['target'].map(
      dict(zip(range(len(target_names)), target_names)))
```

Step 6: Print the DataFrame

```
[In]: print(pandas_df.head())
[Out]:
```

	sepal length (cm)	sepal width (cm)	petal length (cm)	petal width (cm)
0	5.1	3.5	1.4	0.2
1	4.9	3	1.4	0.2
2	4.7	3.2	1.3	0.2
3	4.6	3.1	1.5	0.2
4	5	3.6	1.4	0.2

To get the output, we loaded the Iris dataset into a Pandas DataFrame. First, we import the necessary libraries, including `load_iris` from Scikit-Learn (to load the Iris dataset) and Pandas (for DataFrame handling). We then use the `load_iris()` method to load the Iris dataset, which returns an object containing feature data (X) and target labels (y). We define column names for the DataFrame using the feature_names and target_names attributes. Next, we create a Pandas DataFrame (pd.DataFrame()) with the feature data, adding target and target_names columns using the `map()` method and a dictionary. Finally, we print the first few rows of the DataFrame to inspect the loaded data's structure and content using the `head()` method.

One important feature of this dataset is that all features (sepal length, sepal width, petal length, and petal width) are on the same scale (centimeters). This suggests that scaling will not be necessary for applying the k-means model to the Iris dataset.

To achieve the same results with PySpark code, we first convert the pandas_df DataFrame to a PySpark DataFrame and then display the top five rows as follows:

Step 1: Import the SparkSession

```
[In]: from pyspark.sql import SparkSession
```

Step 2: Create a Spark object named spark

```
[In]: spark = SparkSession.builder.getOrCreate()
```

Step 3: Convert the pandas_df DataFrame to a PySpark DataFrame

```
[In]: spark_df = spark.createDataFrame(pandas_df)
```

Step 4: Show the first five rows of the Spark DataFrame

```
[In]: spark_df.show(5)
[Out]:
```

sepal length (cm)	sepal width (cm)	petal length (cm)	petal width (cm)
5.1	3.5	1.4	0.2
4.9	3	1.4	0.2
4.7	3.2	1.3	0.2
4.6	3.1	1.5	0.2
5	3.6	1.4	0.2

To get this output, we followed a number of steps.

In the first step, we import the `SparkSession` class, which is necessary for creating Spark applications. In the second step, we create a Spark object named spark using the `SparkSession.builder.getOrCreate()` method. This allows us to access the Spark functionality. In the third step, we convert pandas_df DataFrame into spark_df DataFrame by utilizing the `createDataFrame()` method. This transformation enables us to leverage the distributed computing capabilities of Spark for large-scale data analysis. In the final step, we display the first five rows of the newly created PySpark DataFrame using the `show()` method.

Let's explore both pandas_df and spark_df a little bit further. We can print the shape of the pandas_df (number of rows, number of columns) using the Pandas shape attribute as follows:

```
[In]: print(pandas_df.shape)
[Out]: (150, 6)
```

We can achieve the same results with the following PySpark code:

```
[In]: print((spark_df.count(), len(spark_df.columns)))
[Out]: (150, 6)
```

In this PySpark code, we print the shape of the Spark DataFrame by utilizing the count() method to retrieve the number of rows and the len() function to determine the number of columns.

Let's count the number of species within each target variable using the Pandas value_counts() method:

```
[In]: species_count = pandas_df['target_names'].value_counts()
[In]: print(species_count)
[Out]:
setosa        50
versicolor    50
virginica     50
Name: target_names, dtype: int64
```

We can achieve similar results with the following PySpark code:

```
[In]: species_count = spark_df.groupBy('target_names').count()
[In]: species_count.show()
[Out]:
+------------+-----+
|target_names|count|
+------------+-----+
|      setosa|   50|
|  versicolor|   50|
|   virginica|   50|
+------------+-----+
```

This PySpark code performs a groupBy operation on the spark_df using the column target_names. By applying the groupBy() method, the data is grouped based on unique values in the target_names column. The count() method is then applied to the grouped DataFrame to calculate the number of occurrences for each flower species.

We can see from the output of both Pandas and PySpark that each target has 50 species. This suggests that an accurate k-means algorithm should group the Iris dataset into three clusters, each with 50 observations. We will see if this is the case in the next section.

Machine Learning with k-Means

In this section, we build, train, and evaluate a k-means clustering algorithm in both Scikit-Learn and PySpark and use it to group the Iris flowers data into groups or clusters. We know from the data analysis in the Dataset section that there are three clusters (setosa, versicolor, and virginica), so if our clustering algorithm is 100% accurate, it should group the data into those clusters, each having 50 species.

k-Means Clustering with Scikit-Learn

We begin by building a k-means clustering model using the Scikit-Learn library. These are the steps to be followed:

Step 1: Import necessary libraries

```
[In]: from sklearn.datasets import load_iris
[In]: from sklearn.cluster import KMeans
[In]: from sklearn.metrics import silhouette_score
[In]: import matplotlib.pyplot as plt
```

This code imports several modules from the Scikit-Learn and Matplotlib libraries. The first line imports the load_iris function, which is used to load the Iris dataset. The second line imports the KMeans class, which is an implementation of the k-means clustering algorithm used to partition data points into clusters based on their similarity. The third line imports the silhouette_score function, which calculates the silhouette metric that measures how well the data points are separated into different clusters. The last line imports the pyplot module, which will be used to plot SSE (Sum of Squared Errors) scores against k values to decide on the optimal number of clusters using the elbow method.

Step 2: Load the Iris dataset

```
[In]: iris = load_iris()
[In]: X = iris.data
[In]: feature_names = iris.feature_names
[In]: pandas_df = pd.DataFrame(X, columns=feature_names)
```

This code loads the Iris dataset and assigns it to the variable iris. The dataset is then assigned to the variable X, which will be used to store the feature data of the dataset. The code then defines column names for the feature data of the Iris dataset and creates a DataFrame using the Pandas library.

Step 3: Choose the value of k

```
[In]: k_values = range(2, 10)
[In]: sse_values = []
[In]: for k in k_values:
          kmeans = KMeans(n_clusters=k, random_state=42)
          kmeans.fit(X)
          sse_values.append(kmeans.inertia_)
```

The algorithm needs a specific value of k to start with. We know that the correct value is 3 since we know that there are three classes (setosa, versicolor, virginica), but in many real-world applications, we wouldn't know the number of efficient groups or clusters. Therefore, we need a method to choose the number of clusters, k.

The elbow method is a widely used technique for determining the optimal number of clusters in a dataset. It involves selecting different values of k and training corresponding clustering models. For each model, the Sum of Squared Errors (SSE) or sum of within-cluster distances of centroids is computed and plotted against k. The optimal k is determined at the point where the SSE shows a significant decrease, resembling the shape of an elbow. This indicates that additional clusters beyond that point do not contribute significantly to improving the clustering quality.

The code provided previously calculates the SSE values for different values of k using the k-means clustering algorithm. The following is what the code does line by line:

- k_values = range(2, 10) defines a range of values for k (the number of clusters) from 2 to 9 (inclusive).
- sse_values = [] initializes an empty list called sse_values. This list will be used to store the SSE values obtained for each value of k.

- for k in k_values starts a loop that iterates over each value of k in the k_values range.
- kmeans = KMeans(n_clusters=k, random_state=42) creates an instance of the KMeans class. The n_clusters parameter is set to the current value of k, indicating the number of clusters to be created. The random_state parameter is set to 42 to ensure reproducibility of the results.
- kmeans.fit(X) fits the k-means model to the feature data (X) using the fit() method. The k-means algorithm will partition the data into k clusters based on their similarity.
- sse_values.append(kmeans.inertia_) appends the SSE value of the k-means model to the sse_values list. The SSE value is a measure of how far the samples within each cluster are from the centroid of that cluster. Lower SSE values indicate better clustering.

After executing these lines of code, the sse_values list will contain the SSE values for each value of k specified in the k_values range. These values can be used to generate a plot to decide on the optimum number of clusters by identifying the elbow point.

Step 4: Plot the SSE values to identify the elbow point

```
[In]: plt.plot(k_values, inertia_values, 'bx-')
[In]: plt.xlabel('Number of Clusters (k)')
[In]: plt.ylabel('Inertia')
[In]: plt.title('The Elbow Method')
[In]: plt.show()
[Out]:
```

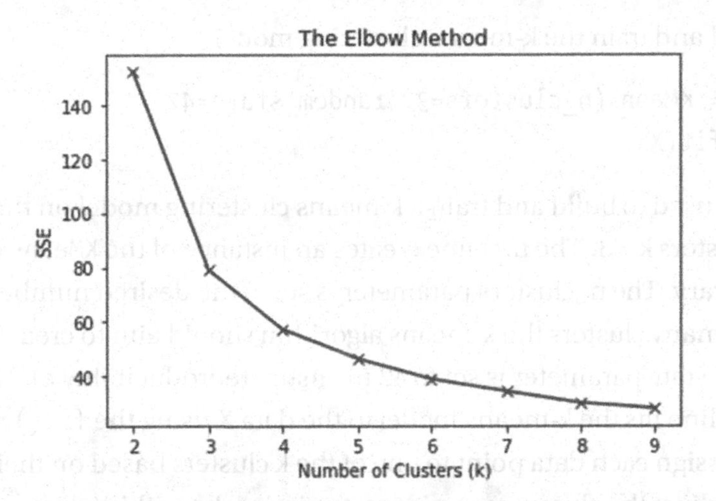

Let's first explain the code and then the plot. The code is used to plot the SSE values against the number of clusters (k) in order to identify the elbow point, which helps determine the optimal number of clusters for the dataset. The following is a breakdown of what the code does:

- plt.plot(k_values, sse_values, 'bx-') creates a line plot using the plot() function from the Matplotlib library (plt). It plots the values of k on the x axis (k_values) and the corresponding SSE values on the y axis (sse_values). The 'bx-' argument specifies the line style and marker style for the plot. In this case, 'bx-' represents blue markers ('b') connected by a solid line ('-').
- plt.xlabel('Number of Clusters (k)') sets the label for the x axis of the plot to "Number of Clusters (k)" using the xlabel() function.
- plt.ylabel('SSE') sets the label for the y axis of the plot to "SSE" using the ylabel() function.
- plt.title('The Elbow Method') sets the title of the plot to "The Elbow Method" using the title() function.
- plt.show() displays the plot on the screen using the show() function.

The plot indicates that the elbow point, at which the number of clusters is optimal, is k=3, as this is the point located at the bend in the plot.

Step 5: Build and train the k-means clustering model

```
[In]: kmeans = KMeans(n_clusters=3, random_state=42)
[In]: kmeans.fit(X)
```

This code is used to build and train a k-means clustering model on the dataset X with a number of clusters k = 3. The first line creates an instance of the `KMeans` class from the Scikit-Learn library. The n_clusters parameter is set to the desired number of clusters k, indicating how many clusters the k-means algorithm should aim to create (in this case, 3). The random_state parameter is set to 42 to ensure reproducibility of the results.

The second line fits the k-means model to the data X using the `fit()` method. The algorithm will assign each data point to one of the k clusters based on their similarity and iteratively adjust the cluster centroids to minimize the within-cluster sum of squares.

After executing these lines of code, the kmeans object will contain the trained k-means clustering model, which will have learned the cluster assignments and the final centroids for the specified number of clusters.

Step 6: Evaluate the clustering model

```
[In]: labels = kmeans.labels_
[In]: silhouette = silhouette_score(X, labels)
[In]: print("Silhouette Score:", silhouette)
[Out]: Silhouette Score: 0.55
```

In this final step, we evaluate the clustering model by calculating the Silhouette Score. The first line retrieves the cluster labels assigned to each data point by the k-means clustering model. The `labels_` attribute of the `kmeans` object contains an array-like object where each element represents the cluster label assigned to the corresponding data point.

The second line calculates the Silhouette Score using the `silhouette_score()` function. This takes two parameters: X, which is the feature data used for clustering, and labels, which are the cluster labels assigned by the clustering model to each data point. The function returns a single value representing the average Silhouette Score for all data points. The last line prints the score, which is approximately 0.55.

The Silhouette Score ranges from -1 to 1, with values close to 1 indicating that the data points are well clustered, with each point being closer to the data points within its own cluster than to the points in other clusters. If the Silhouette Score is close to 0, it suggests that the data point is on or very close to the decision boundary between two neighboring clusters, while if the score is negative (closer to -1), it indicates that the data point might have been assigned to the wrong cluster, as it is closer to the points in a different cluster than to the points in its own cluster.

The Silhouette Score of 0.55 is positive and relatively closer to 1. This suggests that the clustering results are reasonably good. However, to determine the actual quality, we need to compare the predicted counts with the ground truth.

The following code calculates and prints the predicted cluster counts:

```
[In]: cluster_counts = {}
[In]: for label in labels:
          if label in cluster_counts:
              cluster_counts[label] += 1
          else:
              cluster_counts[label] = 1
[In]: for cluster, count in cluster_counts.items():
          print("Cluster {}: {}".format(cluster, count))
[Out]: Cluster Counts: Cluster 0: 62 Cluster 1: 50 Cluster 2: 38
```

We observe from this output that cluster 1 has 50 counts, as expected, while the counts for the other two clusters deviate from 50, either higher (cluster 0: 62) or lower (cluster 2: 38).

Seaborn pair scatterplots can enhance our understanding of these results:

```
[In]: import seaborn as sns
[In]: pandas_df['Cluster'] = kmeans.labels_
[In]: sns.pairplot(pandas_df, hue='Cluster', palette='viridis',
      diag_kind='kde', markers=['o', 's', 'D'])
[In]: plt.suptitle('Pair Plot of Features with Cluster Assignments')
[In]: plt.show()
```

In this code, we are using the seaborn library to create a pair plot for visualizing relationships between pairs of features in the Iris dataset, with a focus on how the data points are clustered according to a previously fitted k-means clustering model (kmeans).

The first line imports the seaborn library and names it sns. The next step adds cluster labels to the pandas_df DataFrame. This is done by assigning the cluster labels obtained from the k-means model to a new column in the DataFrame named Cluster.

The next line creates the pair plot using seaborn. The pairplot function generates a grid of scatterplots for each pair of features in the DataFrame, and it color-codes the points based on their cluster assignments. The following is a breakdown of the parameters used:

- pandas_df: The DataFrame containing the data and cluster labels.
- hue='Cluster': This parameter specifies that we want to color-code the data points by the Cluster column, allowing us to distinguish between clusters visually.
- palette='viridis': This sets the color palette to the viridis color scheme.
- diag_kind='kde': The diag_kind parameter specifies that we want kernel density estimate (KDE) plots along the diagonal cells, showing the distribution of each feature.
- markers=['o', 's', 'D']: This parameter defines the markers used for data points in each cluster. Here, 'o', 's', and 'D' represent different marker shapes for clarity.
- plt.suptitle('Pair Plot of Features with Cluster Assignments'): This line sets the super title for the pair plot, providing a descriptive title for the visualization.
- plt.show(): Finally, this command displays the pair plot with all the specified visualizations.

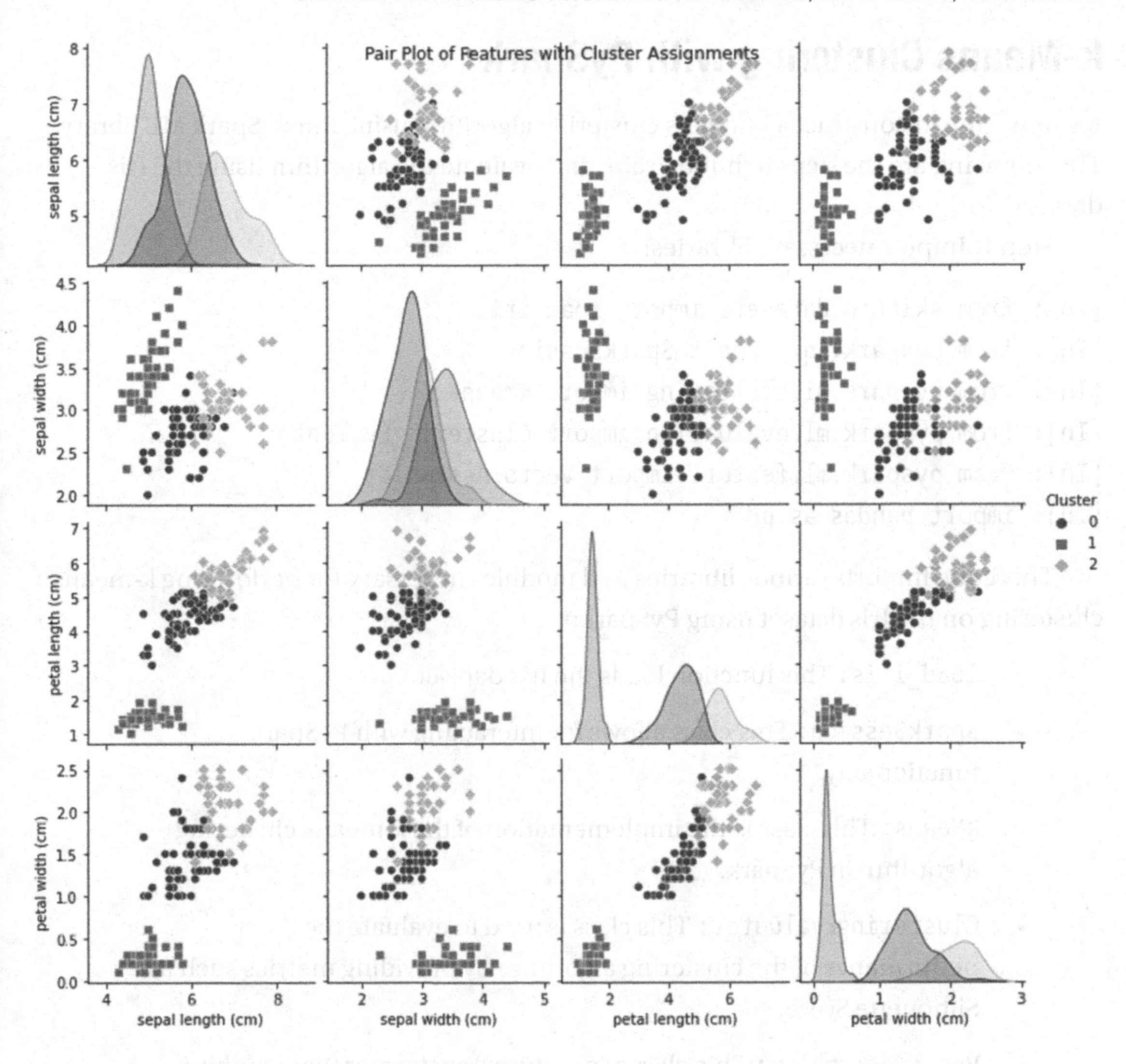

The pair plots reveal noticeable differences in the relationships between pairs of features for cluster 1 when compared to the other two clusters. In other words, cluster 1 can be easily distinguished from the other two clusters (0 and 2). This separation suggests that cluster 1 represents a distinct subgroup within the Iris dataset. In contrast, there is an overlap between the other two groups, which explains why the counts for clusters 0 and 2 deviate from the expected count of 50 each. More specifically, clusters 0 and 2 have 24 data points (62-38) that are misclassified due to the overlap.

k-Means Clustering with PySpark

It's now time to construct a k-means clustering algorithm using the PySpark ML library. The following are the steps to build, train, and evaluate the algorithm using the Iris dataset:

Step 1: Import necessary libraries:

```
[In]: from sklearn.datasets import load_iris
[In]: from pyspark.sql import SparkSession
[In]: from pyspark.ml.clustering import KMeans
[In]: from pyspark.ml.evaluation import ClusteringEvaluator
[In]: from pyspark.ml.feature import VectorAssembler
[In]: import pandas as pd
```

This code imports various libraries and modules necessary for performing k-means clustering on the Iris dataset using PySpark:

- `load_iris`: This function loads the Iris dataset.
- `SparkSession`: This class allows for interacting with PySpark functionality.
- `KMeans`: This class is the implementation of the k-means clustering algorithm in PySpark.
- `ClusteringEvaluator`: This class is used to evaluate the performance of the clustering algorithm by providing metrics such as Silhouette Score.
- `VectorAssembler`: This class is a feature transformer that combines multiple columns into a vector column in PySpark.

Step 2: Load the Iris dataset (same as step 2 in Scikit-Learn)

```
[In]: iris = load_iris()
[In]: X = iris.data
[In]: feature_names = iris.feature_names
[In]: pandas_df = pd.DataFrame(X, columns=feature_names)
```

This code is the same as step 2 in Scikit-Learn: it loads the Iris dataset and assigns it to the variable iris. The dataset is then assigned to the variable X, which stores the feature data of the dataset. The code then defines column names for the feature data of the Iris dataset and creates a DataFrame using the Pandas library.

Step 3: Convert the Pandas DataFrame to a PySpark DataFrame

```
[In]: spark = SparkSession.builder.getOrCreate()
[In]: spark_df = spark.createDataFrame(pandas_df)
```

This code first creates a `SparkSession` as an entry point to Spark functionality and then converts the pandas_df DataFrame to spark_df using the spark object's `createDataFrame()` method.

Step 4: Assemble the input features into a vector column

The following code defines the input features, creates a vector column, and transforms a Spark DataFrame.

```
[In]: input_features = feature_names
[In]: assembler = VectorAssembler(inputCols=input_features,
      outputCol="features")
[In]: spark_df = assembler.transform(spark_df)
```

In the first line, the feature_names is assigned to the variable input_features. This represents the names of the columns that will be used as input features for the vector column. In the second line, an instance of the `VectorAssembler` class, which is a feature transformer in PySpark that combines multiple columns into a single vector column, is created and assigned to the variable assembler. The inputCols parameter is set to input_features, specifying the names of the columns to be assembled. The outputCol parameter is set to features, indicating the name of the resulting vector column. In the last line, the `transform` method is called on the assembler object, passing the spark_df DataFrame as the argument. This performs the transformation and creates a new DataFrame called spark_df with an additional column named features that contains the assembled vector of the input features.

Step 5: Choose the optimal value of k

In this step, the desired number of clusters is 3, which was determined earlier by the elbow method:

```
[In]: k = 3
```

Step 6: Build and train the k-means clustering model

```
[In]: kmeans = KMeans().setK(k).setSeed(42)
[In]: model = kmeans.fit(spark_df)
```

This code creates a k-means clustering model and fits it to a Spark DataFrame. The first line creates a `KMeans` object and assigns it to the variable kmeans. The `setK()` method is called on `kmeans` and sets the number of clusters to k. The value of k represents the desired number of clusters in the k-means model. The `setSeed()` method is called to set the random seed for reproducibility.

The second line fits the k-means model to the spark_df DataFrame. The `fit()` method is called on the kmeans object, passing spark_df as the argument. This trains the k-means model on the provided DataFrame. The resulting trained model is assigned to the variable model.

Step 7: Make predictions

```
[In]: predictions = model.transform(spark_df)
```

This code performs predictions using the trained k-means clustering model on the Spark DataFrame. The `transform()` method is called on the trained k-means model (model) with the spark_df DataFrame as the argument. This applies the trained model to the spark_df DataFrame and generates predictions based on the input data. The resulting predictions, along with the original columns of the DataFrame, are stored in a new DataFrame called predictions.

Step 8: Evaluate the clustering model

```
[In]: evaluator = ClusteringEvaluator()
[In]: silhouette = evaluator.evaluate(predictions)
[In]: print("Silhouette Score:", silhouette)
[Out]: Silhouette Score: 0.74
```

This code calculates and prints the Silhouette Score for the clustering results obtained from the k-means model. The first line creates an instance of the `ClusteringEvaluator`, which is a class in PySpark that provides methods to evaluate the quality of clustering results. The created object is assigned to the variable evaluator.

In the second line, the `evaluate()` method is called on the evaluator object, passing the predictions DataFrame as the argument. This computes the Silhouette Score for the clustering results stored in the predictions DataFrame. The computed score is assigned to the variable silhouette. Finally, the code prints the Silhouette Score using the `print()` function.

The Silhouette Score is approximately 0.74, which is substantially higher than the score from Scikit-Learn (0.55). This difference can be attributed to the choice of distance measure used by each library: Spark uses the squared Euclidean distance as the default distance measure for calculating the Silhouette Score, while Scikit-Learn uses the normal (non-squared) Euclidean distance as the default distance measure. The choice of distance measure has implications for how the Silhouette Score is computed and can lead to variations in the score for the same data and clustering.

Spark's choice of the squared Euclidean distance is primarily motivated by efficiency and parallelism. Squaring the Euclidean distance allows certain parts of the equation to be precomputed, reducing the computational complexity when dealing with large datasets and parallel processing. This design is advantageous in distributed computing environments where Spark excels.

While both squared and normal Euclidean distances are measures of dissimilarity, they can yield different results, especially when comparing distances of data points to cluster centroids. Squaring the distances emphasizes larger differences more, which can affect the assignment of data points to clusters and, subsequently, the Silhouette Scores.

Step 9: Count and print the occurrences for each predicted cluster

```
[In]: cluster_counts = predictions.groupBy("prediction").count()
[In]: cluster_counts.show()
[Out]:

+----------+-----+
|prediction|count|
+----------+-----+
|         0|   50|
|         1|   38|
|         2|   62|
+----------+-----+
```

This code calculates and displays the counts of data points in each cluster from the clustering predictions obtained from the k-means model. In the first line, the `groupBy()` method is called on the predictions DataFrame, specifying the column prediction as the grouping criterion. The `count()` method is applied to the grouped DataFrame to calculate the number of occurrences of each unique value in the prediction column. The resulting DataFrame, cluster_counts, contains two columns: prediction (representing the cluster label) and count (representing the number of data points in that cluster). In the last line, the `show()` method is called on the cluster_counts DataFrame to display its contents.

Even though a Silhouette Score of 0.74 is relatively high (both positive and close to 1), we observe from the cluster counts that only cluster 0 has the expected 50 counts, while the counts for the other two clusters deviate from 50, either higher (cluster 2: 62) or lower (cluster 1: 38). In other words, the model's accuracy is not as high as one would have hoped.

Bringing It All Together

Here, we consolidate all the relevant code from the previous steps into a single code block. This way, the reader can execute the code as a single block in both Scikit-Learn and PySpark.

Scikit-Learn

Step 1: Import necessary libraries

```
[In]: from sklearn.datasets import load_iris
[In]: from sklearn.cluster import KMeans
[In]: from sklearn.metrics import silhouette_score
[In]: import matplotlib.pyplot as plt
[In]: import pandas as pd
```

Step 2: Load the Iris dataset

```
[In]: iris = load_iris()
[In]: X = iris.data
```

Step 3: Define column names

```
[In]: feature_names = iris.feature_names
```

Step 4: Create a DataFrame with column names

```
[In]: pandas_df = pd.DataFrame(X, columns=feature_names)
```

Step 5: Generate multiple SSE values for corresponding cluster values

```
[In]: k_values = range(2, 10)
[In]: sse_values = []
[In]: for k in k_values:
        kmeans = KMeans(n_clusters=k, random_state=42)
        kmeans.fit(X)
        sse_values.append(kmeans.inertia_)
```

Step 6: Plot the SSE values to identify the elbow point

```
[In]: plt.plot(k_values, sse_values, 'bx-')
[In]: plt.xlabel('Number of Clusters (k)')
[In]: plt.ylabel('SSE')
[In]: plt.title('The Elbow Method')
[In]: plt.show()
```

Step 7: Choose the optimal value of k based on the elbow point

```
[In]: k = 3
```

Step 8: Build and train the k-means clustering model

```
[In]: kmeans = KMeans(n_clusters=k, random_state=42)
[In]: kmeans.fit(X)
```

Step 9: Evaluate the clustering model

```
# a) Get the cluster labels for each data point
[In]: labels = kmeans.labels_
# b) Calculate the silhouette score
[In]: silhouette = silhouette_score(X, labels)
# c) Print the Silhouette Score
[In]: print("Silhouette Score:", silhouette)
```

Step 10: Calculate cluster counts

```
# a) Count the occurrences for each predicted cluster
[In]: cluster_counts = {}
[In]: for label in labels:
          if label in cluster_counts:
              cluster_counts[label] += 1
          else:
              cluster_counts[label] = 1
# b) Print the cluster counts
[In]: print("Cluster Counts:")
[In]: for cluster, count in cluster_counts.items():
          print("Cluster {}: {}".format(cluster, count))
```

PySpark

Step 1: Import necessary libraries

```
[In]: from sklearn.datasets import load_iris
[In]: from pyspark.sql import SparkSession
[In]: from pyspark.ml.clustering import KMeans
[In]: from pyspark.ml.evaluation import ClusteringEvaluator
[In]: from pyspark.ml.feature import VectorAssembler
[In]: import pandas as pd
```

Step 2: Load the Iris dataset

```
[In]: iris = load_iris()
[In]: X = iris.data
```

Step 3: Define column names

```
[In]: feature_names = iris.feature_names
```

Step 4: Create a DataFrame with column names

```
[In]: pandas_df = pd.DataFrame(X, columns=feature_names)
```

Step 5: Create a SparkSession

```
[In]: spark = SparkSession.builder.getOrCreate()
```

Step 6: Convert the Pandas DataFrame to a PySpark DataFrame

```
[In]: spark_df = spark.createDataFrame(pandas_df)
```

Step 7: Define the input features

```
[In]: input_features = feature_names
```

Step 8: Assemble the input features into a vector column

```
[In]: assembler = VectorAssembler(inputCols=input_features,
      outputCol="features")
[In]: spark_df = assembler.transform(spark_df)
```

Step 9: Choose the optimal value of k

```
[In]: k = 3
```

Step 10: Build and train the k-means clustering model

```
[In]: kmeans = KMeans().setK(k).setSeed(42)
[In]: model = kmeans.fit(spark_df)
```

Step 11: Make predictions

```
[In]: predictions = model.transform(spark_df)
```

Step 12: Evaluate the clustering model

```
[In]: evaluator = ClusteringEvaluator()
[In]: silhouette = evaluator.evaluate(predictions)
[In]: print("Silhouette Score:", silhouette)
```

Step 13: Count the occurrences for each predicted cluster

```
[In]: cluster_counts = predictions.groupBy("prediction").count()
```

Step 14: Print the cluster counts

```
[In]: print("Cluster Counts:")
[In]: cluster_counts.show()
```

Summary

In this chapter, we delved into the area of unsupervised learning with k-means clustering. We demonstrated how to build, train, and evaluate a k-means algorithm using the open source Iris dataset. The purpose of this exercise is to group the species of Iris flowers into clusters based on their characteristics.

We leveraged the power of Pandas, Scikit-Learn, and PySpark. The main objective of this approach is to assist data scientists who are familiar with Pandas and Scikit-Learn transition smoothly to PySpark. By exploring the similarities in steps between the platforms, our goal was to provide these machine learning practitioners with a seamless transition to PySpark, enabling them to harness its distributed computing capabilities for enhanced scalability and performance.

In the next chapter, we delve into a new area of machine learning: hyperparameter tuning. Up to this point, we have built algorithms with default parameters. The next chapter demonstrates how to fine-tune those parameters to customize the default values.

CHAPTER 16

Hyperparameter Tuning with Scikit-Learn and PySpark

In this chapter, we investigate the subject of hyperparameter tuning. This is a critical step in machine learning that involves finding the optimal set of hyperparameters for a given algorithm. Hyperparameters are parameters that are set before the learning process begins and affect the behavior and performance of the model.

The process of hyperparameter tuning aims to find the best combination of hyperparameters that results in the highest performance of the model on a given task or dataset. This process is important because different hyperparameter values can significantly impact the model's ability to learn and generalize from the data.

In this chapter, we provide examples of such parameters in both Scikit-Learn and PySpark. We also look at a detailed example of how to fine-tune key hyperparameters in random forests. For this, we use the Iris dataset, which is the same dataset we used in Chapter 9 to classify the Iris flowers into different species.

Examples of Hyperparameters

In this section, we provide examples of the key hyperparameters for various popular machine learning algorithms examined in this book. The hyperparameters for a specific algorithm generally remain the same regardless of whether it is used for regression or classification tasks. For example, if we consider the random forest algorithm, hyperparameters such as the number of trees, maximum depth, and the number of features to consider for each split would be the same, whether used for regression or classification.

A. Testas, *Distributed Machine Learning with PySpark*, https://doi.org/10.1007/978-1-4842-9751-3_16

The following table compares the key parameters of decision trees, random forests, gradient-boosted trees, neural networks, and k-means clustering in both Scikit-Learn and PySpark, along with a description of what each parameter means.

Model	Scikit-Learn	PySpark	Description
Decision trees	max_depth	maxDepth	Maximum depth of the decision tree
	min_samples_split	minInstancesPerNode	Minimum samples required to split an internal node
	min_samples_leaf	minInfoGain	Minimum samples required to be at a leaf node
	max_features	maxBins	Maximum number of features for the best split
Random forests	n_estimators	numTrees	Number of decision trees in the random forest
	max_depth	maxDepth	Maximum depth of each tree in the forest
	min_samples_split	minInstancesPerNode	Minimum samples required to split an internal node
	min_samples_leaf	minInfoGain	Minimum samples required to be at a leaf node
	max_features	featureSubsetStrategy	Maximum features for the best split
Gradient-boosted trees	n_estimators	numTrees	Number of decision trees in the boosting process
	learning_rate	stepSize	Contribution of each tree to the final prediction
	max_depth	maxDepth	Maximum depth of each decision tree
	min_samples_split	minInstancesPerNode	Minimum samples required to split an internal node
	min_samples_leaf	minInfoGain	Minimum samples required to be at a leaf node

(*continued*)

Model	Scikit-Learn	PySpark	Description
Neural networks	hidden_layers	numLayers	Number of hidden layers in the neural network
	hidden_units	layerSizes	Neurons in each hidden layer
	learning_rate	stepSize	Step size in updating network weights during training
k-Means clustering	n_clusters	k	Desired number of clusters
	max_iter	maxIterations	Maximum number of algorithm iterations or epochs

Tuning the Parameters of a Random Forest

Hyperparameter tuning plays a crucial role in optimizing the performance of machine learning algorithms, and random forests are no exception. Random forests are a robust ensemble learning method known for their robustness and effectiveness in handling complex data tasks. However, the performance of a random forest model heavily depends on the values assigned to its hyperparameters.

In this section, we demonstrate how to fine-tune four key hyperparameters in random forests in both Scikit-Learn and PySpark:

- Number of decision trees (n_estimators / numTrees)
- Maximum depth of each tree (max_depth / maxDepth)
- Minimum number of samples required to split an internal node (min_samples_split / minInstancesPerNode)
- Minimum samples required to be at a leaf node (min_samples_leaf / minInfoGain)

By carefully adjusting these hyperparameters, we aim to find the best configuration that maximizes the model's predictive power while avoiding overfitting or underfitting.

We use the same Iris dataset we used in Chapter 9 to classify the Iris flowers into different species. By leveraging hyperparameter tuning techniques in random forests, we aim to build a highly accurate classification model that can effectively distinguish between the three species of Iris flowers: setosa, versicolor, and virginica.

Hyperparameter Tuning in Scikit-Learn

The following code follows the standard process of hyperparameter tuning using Scikit-Learn's `GridSearchCV` with a random forest classifier. It loads the Iris dataset, splits it into training and testing sets, defines the parameter grid for tuning, performs grid search, retrieves the best model and its parameters, makes predictions on the test data, evaluates the model's performance, and finally prints the best parameter values.

We start the process by importing the necessary libraries. These are required to provide the essential functionality and tools necessary for our analysis.

Step 1: Import necessary libraries

```
[In]: from sklearn.datasets import load_iris
[In]: from sklearn.ensemble import RandomForestClassifier
[In]: from sklearn.model_selection import train_test_split, GridSearchCV
[In]: from sklearn.metrics import accuracy_score
```

Here is what each import does:

- `load_iris` loads the Iris dataset.
- `RandomForestClassifier` defines the Random Forest classifier.
- `train_test_split` splits the data into training and testing sets.
- `GridSearchCV`, which stands for Grid Search Cross-Validation, searches for the best combination of hyperparameters for the machine learning model.
- `accuracy_score` calculates the accuracy of the model.

Step 2: Load the Iris dataset

```
[In]: iris_sklearn = load_iris()
```

This line loads the Iris dataset using the `load_iris()` function.

Step 3: Split the data into features and labels

```
[In]: X = iris_sklearn.data
[In]: y = iris_sklearn.target
```

This code separates the features and labels from the loaded dataset.

Step 4: Split the data into training and testing sets

```
[In]: X_train, X_test, y_train, y_test = train_test_split(X, y, test_
      size=0.2, random_state=42)
```

In this step, we split the data into training and testing sets, with 80% for training and 20% for testing.

Step 5: Define the random forest classifier

```
[In]: rf = RandomForestClassifier(random_state=42)
```

This line defines an instance of the random forest classifier. The random seed ensures reproducibility of the results.

Step 6: Define the parameter grid for tuning

```
[In]: param_grid = {
    'max_depth': [2, 5, 7, 10],
    'n_estimators': [20, 50, 75, 100],
    'min_samples_split': [2, 3, 5, 10],
    'min_samples_leaf': [1, 2, 3, 4],
}
```

In this code, we define the param_grid so that the grid search algorithm will systematically evaluate various combinations of these hyperparameters to find the best combination that results in the highest model performance.

Given the relatively small size of the Iris dataset (150 records), we have made conservative parameter choices. This is to prevent overfitting and to ensure that the model does not become overly complex for the given dataset size.

For reference, the default parameters for the Scikit-Learn random forest classifier are as follows:

- max_depth: None (unlimited depth by default)
- n_estimators: 100
- min_samples_split: 2
- min_samples_leaf: 1

The following are the details of the parameters we used to customize the default settings:

- `'max_depth': [2, 5, 7, 10]` specifies the max_depth hyperparameter of the random forest classifier. The grid search will consider four values for max_depth: 2, 5, 7, and 10. The model will be trained and evaluated with each of these values to determine the best max_depth value.
- `'n_estimators': [20, 50, 75, 100]` defines the n_estimators hyperparameter, which determines the number of decision trees in the random forest. The grid search will explore three values for n_estimators: 20, 50, 75, and 100.
- `'min_samples_split': [2, 3, 5, 10]` specifies the min_samples_split hyperparameter, which determines the minimum number of samples required to split an internal node in a decision tree. The grid search will evaluate the model with four different values for min_samples_split: 2, 3, 5, and 10.
- `'min_samples_leaf': [1, 2, 3, 4]` defines the min_samples_leaf hyperparameter, which determines the minimum number of samples required to be present in a leaf node of a decision tree in the random forest. The grid search will assess the model's performance using four distinct values for min_samples_leaf: 1, 2, 3, and 4.

Step 7: Perform grid search using GridSearchCV

```
[In]: grid_search = GridSearchCV(estimator=rf, param_grid=param_grid,
    cv=5)
[In]: grid_search.fit(X_train, y_train)
```

In these two lines of code, the grid search is performed using the `GridSearchCV` class from the Scikit-Learn library. Here is what the code does:

- The first line creates an instance of GridSearchCV and initializes it with the following three parameters:
    - The estimator parameter specifies the machine learning model or estimator that will be used. In this case, it is set to rf, which is an instance of the RandomForestClassifier.

- The param_grid parameter specifies the grid of hyperparameters and their possible values that will be explored during the grid search. In this case, it is set to the previously defined param_grid dictionary.
- The cv parameter determines the number of folds in the cross-validation process. It is set to 5, which means the dataset will be split into five folds, and the grid search will perform five-fold cross-validation during the evaluation of different hyperparameter combinations. With a small dataset like the Iris dataset, a five-fold cross-validation is sufficient. It strikes a balance between training the model on a substantial portion of the data while still having multiple validation sets for robust evaluation.

- The second line of code initiates the grid search process. The `fit()` method of the `GridSearchCV` object is called with the training data X_train and the corresponding labels y_train. The grid search algorithm then performs the search, training and evaluating the model with different hyperparameter combinations using cross-validation.

The grid search algorithm exhaustively searches through all possible combinations of hyperparameters specified in the param_grid dictionary and evaluates the model's performance using the chosen evaluation metric. It keeps track of the best-performing model based on this metric. After the grid search is completed, the best model can be accessed using the `best_estimator_` attribute of the `GridSearchCV` object.

Step 8: Get the best model and its parameters

```
[In]: best_model = grid_search.best_estimator_
[In]: best_max_depth = best_model.max_depth
[In]: best_num_estimators = best_model.n_estimators
[In]: best_min_samples_split = best_model.min_samples_split
[In]: best_min_samples_leaf = best_model.min_samples_leaf
```

In this code, after the grid search is performed, the best model and its corresponding hyperparameters are extracted for further analysis. Here are the steps:

1. In the first line, the `best_estimator_` attribute of the `GridSearchCV` object stores the best model found during the grid search process. This model represents the combination of hyperparameters that achieved the highest performance according to the chosen evaluation metric.
2. In the second line, the `max_depth` attribute of the `best_model` object retrieves the value of the max_depth hyperparameter of the best model. It represents the maximum depth of each individual tree in the random forest classifier.
3. In the third line, the `n_estimators` attribute of the `best_model` object retrieves the value of the n_estimators hyperparameter of the best model. It represents the number of decision trees in the random forest classifier.
4. In the fourth line, the `min_samples_split` attribute of the `best_model` object retrieves the value of the min_samples_split hyperparameter of the best model. It represents the minimum number of samples required to split an internal node in the decision trees of the random forest classifier.
5. In the fifth line, the `min_samples_leaf` attribute of the best_model object retrieves the value of the min_samples_leaf hyperparameter of the best model. This hyperparameter defines the minimum number of samples required to be at a leaf node in the decision trees of the random forest classifier.

By accessing these attributes of the `best_model` object, we can obtain the optimal hyperparameter values that yielded the best performance during the grid search.

Step 9: Make predictions on the test data using the best model

```
[In]: y_pred = best_model.predict(X_test)
```

In this line of code, we use the best model to predict the labels for the test data.

Step 10: Evaluate the model's performance

```
[In]: accuracy = accuracy_score(y_test, y_pred)
[In]: print("Test Accuracy:", accuracy)
[Out]: 1.0
```

This step evaluates the model's performance using the accuracy metric. The first line calculates the accuracy of the model by comparing the predicted labels with the true labels. The second line prints the accuracy of the model on the test data.

Step 11: Print the best parameter values

```
[In]: print("Best max_depth:", best_max_depth)
[In]: print("Best n_estimators:", best_num_estimators)
[In]: print("Best min_samples_split:", best_min_samples_split)
[In]: print("Best min_samples_leaf:", best_min_samples_leaf)

[Out]:
Best max_depth: 5
Best n_estimators: 50
Best min_samples_split: 3
Best min_samples_leaf: 1
```

In this final step, after obtaining the best model and its corresponding hyperparameters, the code prints the values of these hyperparameters.

Here are the details of this process:

1. The first line prints the value of the max_depth hyperparameter of the best model, which represents the maximum depth of each individual tree in the random forest classifier.

2. The second line prints the value of the n_estimators hyperparameter of the best model, which represents the number of decision trees in the random forest classifier.

3. The third line prints the value of the min_samples_split hyperparameter of the best model, which represents the minimum number of samples required to split an internal node in the decision trees of the random forest classifier.

4. The last line prints the value of the min_samples_leaf hyperparameter of the best model, which represents the minimum number of samples required to be present at a leaf node in the decision trees of the random forest classifier.

Here is an interpretation of each line of the output:

- Test Accuracy: The model achieved a perfect accuracy of 1.0 on the test dataset, meaning that 100% of the samples were correctly predicted.
- Best max_depth: The optimal value for the maximum depth of each decision tree in the random forest classifier is 5.
- Best n_estimators: The optimal number of decision trees in the random forest classifier is 50.
- Best min_samples_split: The optimal minimum number of samples required to split an internal node in the decision trees of the random forest classifier is 3.
- Best min_samples_leaf: The optimal minimum number of samples required to be present at a leaf node in the decision trees of the random forest classifier is 1. This hyperparameter defines the granularity of splitting the trees in the ensemble, contributing to the model's ability to generalize effectively and make accurate predictions on unseen data.

Hyperparameter Tuning in PySpark

In this subsection, we replicate Scikit-Learn functionality using PySpark code. Like Scikit-Learn, we initiate the process of fine-tuning the hyperparameters of a random forest by first importing the necessary libraries:

Step 1: Import necessary libraries

```
[In]: from pyspark.sql import SparkSession
[In]: from pyspark.ml.feature import VectorAssembler
[In]: from pyspark.ml.classification import RandomForestClassifier
[In]: from pyspark.ml.evaluation import
      MulticlassClassificationEvaluator
```

```
[In]: from pyspark.ml.tuning import ParamGridBuilder, CrossValidator
[In]: from sklearn.datasets import load_iris
[In]: from sklearn.model_selection import train_test_split
[In]: from sklearn.metrics import accuracy_score
[In]: import pandas as pd
```

In this first step, we import the necessary libraries. Here is what each import does:

- `SparkSession` is the entry point for interacting with Spark.
- `VectorAssembler` is used to transform the input features into a single vector column.
- `RandomForestClassifier` is an implementation of the random forest classifier in PySpark.
- `MulticlassClassificationEvaluator` is used to evaluate the performance of the model by computing the accuracy metric.
- `ParamGridBuilder` and `CrossValidator` are used for performing hyperparameter tuning with cross-validation in PySpark.
- `load_iris` is used to load the Iris dataset.
- `train_test_split` is used to split the dataset into training and testing subsets for model evaluation.
- `accuracy_score` is used to calculate the accuracy of the model by comparing the predicted and true labels.
- `pandas as pd` is used to create a Pandas DataFrame.

Step 2: Load the Iris dataset

```
[In]: iris_data = load_iris()
```

This code simply loads the Iris dataset using the `load_iris()` function.

Step 3: Convert the data and target arrays to a Pandas DataFrame

```
[In]: X = iris_data.data
[In]: y = iris_data.target
```

This code converts the data and target arrays from the `iris_data` object into a Pandas DataFrame. The first line assigns the feature data to the variable X, which contains the input features of the Iris dataset. The second line assigns the target labels from the same `iris_data` object to the variable y. The target attribute contains the corresponding target labels for the Iris dataset, which represents the species of the flowers.

Step 4: Create a SparkSession

```
[In]: spark = SparkSession.builder.getOrCreate()
```

This line creates a Spark Session to enable interaction with Spark.

Step 5: Create a Pandas DataFrame from the Iris dataset

```
[In]: iris_pandas = pd.DataFrame(data=X,
      columns=iris_data.feature_names)
[In]: iris_pandas["label"] = y
```

This code creates a Pandas DataFrame named iris_pandas from the Iris dataset, which has been loaded into the variables X (feature data) and y (target labels) in step 3. More precisely, the first line creates a new DataFrame called iris_pandas using the Pandas DataFrame constructor. It takes two arguments: data and columns. The data argument (X) is the feature data array, and the columns argument specifies the column names for the DataFrame, which are extracted from the `feature_names` attribute of the iris_data object.

The second line adds a new column named label to the iris_pandas DataFrame and assigns it the values from the y array (which represents the target labels of the Iris dataset).

Step 6: Create a PySpark DataFrame from the Pandas DataFrame

```
[In]: iris_df = spark.createDataFrame(iris_pandas)
```

The aim of this step is to convert the Pandas DataFrame to a PySpark DataFrame so that we can apply PySpark code to it. The `createDataFrame()` function in Spark is used to perform this conversion.

Step 7: Split the data into features and labels and assemble the features

```
[In]: feature_columns = iris_data.feature_names
[In]: assembler = VectorAssembler(inputCols=feature_columns,
      outputCol="features")
[In]: iris_df = assembler.transform(iris_df)
```

This code splits the data in the PySpark DataFrame into features and labels. The first line assigns the list of features to the variable feature_columns. The second line creates a `VectorAssembler` object named assembler. This is a transformer in PySpark that combines multiple columns into a single vector column. It takes two arguments: inputCols specifies the names of the input feature columns (feature_columns), and outputCol specifies the name of the output vector column (features).

The last line applies the `transform` method of the assembler object to the PySpark iris_df DataFrame. The method takes the input DataFrame and adds a new column named features to it. This new column contains the combined vector representation of the input features specified by inputCols.

Step 8: Split the data into training and testing sets

```
[In]: X_train, X_test = iris_df.randomSplit([0.8, 0.2], seed=42)
```

This code splits the data in the iris_df into training and testing sets. The `randomSplit` function takes two arguments. The first argument, [0.8, 0.2], represents the proportions for splitting the data: 80% of the data will be allocated to the training set, and 20% will be allocated to the test set. The second argument, seed=42, sets the random seed to ensure reproducibility of the splits.

The result is a list containing two PySpark DataFrames: X_train and X_test. The X_train DataFrame contains 80% of the data, randomly sampled, which will be used for training the classifier. The X_test DataFrame contains the remaining 20% of the data, which will be used for evaluating the performance of the trained model on unseen data.

Step 9: Define the random forest classifier

```
[In]: spark.conf.set("spark.seed", 42)
[In]: rf = RandomForestClassifier(featuresCol="features",
         labelCol="label")
```

This code creates an instance of the `RandomForestClassifier` class named rf, which is an algorithm for classification tasks based on the random forest ensemble method. It takes two arguments: featuresCol specifies the name of the input feature column in the PySpark DataFrame (featuresCol="features"), and labelCol specifies the name of the target label column (labelCol="label").

Note that in PySpark's RandomForestClassifier, there is no direct parameter called seed that we can set inside the classifier constructor to control the randomness. Unlike Scikit-Learn's RandomForestClassifier, where we can set the random_state parameter inside the constructor, PySpark's RandomForestClassifier relies on a global Spark Session configuration for controlling randomness.

Step 10: Define the parameter grid for tuning

```
[In]: param_grid = ParamGridBuilder() \
      .addGrid(rf.maxDepth, [2, 5, 7, 10]) \
      .addGrid(rf.numTrees, [20, 50, 75, 100]) \
      .addGrid(rf.minInstancesPerNode, [2, 3, 5, 10]) \
      .addGrid(rf.minInfoGain, [0.0, 0.01, 0.02, 0.03]) \
      .build()
```

This code defines a parameter grid for tuning the hyperparameters of the random forest classifier. By defining this grid, the code sets up a range of values for each hyperparameter that will be tested during the hyperparameter tuning process. The grid builder allows for systematically evaluating different combinations of hyperparameters to find the optimal configuration for the random forest classifier.

For reference, the default hyperparameters for the PySpark random forest classifier are as follows:

- maxDepth: 5
- numTrees: 20
- minInstancesPerNode: 1
- minInfoGain: 0.0

Now, let's explain the parameter grid we've set up to customize the default values:

- `param_grid = ParamGridBuilder()` creates an instance of the ParamGridBuilder class, which is used to build a grid of hyperparameter combinations.
- `.addGrid(rf.maxDepth, [2, 5, 7, 10])` adds a grid of values for the maxDepth hyperparameter of the random forest classifier. It specifies a range of values [2, 5, 7, 10] to be tested for the maxDepth parameter.

- `.addGrid(rf.numTrees, [20, 50, 75, 100])` adds a grid of values for the numTrees hyperparameter of the model. It specifies a range of values [20, 50, 75, 100] to be tested for the numTrees parameter.
- `.addGrid(rf.minInstancesPerNode, [2, 3, 5, 10])` adds a grid of values for the minInstancesPerNode hyperparameter of the model. It specifies a range of values [2, 3, 5, 10] to be tested for the minInstancesPerNode parameter.
- `.addGrid(rf.minInfoGain, [0.0, 0.01, 0.02, 0.03])` adds a grid of values for the minInfoGain hyperparameter of the random forest classifier. It specifies a range of values [0.0, 0.01, 0.02, 0.03] to be tested for the minInfoGain parameter. This hyperparameter controls the minimum information gain required for a split to happen in the decision trees. A higher value implies stricter splitting criteria, and a lower value allows for more splits, potentially leading to a more complex model. Notice how PySpark uses decimals for the minInfoGain hyperparameter, unlike Scikit-Learn, where we used integers [1, 2, 3, 4] for the min_samples_leaf. This difference in representation reflects variations in implementation between the two platforms, which may impact accuracy and the best parameter combinations.
- Finally, `.build()` builds the parameter grid by combining all the specified hyperparameter grids.

Step 11: Create the cross-validator

```
[In]: evaluator =
      MulticlassClassificationEvaluator(metricName="accuracy")
[In]: cross_validator = CrossValidator(estimator=rf,
      estimatorParamMaps=param_grid, evaluator=evaluator, numFolds=5)
```

This code creates a cross-validator object for performing model selection and hyperparameter tuning using k-fold cross-validation. It utilizes the random forest classifier, the parameter grid, and the evaluator to evaluate and select the best model configuration based on the specified metric (accuracy).

Here are the code details:

- The first line creates an instance of the `MulticlassClassificationEvaluator` class named evaluator. This is used to evaluate the performance of the classification model based on the specified metric (accuracy). This measures the proportion of correctly classified instances.
- The second line creates an instance of the `CrossValidator` class named cross_validator. This is used for model selection and hyperparameter tuning using k-fold cross-validation, and has the following parameters:
    - estimator=rf specifies the model to be evaluated, which is the random forest classifier (rf).
    - estimatorParamMaps=param_grid specifies the parameter grid that contains different combinations of hyperparameters to be tested.
    - evaluator=evaluator specifies the evaluator object to be used for evaluating the model's performance.
    - numFolds=5 specifies the number of folds or partitions to be used in the cross-validation process. In this case, five-fold cross-validation will be performed.

Step 12: Fit the cross-validator to the training data

```
[In]: cv_model = cross_validator.fit(X_train)
```

This line fits the cross-validator (cross_validator) to the training data (X_train). It trains and evaluates multiple random forest models with different hyperparameter configurations. This helps to identify the best combination of hyperparameters that yields the highest accuracy on the training data. The resulting cv_model object can then be used to make predictions on new, unseen data.

Step 13: Get the best model and its parameters

```
[In]: best_model = cv_model.bestModel
[In]: best_max_depth = best_model.getMaxDepth()
[In]: best_num_estimators = best_model.getNumTrees
```

```
[In]: best_min_samples_split = best_model.getMinInstancesPerNode()
[In]: best_min_info_gain = best_model.getMinInfoGain()
```

This code retrieves the best model and its corresponding hyperparameters from the trained cross-validated model. This allows us to access and use the optimal configuration found through the hyperparameter tuning process.

Here are the explanations for each step:

- The first line assigns the best model found during the cross-validation process to the variable best_model. The cv_model object contains information about all the models trained during cross-validation, and bestModel retrieves the model with the highest performance based on the evaluation metric (accuracy).
- The second line retrieves the value of the maxDepth hyperparameter from the best_model object. It uses the `getMaxDepth()` method to obtain the maximum depth setting of the decision trees in the random forest.
- The third line retrieves the value of the numTrees hyperparameter from the best_model object. It uses the `getNumTrees` attribute to obtain the number of decision trees in the random forest.
- The fourth line retrieves the value of the minInstancesPerNode hyperparameter from the best_model object. It uses the `getMinInstancesPerNode()` method to obtain the minimum number of samples required to split an internal node in the decision trees.
- The last line retrieves the value of the minInfoGain hyperparameter from the best_model object. It uses the `getMinInfoGain()` method to obtain the minimum information gain required for a split to occur in the decision trees. This parameter controls the strictness of the splitting criteria, where a higher value implies a more stringent criterion for splitting, while a lower value allows for more splits, potentially resulting in a more complex model.

Step 14: Make predictions on the test data using the best model

```
[In]: predictions = best_model.transform(X_test)
```

This code makes predictions on the test data using the best model obtained from the hyperparameter tuning process. This allows us to evaluate the model's performance on unseen data. The predictions object can be further analyzed and compared to the true labels to assess the accuracy and effectiveness of the trained model.

The code applies the `transform` method to the best_model object, passing in the test data (X_test). This method applies the trained model to the test data and generates predictions, which are stored in the predictions object.

Step 15: Evaluate the model's performance

```
[In]: accuracy = evaluator.evaluate(predictions)
```

This code calculates the accuracy of the predictions made by the model on the test data. This is done by comparing the predicted labels to the true labels in the test data. The resulting accuracy value is assigned to the variable accuracy, which can be used to assess the performance of the model on the test data.

Step 16: Print the test accuracy and best parameter values

```
[In]: print("Test Accuracy:", accuracy)
[In]: print("Best maxDepth:", best_max_depth)
[In]: print("Best numTrees:", best_num_estimators)
[In]: print("Best minInstancesPerNode:", best_min_samples_split)
[In]: print("Best minInfoGain:", best_min_info_gain)
```

```
[Out]:
Test Accuracy: 0.97
Best maxDepth: 5
Best numTrees: 50
Best minInstancesPerNode: 5
Best minInfoGain: 0.0
```

In this step, the code prints out various metrics and hyperparameters related to the trained model and its performance on the test data. This provides an overview of the model's performance on the test data and highlights the optimal settings for the hyperparameters that were tuned during the training process.

The following are explanations for each line of the code:

- print("Test Accuracy:", accuracy) prints the value of the test accuracy. This value is represented by the variable accuracy, which was computed earlier using the evaluation metric.
- print("Best maxDepth:", best_max_depth) prints the value of the best maximum depth hyperparameter.
- print("Best numTrees:", best_num_estimators) prints the value of the best number of trees hyperparameter.
- print("Best minInstancesPerNode:", best_min_samples_split) prints the value of the best minimum instances per node hyperparameter.
- print("Best minInfoGain:", best_min_info_gain) prints the value of the best minimum information gain hyperparameter.

Turning to the output, here is what the results mean:

- Test Accuracy (approximately 0.97): This indicates the test accuracy achieved by the random forest classifier on the given test dataset. The value 0.97 represents the accuracy score, which is a measure of how well the model's predictions align with the true labels in the test data. The model achieved an accuracy of 97%.
- Best maxDepth (5): This reveals the best value found for the maximum depth hyperparameter during the hyperparameter tuning process. The value 5 indicates that the optimal maximum depth setting for the decision trees in the random forest is 5.
- Best numTrees (50): This indicates the best number of trees chosen for the random forest classifier. The value 50 indicates that the optimal number of trees in the forest is set to 50.
- Best minInstancesPerNode (5): This shows the best value for the minimum number of instances per tree node, which is a hyperparameter that controls the splitting process in the decision trees. The value 5 indicates that the optimal setting for the minimum instances per node is 5, meaning that a node will only be split further if it contains at least 5 instances.

- Best minInfoGain (0.0): This indicates the best value for the minimum information gain hyperparameter. In this case, the optimal setting for minInfoGain is 0.0, signifying that even a small improvement in information gain is sufficient to trigger a split in the decision trees. A value of 0.0 means that the splitting criteria are not very strict, potentially leading to a more complex model with finer-grained splits.

Finally, it's worth noting that the results from PySpark differ slightly from those of Scikit-Learn. Both algorithms agreed on the Best maxDepth and Best numTrees, which were 5 and 50, respectively, in both. However, there are differences in Best minInstancesPerNode (3 in Scikit-Learn vs. 5 in PySpark) and Best minInfoGain (1 in Scikit-Learn vs. 0 in PySpark). The accuracy was also slightly lower in PySpark (97% vs. 100%). These variations reflect some subtle implementation differences between the two frameworks.

Bringing It All Together

Now, let's compile all the relevant code from the previous steps into a unified code block. This will allow the reader to execute the code as a cohesive unit in both Scikit-Learn and PySpark.

Scikit-Learn

Step 1: Import necessary libraries

```
[In]: from sklearn.datasets import load_iris
[In]: from sklearn.ensemble import RandomForestClassifier
[In]: from sklearn.model_selection import train_test_split, GridSearchCV
[In]: from sklearn.metrics import accuracy_score
```

Step 2: Load the Iris dataset using the load_iris() function

```
[In]: iris_sklearn = load_iris()
```

Step 3: Split the data into features and labels

```
[In]: X = iris_sklearn.data
[In]: y = iris_sklearn.target
```

Step 4: Split the data into training and testing sets

```
[In]: X_train, X_test, y_train, y_test = train_test_split(X, y, test_
      size=0.2, random_state=42)
```

Step 5: Define the random forest classifier

```
[In]: rf = RandomForestClassifier(random_state=42)
```

Step 6: Define the parameter grid for tuning

```
[In]: param_grid = {
    'max_depth': [2, 5, 7, 10],
    'n_estimators': [20, 50, 75, 100],
    'min_samples_split': [2, 3, 5, 10],
    'min_samples_leaf': [1, 2, 3, 4],
}
```

Step 7: Perform grid search using GridSearchCV

```
[In]: grid_search = GridSearchCV(estimator=rf, param_grid=param_grid,
      cv=5)
[In]: grid_search.fit(X_train, y_train)
```

Step 8: Get the best model and its parameters

```
[In]: best_model = grid_search.best_estimator_
[In]: best_max_depth = best_model.max_depth
[In]: best_num_estimators = best_model.n_estimators
[In]: best_min_samples_split = best_model.min_samples_split
[In]: best_min_samples_leaf = best_model.min_samples_leaf
```

Step 9: Make predictions on the test data using the best model

```
[In]: y_pred = best_model.predict(X_test)
```

Step 10: Evaluate the model's performance

```
[In]: accuracy = accuracy_score(y_test, y_pred)
[In]: print("Test Accuracy:", accuracy)
```

Step 11: Print the best parameter values

```
[In]: print("Best max_depth:", best_max_depth)
[In]: print("Best n_estimators:", best_num_estimators)
[In]: print("Best min_samples_split:", best_min_samples_split)
[In]: print("Best min_samples_leaf:", best_min_samples_leaf)
```

PySpark

Step 1: Import necessary libraries

```
[In]: from pyspark.sql import SparkSession
[In]: from pyspark.ml.feature import VectorAssembler
[In]: from pyspark.ml.classification import RandomForestClassifier
[In]: from pyspark.ml.evaluation import
      MulticlassClassificationEvaluator
[In]: from pyspark.ml.tuning import ParamGridBuilder, CrossValidator
[In]: from sklearn.datasets import load_iris
[In]: from sklearn.model_selection import train_test_split
[In]: from sklearn.metrics import accuracy_score
[In]: import pandas as pd
```

Step 2: Load the Iris dataset using the load_iris() function

```
[In]: iris_data = load_iris()
```

Step 3: Convert the data and target arrays to a Pandas DataFrame

```
[In]: X = iris_data.data
[In]: y = iris_data.target
```

Step 4: Create a SparkSession

```
[In]: spark = SparkSession.builder.getOrCreate()
```

Step 5: Create a Pandas DataFrame from the Iris dataset

```
[In]: iris_pandas = pd.DataFrame(data=X,
          columns=iris_data.feature_names)
[In]: iris_pandas["label"] = y
```

Step 6: Create a PySpark DataFrame from the Pandas DataFrame

```
[In]: iris_df = spark.createDataFrame(iris_pandas)
```

Step 7: Split the data into features and labels and assemble the features

```
[In]: feature_columns = iris_data.feature_names
[In]: assembler = VectorAssembler(inputCols=feature_columns,
      outputCol="features")
[In]: iris_df = assembler.transform(iris_df)
```

Step 8: Split the data into training and testing sets

```
[In]: X_train, X_test = iris_df.randomSplit([0.8, 0.2], seed=42)
```

Step 9: Define the random forest classifier

```
[In]: spark.conf.set("spark.seed", 42)
[In]: rf = RandomForestClassifier(featuresCol="features",
          labelCol="label")
```

Step 10: Define the parameter grid for tuning

```
[In]: param_grid = ParamGridBuilder() \
          .addGrid(rf.maxDepth, [2, 5, 7, 10]) \
          .addGrid(rf.numTrees, [20, 50, 75, 100]) \
          .addGrid(rf.minInstancesPerNode, [2, 3, 5, 10]) \
          .addGrid(rf.minInfoGain, [0.0, 0.01, 0.02, 0.03]) \
          .build()
```

Step 11: Create the cross-validator

```
[In]: evaluator =
      MulticlassClassificationEvaluator(metricName="accuracy")
[In]: cross_validator = CrossValidator(estimator=rf,
      estimatorParamMaps=param_grid, evaluator=evaluator, numFolds=5)
```

Step 12: Fit the cross-validator to the training data

```
[In]: cv_model = cross_validator.fit(X_train)
```

Step 13: Get the best model and its parameters

```
[In]: best_model = cv_model.bestModel
[In]: best_max_depth = best_model.getMaxDepth()
[In]: best_num_estimators = best_model.getNumTrees
[In]: best_min_samples_split = best_model.getMinInstancesPerNode()
[In]: best_min_info_gain = best_model.getMinInfoGain()
```

Step 14: Make predictions on the test data using the best model

```
[In]: predictions = best_model.transform(X_test)
```

Step 15: Evaluate the model's performance

```
[In]: accuracy = evaluator.evaluate(predictions)
```

Step 16: Print the test accuracy

```
[In]: print("Test Accuracy:", accuracy)
[In]: print("Best maxDepth:", best_max_depth)
[In]: print("Best numTrees:", best_num_estimators)
[In]: print("Best minInstancesPerNode:", best_min_samples_split)
[In]: print("Best minInfoGain:", best_min_info_gain)
```

Summary

In this chapter, we investigated an important topic in machine learning: hyperparameter tuning. This is a critical step in model building that involves finding the optimal set of hyperparameters for a given algorithm. We also provided examples of such parameters in both Scikit-Learn and PySpark and delved into a detailed example of how to fine-tune key hyperparameters in random forests using the Iris dataset.

In the next chapter, we explore an equally important topic: pipelines. By harnessing the power of pipelines, data scientists can automate and standardize the steps involved in the modeling workflow. This enables the building of robust and scalable models, enhances model interpretability, and facilitates the integration of additional preprocessing steps and feature engineering techniques.

CHAPTER 17

Pipelines with Scikit-Learn and PySpark

In this chapter, we explore the topic of pipeline techniques in both Scikit-Learn and PySpark. By harnessing the power of pipelines, data scientists can automate and standardize the steps involved in the modeling workflow. This enables the building of robust and scalable models, enhances model interpretability, and facilitates the integration of additional preprocessing steps and feature engineering techniques.

To illustrate how pipelines can streamline the modeling process and improve efficiency, we implement regression pipelines using a simulated dataset as an example. We generate simulated data that reflects a regression problem, consisting of a set of input features and a continuous target variable. Through the use of pipeline techniques, we demonstrate how to preprocess the data, apply feature scaling, and build regression models in both Scikit-Learn and PySpark. While we demonstrate the pipeline method using regression, this technique is transferable to any other algorithm, whether it involves regression or classification tasks.

To begin, we introduce the concept of pipelines and discuss their significance in the context of machine learning. We explore how pipelines enable us to encapsulate and sequence various data preprocessing and modeling steps into a coherent workflow. By structuring our analysis in this way, we can ensure reproducibility and maintainability and easily experiment with different combinations of preprocessing and modeling techniques.

We then explore the pipeline capabilities in Scikit-Learn. We demonstrate how to construct a pipeline that incorporates a feature scaling step using the `StandardScaler` and a regression step using the `LinearRegression` class. We walk through the process of fitting the pipeline to the training data, making predictions on test data, and evaluating the model's performance.

A. Testas, *Distributed Machine Learning with PySpark*, https://doi.org/10.1007/978-1-4842-9751-3_17

In the final part of the chapter, we shift our focus to PySpark, allowing data scientists planning to migrate from Scikit-Learn to reap the benefits of a powerful distributed data processing framework. PySpark offers a scalable and efficient solution for working with large-scale datasets. We illustrate how to build a regression pipeline using PySpark's MLlib library. Our pipeline incorporates the `VectorAssembler` to assemble input features, the `StandardScaler` for feature scaling, and the `LinearRegression` class for regression modeling. We explore the process of fitting the pipeline to the training data, making predictions on test data, and evaluating the model's performance.

By the end of this chapter, the reader will have a solid understanding of how to leverage pipeline techniques to streamline their machine learning workflow in these two widely used frameworks.

The Significance of Pipelines

In the domain of machine learning, managing and organizing the various stages of data preprocessing and modeling can be a complex task. It often involves multiple steps such as data cleaning, feature scaling, feature engineering, and model fitting. Keeping track of these steps and ensuring their proper sequencing can be challenging, especially when experimenting with different combinations of techniques or working with large datasets.

This is where the concept of pipelines comes into play. A pipeline is a powerful tool that allows us to encapsulate and sequence these preprocessing and modeling steps into a coherent and streamlined workflow. By structuring our analysis using pipelines, we can achieve several key benefits.

One key benefit is encapsulating preprocessing and modeling steps. Pipelines provide a framework for encapsulating individual steps of the machine learning process. Each step, such as feature scaling or feature engineering, can be defined as a separate component within the pipeline. This modular approach enhances code organization and readability. Moreover, it enables us to easily modify or replace specific steps without impacting the rest of the pipeline. By encapsulating these steps, we can maintain a clear separation of concerns and ensure that each step performs its designated task effectively.

A second benefit is seamless workflow and reproducibility. Pipelines enable us to define a clear and structured workflow for machine learning. Each step is executed in a sequential manner, ensuring that the data flows seamlessly from one stage to another. This ensures reproducibility, as the entire workflow can be easily replicated on new datasets or shared with colleagues. By providing a well-defined structure, pipelines reduce the chances of errors and inconsistencies that may arise from manual intervention or ad hoc data processing.

A third benefit is experimentation and flexibility. By encapsulating preprocessing and modeling steps within a pipeline, we can easily modify and experiment with different combinations of techniques. For instance, we can switch between different feature scaling methods or try out alternative feature engineering approaches with minimal code changes. This flexibility empowers us to explore various avenues and refine our regression models efficiently.

Finally, pipelines enhance maintainability. Maintaining and updating machine learning models can become increasingly challenging as projects evolve. With pipelines, we can ensure maintainability by having a single code base that encapsulates the entire workflow. This allows us to make modifications or add new preprocessing or modeling steps without disrupting the overall structure. Pipelines also contribute to code reusability, as individual steps can be shared across multiple machine learning projects. By adopting a modular and maintainable approach, pipelines help us manage the complexity of machine learning projects effectively.

In the upcoming sections of this chapter, we explore the implementation of pipelines in two widely used frameworks: Scikit-Learn and PySpark. We dive into the details of constructing pipelines and demonstrate their practical applications in regression analysis. Through clear examples and step-by-step explanations, we highlight the benefits of using pipelines to streamline the modeling process and improve the efficiency of machine learning workflows.

Pipelines with Scikit-Learn

Scikit-Learn leverages the power of the Pipeline concept. By utilizing the `Pipeline` class, you can seamlessly link transformers and models, treating the entire process as a single Scikit-Learn model. This approach simplifies the workflow and allows for efficient and intuitive data transformation and modeling.

The first step in the process is importing the necessary libraries:

Step 1: Import necessary libraries

The following code imports several functions and classes necessary for the pipeline process:

```
[In]: from sklearn.model_selection import train_test_split
[In]: from sklearn.pipeline import Pipeline
[In]: from sklearn.preprocessing import StandardScaler
[In]: from sklearn.linear_model import LinearRegression
[In]: from sklearn.metrics import mean_squared_error
[In]: import numpy as np
```

Here is a description of what each of these functions and classes does:

- `train_test_split`: This function is used to split the dataset into training and testing subsets. It randomly divides the data into two portions: one for training the regression model and the other for evaluating its performance.
- `Pipeline`: This class provides a convenient way to chain multiple data processing steps together into a single object. It allows us to define a sequence of transformers (data preprocessing steps) and an estimator (machine learning model) as a single unit.
- `StandardScaler`: This class is used to scale each feature by subtracting the mean and dividing by the standard deviation, resulting in a transformed feature with a mean of 0 and a standard deviation of 1.
- `LinearRegression`: This class represents a linear regression model, which models the relationship between a dependent variable and one or more independent variables.
- `mean_squared_error`: This function is a common evaluation metric used to measure the average squared difference between the predicted and actual values in a regression problem. It provides a way to assess the performance of a regression model.
- `numpy as np`: This is a computing library in Python that provides support for large, multidimensional arrays and a collection of mathematical functions to operate on these arrays. We use it to generate a simulated dataset.

Step 2: Generate data

```
[In]: np.random.seed(0)
[In]: num_samples = 1000
[In]: X = np.random.rand(num_samples, 2) * 10
[In]: y = np.random.rand(num_samples) * 20 - 10
```

In this step, we generate some random data to illustrate the pipeline process. The first line sets the seed value for the random number generator in NumPy. By setting a specific seed, such as 0, we ensure that the same sequence of random numbers will be generated each time we run the code.

The second line assigns the value 1000 to the variable num_samples. This represents the number of samples (data points) we want to generate.

The third line generates a two-dimensional array (X) of random numbers between 0 and 1 using the rand function. The shape of X is (num_samples, 2), indicating that there will be num_samples rows and 2 columns. Multiplying the random numbers by 10 scales the range from [0, 1) to [0, 10).

The last line generates a one-dimensional array (y) of random numbers between 0 and 1 using the rand function. The shape of y is (num_samples), indicating that it will have num_samples elements. Multiplying the random numbers by 20 expands the range to [0, 20), and subtracting 10 shifts the range to [-10, 10).

We can take a look at the top five rows of X and y as follows:

Top five rows of X:

```
[In]: print(X[0:5])
[Out]:
[[5.48813504 7.15189366]
 [6.02763376 5.44883183]
 [4.23654799 6.45894113]
 [4.37587211 8.91773001]
 [9.63662761 3.83441519]]
```

Top five elements of y:

```
[In]: print(y[0:5])
[Out]: [ 6.23036941 -0.47832028  0.4631198  -4.98958827  2.10086034]
```

In the code that generates the preceding output, the X[0:5] line slices the X variable to retrieve the rows from index 0 to index 4 (inclusive), which correspond to the first five rows of X. The resulting rows are then printed, displaying the top five rows of X. Similarly, the y[0:5] slices the y variable to retrieve the elements from index 0 to index 4 (inclusive), which correspond to the first five elements of y. The resulting elements are then printed, displaying the top five elements of y.

Step 3: Split the data into training and testing sets

The following code uses the `train_test_split` function to split the dataset into training and testing subsets:

```
[In]: X_train, X_test, y_train, y_test = train_test_split(X, y,
      test_size=0.2, random_state=0)
```

In this code, X represents the input features or independent variables, and y represents the corresponding target variable or dependent variable. The test_size=0.2 parameter specifies that 20% of the data will be reserved for testing, while the remaining 80% will be used for training the machine learning model. The random_state=0 parameter sets the random seed to ensure reproducibility. The X_train, X_test, y_train, y_test variables represent the resulting subsets of the data after splitting. X_train and y_train will contain the training data, while X_test and y_test will contain the testing data.

Step 4: Define the pipeline stages

This is an important step in the pipeline process. In the following code, two stages of the machine learning pipeline are defined (we could define any number of stages):

```
[In]: scaler = StandardScaler()
[In]: regression = LinearRegression()
```

The first line creates an instance of the `StandardScaler` class and assigns it to the variable scaler. This is a data preprocessing step used to standardize numerical features. It scales the features by subtracting the mean and dividing by the standard deviation, resulting in transformed features with zero mean and unit variance.

The second line creates an instance of the `LinearRegression` class and assigns it to the variable regression. This class is our machine learning model used for linear regression, which aims to model the relationship between a dependent variable and independent variables by fitting a linear equation to the observed data.

In a typical pipeline, as shown in the next step, these stages are combined together using the Pipeline class, where each stage is defined as a tuple consisting of a name (optional) and the corresponding object.

Step 5: Create the pipeline

In the following code, a pipeline is created using the Pipeline class:

```
[In]: pipeline = Pipeline(steps=[('scaler', scaler), ('regression',
      regression)])
```

This line of code creates an instance of the `Pipeline` class and assigns it to the variable pipeline. This class allows us to define a sequence of data processing steps and an estimator as a single unit.

The `steps` parameter of the `Pipeline` constructor accepts a list of tuples, where each tuple represents a step in the pipeline. In this case, the list contains two tuples:

1) ('scaler', scaler): This defines the name 'scaler' for the step and assigns the scaler object to it. This corresponds to the `StandardScaler` stage in the pipeline, which is responsible for scaling the input features.
2) ('regression', regression): This defines the name regression for the step and assigns the regression object to it. This corresponds to the `LinearRegression` stage in the pipeline, which represents the linear regression model.

It is important to note that the order of the tuples in the list determines the order in which the steps will be executed.

Step 6: Fit the pipeline to the training data

In this step, we fit the previously defined pipeline to the training data:

```
[In]: pipeline.fit(X_train, y_train)
```

This line of code calls the `fit` method on the pipeline object, which fits the pipeline to the training data. The X_train and y_train variables represent the training data used for fitting the pipeline. X_train contains the input features, and y_train contains the corresponding target variable or labels. When we call `fit` on the pipeline, it applies the `StandardScaler` to scale the features and then fits the `LinearRegression` model to the transformed data.

The fitting process involves estimating the coefficients of the linear regression model based on the provided training data. This step allows the pipeline to learn from the training data and prepare the model for making predictions.

Step 7: Print the coefficients

This step extracts the coefficients and the intercept from the fitted pipeline's regression model:

```
[In]: coefficients = pipeline.named_steps['regression'].coef_
[In]: intercept = pipeline.named_steps['regression'].intercept_
[In]: print("Coefficients:", coefficients)
[In]: print("Intercept:", intercept)
[Out]:
Coefficients: [0.1555 -0.0093]
Intercept: 0.0938
```

The first line retrieves the coefficients of the regression model that is part of the pipeline. The `named_steps` attribute of the pipeline allows accessing individual steps by their specific names. In this case, regression is the name assigned to the `LinearRegression` step. By using `.coef_` on the `LinearRegression` object, we obtain the weights or parameters of the model.

The second line retrieves the intercept (or bias) term of the regression model. Similar to the previous line, it accesses the `LinearRegression` step using the name regression and retrieves the intercept value using `.intercept_`.

The last two statements print the coefficients and the intercept. The coefficients represent the weights assigned to each feature in the linear regression model, indicating the contribution of each feature to the predicted outcome. The intercept term represents the baseline prediction when all features have a value of zero.

Step 8: Make predictions on the test data

The following code performs predictions on the test data using the fitted pipeline:

```
[In]: y_pred = pipeline.predict(X_test)
```

This line uses the `predict` method of the fitted pipeline object to make predictions on the test data X_test. This method applies the transformations defined in the pipeline's steps to the input data and then uses the trained model to generate predictions. The X_test variable represents the test data, which contains the input features on which we want to make predictions.

By calling `predict` on the pipeline, it applies scaling using the `StandardScaler` step. It then uses the trained `LinearRegression` model to generate predictions based on the scaled data. The resulting predictions are assigned to the variable y_pred, an array or vector that contains the predicted values for the target variable corresponding to the test data.

Step 9: Print top five actual vs. predicted values

In this step, we want to take a look at the top five actual values and corresponding predicted values for the target variable:

```
[In]: import pandas as pd
[In]: data = {'Actual': y_test[:5], 'Predicted': y_pred[:5]}
[In]: df = pd.DataFrame(data)
[In]: print(df.head())
[Out]:
```

	Actual	Predicted
0	7.59	0.03
1	8.62	-0.07
2	6.43	0.02
3	7.07	0.12
4	-1.04	0.19

The first line of the code imports Pandas as pd. In the second line, a Python dictionary called data is created, with the keys representing the names of the columns we want to create in the DataFrame. The values are obtained by slicing the first five elements from y_test and y_pred, which contain the actual and predicted values from the machine learning model. In the next line of code, the Pandas DataFrame df is created by passing the data dictionary as an argument to the `DataFrame()` constructor. Finally, the `head()` method displays the first few rows of the DataFrame.

We can see from the output that there is a big difference between the actual and predicted values, indicating that the performance of the model isn't that great. This is fine for the task at hand since our simulated data is used purely for illustration purposes.

Step 10: Evaluate the model

In this final step, we evaluate the performance of the model by calculating the root mean squared error (RMSE) on the test data:

```
[In]: rmse = np.sqrt(mean_squared_error(y_test, y_pred))
[In]: print("Root Mean Squared Error (RMSE) on test data:", rmse)
[Out]: Root Mean Squared Error (RMSE) on test data: 5.55
```

The first line of the code calculates the RMSE by using the `mean_squared_error` function. This takes two arguments: the actual target values (y_test) and the predicted target values (y_pred). It computes the mean squared error between the two sets of values. The `np.sqrt` function is then applied to the mean squared error to obtain the RMSE. This is a commonly used metric to measure the average difference between the predicted values and the actual values. It represents the square root of the average of the squared differences. The resulting RMSE value is assigned to the variable rmse.

The next line prints the computed RMSE on the test data, which is approximately 5.55. This value provides a measure of the model's accuracy or goodness of fit. A lower RMSE indicates better predictive performance, as it represents a smaller average difference between the predicted and actual values.

Pipelines with PySpark

Similar to Scikit-Learn, PySpark harnesses the capabilities of the Pipeline concept. By employing the `Pipeline` class, we can effortlessly connect transformers and models, treating the entire process as a cohesive PySpark model. This methodology streamlines the workflow, enabling seamless and effective data transformation and modeling with ease.

Similar to what we did with Scikit-Learn, we begin by importing the necessary PySpark libraries:

Step 1: Import necessary libraries

The first step in the pipeline process is to import the necessary classes from the PySpark library:

```
[In]: from pyspark.sql import SparkSession
[In]: from pyspark.ml.feature import VectorAssembler, StandardScaler
[In]: from pyspark.ml.regression import LinearRegression
[In]: from pyspark.ml import Pipeline
[In]: from pyspark.ml.evaluation import RegressionEvaluator
```

Here are descriptions of what these classes are used for:

- `SparkSession`: This is the entry point to Spark functionalities. It allows us to create a Spark application and interact with data in a structured way.
- `VectorAssembler`: This is a feature transformer that combines a given list of columns into a single vector column. It takes a set of input columns and produces a new column that contains a vector of their values.
- `StandardScaler`: This class is a feature transformer that standardizes numerical features by subtracting the mean and scaling to unit variance. It transforms the feature values such that they have zero mean and unit standard deviation.
- `LinearRegression`: This is an algorithm for predicting a continuous target variable based on one or more predictor variables. It fits a linear equation to the input data, where the target variable is assumed to be a linear combination of the predictors.
- `Pipeline`: This is a utility that allows us to chain multiple stages of transformations and estimators together to form a workflow. It helps in organizing and executing a sequence of operations on the data, such as feature engineering, model training, and evaluation.
- `RegressionEvaluator`: This is a class for evaluating regression models in PySpark. It provides various metrics and methods for assessing the performance of regression models, such as root mean squared error (RMSE).

Step 2: Create a SparkSession

```
[In]: spark = SparkSession.builder.getOrCreate()
```

This line of code creates a Spark Session to enable us to interact with Spark.

Step 3: Generate simulated data

In this step, we generate simulated data to demonstrate the concept of pipelining in PySpark:

```
[In]: num_samples = 1000
[In]: simulated_data = spark.range(num_samples).selectExpr(
          "id as id",
          "(RAND() * 10) as feature1",
          "(RAND() * 5) as feature2",
          "(RAND() * 20 - 10) as label")
[Out]:
```

id	feature1	feature2	label
0	3.51	4.01	-2.85
1	1.97	4.49	2.36
2	8.14	3.70	-6.81
3	7.40	2.95	0.93
4	3.64	1.12	-3.44

In the preceding code, we first assign the value 1000 to the variable num_samples, which represents the number of rows or samples that we want to generate for our simulated dataset. We then create a DataFrame called simulated_data using the `spark.range()` function. This generates a DataFrame with a single column named id containing numbers from 0 to num_samples-1. This acts as an index column for the simulated data.

The `selectExpr()` method in the second line is used to select and transform columns in the DataFrame. It takes multiple arguments, where each argument specifies a transformation expression to be applied to a column:

- The expression "(RAND() * 10) as feature1" generates a random value between 0 and 10 for each row and assigns it to a new column named feature1. The `RAND()` function generates a random number between 0 and 1, and multiplying it by 10 scales the range to 0–10.

- The “(RAND() * 5) as feature2” expression generates a random value between 0 and 5 for each row and assigns it to a new column named feature2. Similar to the previous expression, it uses the `RAND()` function to generate a random number between 0 and 1 and scales it to 0–5.
- The “(RAND() * 20 - 10) as label” expression generates a random value between -10 and 10 for each row and assigns it to a new column named label. The `RAND()` function generates a random number between 0 and 1, and by multiplying it by 20 and subtracting 10, the range is scaled to -10 to 10.

As shown in the output, the resulting simulated_data DataFrame has four columns: id (generated by `spark.range()`), feature1 (random values between 0 and 10), feature2 (random values between 0 and 5), and label (random values between -10 and 10).

Step 4: Split the data into training and testing sets

```
[In]: train_ratio = 0.8
[In]: test_ratio = 1 - train_ratio
[In]: training_data, testing_data = simulated_data.randomSplit([train_
      ratio, test_ratio], seed=42)
```

In this code, we split the simulated_data DataFrame into training and testing sets using the `randomSplit()` function. The first line assigns the value 0.8 to the variable train_ratio, indicating that 80% of the data will be used for training. The second line calculates the ratio for the testing set by subtracting the train_ratio from 1. Since the sum of the training and testing ratios should equal 1, subtracting the train_ratio from 1 gives the remaining portion, which corresponds to the testing ratio.

The last line performs the actual split of the data into training and testing sets. The `randomSplit()` function takes two arguments: an array of ratios for the training and testing sets and a seed parameter to ensure that the random splitting of the data is reproducible.

Step 5: Define the pipeline stages

In this step, the pipeline stages for the machine learning workflow are defined using the PySpark ML library:

```
[In]: assembler = VectorAssembler(
      inputCols=["feature1", "feature2"], outputCol="features")
[In]: scaler = StandardScaler(
```

```
    inputCol="features", outputCol="scaledFeatures")
[In]: regression = LinearRegression(
    featuresCol="ScaledFeatures", labelCol="label")
```

The first line creates a `VectorAssembler` object, which is a transformation stage in the pipeline. The assembler combines multiple input columns (feature1 and feature2) into a single output column (features). The `inputCols` parameter specifies the names of the input columns to be assembled into a vector. In this case, it includes the columns feature1 and feature2. The `outputCol` parameter specifies the name of the output column, which will contain the assembled vector of features.

The second line creates a `StandardScaler` object to standardize the feature vector column obtained from the previous stage. It takes the feature vector column features as input, specified by the `inputCol` parameter, and creates a new column named scaledFeatures that contains the standardized feature values. The scaling is performed by subtracting the mean and dividing by the standard deviation of each feature.

The third line creates a `LinearRegression` object, which is an estimator stage in the pipeline, and is used to build a linear regression model. The `featuresCol` parameter specifies the name of the input column containing the features. It matches the output column name (scaledFeatures) produced by the `VectorAssembler`. The `labelCol` parameter specifies the name of the input column containing the labels.

By defining these pipeline stages, we are setting up the data transformation and modeling steps for the machine learning pipeline. These stages can be further combined with other stages, such as feature transformations or additional models, to build a comprehensive pipeline for data preprocessing and modeling in PySpark.

Step 6: Create the pipeline

```
[In]: pipeline_stages = [assembler, scaler, regression]
[In]: pipeline = Pipeline(stages=pipeline_stages)
```

In this code, the actual pipeline is created using the `Pipeline` class from PySpark. It is no coincidence that PySpark utilizes the same name and concept as Scikit-Learn, as it draws inspiration from Scikit-Learn's pipeline implementation.

The pipeline consists of three stages: assembler, scaler, and regression.

- The assembler stage is responsible for assembling the input features into a vector. It takes the columns feature1 and feature2 as input and creates a new column named features that contains a vector representation of the features.

- The scaler stage is a `StandardScaler` that standardizes the feature vector column obtained from the previous stage. It takes the feature vector column features as input and creates a new column named scaledFeatures that contains the standardized feature values.
- The regression stage is a `LinearRegression` estimator that performs linear regression on the scaled feature vector column. It takes the scaled feature vector column scaledFeatures and the label column label as inputs to train the linear regression model.

The pipeline_stages list in the code contains the stages in the order in which they will be executed. The `Pipeline` object is created with pipeline_stages as the input, representing the entire pipeline.

Step 7: Fit the pipeline to the training data

```
[In]: pipeline_model = pipeline.fit(training_data)
```

In this single line of code, the pipeline is fitted to the training data using the `fit()` method. Fitting the pipeline means executing each stage of the pipeline in order, applying the defined transformations, and training the specified model on the training data.

The `fit()` method takes the training data as input and applies each stage's transformation sequentially, starting from the first stage (assembler) and ending with the last stage (regression). The output of each stage is passed as input to the next stage in the pipeline. During the fitting process, the transformations defined in the pipeline stages are applied to the training data, and the model specified in the last stage (regression) is trained using the transformed data.

The result of calling `fit()` is a fitted pipeline model (pipeline_model) that encapsulates the trained model along with the applied transformations. This fitted pipeline model can be used to make predictions on new data or evaluate the model's performance.

Step 8: Print the coefficients

```
[In]: coefficients = pipeline_model.stages[-1].coefficients
[In]: intercept = pipeline_model.stages[-1].intercept
[In]: print("Coefficients:", coefficients)
[In]: print("Intercept:", intercept)
[Out]: Coefficients: [0.2150,-0.2897] Intercept: 0.1336
```

In this code, we retrieve the coefficients and intercept of the trained linear regression model from the fitted pipeline model (pipeline_model). Since the linear regression model is the last stage in the pipeline, we access it using pipeline_model.stages[-1]. The `stages` attribute of the pipeline model contains a list of all stages in the pipeline, and -1 refers to the last stage. Since we have used two features (feature1 and feature2) in the linear regression model, we have two resulting coefficients, with approximate values of 0.22 and -0.29, respectively. The intercept (approximately 0.13) is produced regardless of the number of features. These results differ from those of the Scikit-Learn regression model, suggesting that there are implementation differences between the two platforms.

Step 9: Make predictions on the testing data

```
[In]: predictions = pipeline_model.transform(testing_data)
```

In this single line of code, we use the trained pipeline model (pipeline_model) to make predictions on the test data (testing_data). The `transform` method is applied to the test data using the pipeline_model. This method applies each stage of the pipeline to the test data in a sequential manner. It takes the test data as input and performs all the necessary transformations defined in the pipeline stages.

The output DataFrame predictions contains the original columns from the test data, along with additional columns such as scaledFeatures and prediction. The prediction column contains the predicted values generated by the trained linear regression model for each corresponding row in the test data.

Step 10: Display top five actual vs. predicted values

```
[In]: predictions.select("label", "prediction").show(5)
[Out]:
```

label	prediction
8.46	-0.22
-9.10	-0.16
-7.05	-0.14
7.65	-0.15
5.21	-0.04

In this step, we would like to take a look at the top five rows of actual labels and corresponding predicted values from the predictions DataFrame. The `select` method is used to select the label and prediction columns. The `show()` method is then used to retrieve the first five rows of the selected columns.

As the output shows, there are large differences between the two sets of values, indicating that the model is making prediction errors. This is not surprising given that we are using a simulated dataset, which is fine as our purpose is to illustrate the concept of pipelining.

Step 11: Evaluate the model

In this final step, we evaluate the performance of the model by calculating the root mean squared error (RMSE) metric on the testing data:

```
[In]: evaluator = RegressionEvaluator(labelCol="label",
      predictionCol="prediction", metricName="rmse")
[In]: rmse = evaluator.evaluate(predictions)
[In]: print("Root Mean Squared Error (RMSE) on test data:", rmse)
[Out]: Root Mean Squared Error (RMSE) on test data: 5.88
```

In the first line of code, we create a `RegressionEvaluator` object named evaluator. We specify the label column as label and the prediction column as prediction using the `labelCol` and `predictionCol` parameters, respectively. We also set the metricName parameter to rmse to indicate that we want to calculate the RMSE.

In the next line, we use the `evaluate` method of the evaluator object to calculate the RMSE. We pass the predictions DataFrame, which contains the actual labels and predicted values, as the argument to the evaluate method. The RMSE value is then assigned to the variable rmse.

Finally, we print the RMSE value (approximately 5.88) using a formatted string, along with a descriptive message. The output displays the RMSE on the test data, which gives an indication of how well the model is performing in terms of the average difference between the actual and predicted values. A lower RMSE indicates better model performance.

It is worth noting that the RMSE value from PySpark (5.88) is different from the value obtained by Scikit-Learn (5.55). Machine learning algorithms, including regression, involve randomness during training, which can lead to slightly different results between runs. Additionally, PySpark and Scikit-Learn may use different optimization techniques for their regression models, leading to variations in model predictions and RMSE values.

Bringing It All Together

Now, let's consolidate all the relevant code from the preceding steps into a single code block. This will enable the reader to execute the code as a coherent entity in both Scikit-Learn and PySpark.

Scikit-Learn

Step 1: Import necessary libraries

```
[In]: from sklearn.model_selection import train_test_split
[In]: from sklearn.pipeline import Pipeline
[In]: from sklearn.preprocessing import StandardScaler
[In]: from sklearn.linear_model import LinearRegression
[In]: from sklearn.metrics import mean_squared_error
[In]: import numpy as np
[In]: import pandas as pd
```

Step 2: Generate simulated data

```
[In]: np.random.seed(0)
[In]: num_samples = 1000
[In]: X = np.random.rand(num_samples, 2) * 10
[In]: y = np.random.rand(num_samples) * 20 - 10
```

Step 3: Split the data into training and testing sets

```
[In]: X_train, X_test, y_train, y_test = train_test_split(X, y,
      test_size=0.2, random_state=0)
```

Step 4: Define the pipeline stages

```
[In]: scaler = StandardScaler()
[In]: regression = LinearRegression()
```

Step 5: Create the pipeline

```
[In]: pipeline = Pipeline(steps=[('scaler', scaler), ('regression',
      regression)])
```

Step 6: Fit the pipeline to the training data

```
[In]: pipeline.fit(X_train, y_train)
```

Step 7: Print the coefficients

```
[In]: coefficients = pipeline.named_steps['regression'].coef_
[In]: intercept = pipeline.named_steps['regression'].intercept_
[In]: print("Coefficients:", coefficients)
[In]: print("Intercept:", intercept)
```

Step 8: Make predictions on the testing data

```
[In]: y_pred = pipeline.predict(X_test)
```

Step 9: Print top five actual vs. predicted

```
[In]: data = {'Actual': y_test[:5], 'Predicted': y_pred[:5]}
[In]: df = pd.DataFrame(data)
[In]: print(df.head())
```

Step 10: Evaluate the model

```
[In]: rmse = np.sqrt(mean_squared_error(y_test, y_pred))
[In]: print("Root Mean Squared Error (RMSE) on test data:", rmse)
```

PySpark

Step 1: Import necessary libraries

```
[In]: from pyspark.sql import SparkSession
[In]: from pyspark.ml.feature import VectorAssembler, StandardScaler
[In]: from pyspark.ml.regression import LinearRegression
[In]: from pyspark.ml import Pipeline
[In]: from pyspark.ml.evaluation import RegressionEvaluator
```

Step 2: Create a SparkSession

```
[In]: spark = SparkSession.builder.getOrCreate()
```

Step 3: Generate simulated data

```
[In]: num_samples = 1000
[In]: simulated_data = spark.range(num_samples).selectExpr(
                  "id as id",
      "(RAND() * 10) as feature1",
                  "(RAND() * 5) as feature2",
                  "(RAND() * 20 - 10) as label")
```

Step 4: Split the data into training and testing sets

```
[In]: train_ratio = 0.8
[In]: test_ratio = 1 - train_ratio
[In]: training_data, testing_data =
      simulated_data.randomSplit([train_ratio, test_ratio], seed=42)
```

Step 5: Define the pipeline stages

```
[In]: assembler = VectorAssembler(
      inputCols=["feature1", "feature2"], outputCol="features")
[In]: scaler = StandardScaler(
      inputCol="features", outputCol="scaledFeatures")
[In]: regression = LinearRegression(
      featuresCol="scaledFeatures", labelCol="label")
```

Step 6: Create the pipeline

```
[In]: pipeline_stages = [assembler, scaler, regression]
[In]: pipeline = Pipeline(stages=pipeline_stages)
```

Step 7: Fit the pipeline to the training data

```
[In]: pipeline_model = pipeline.fit(training_data)
```

Step 8: Print the coefficients

```
[In]: coefficients = pipeline_model.stages[-1].coefficients
[In]: intercept = pipeline_model.stages[-1].intercept
[In]: print("Coefficients:", coefficients)
[In]: print("Intercept:", intercept)
```

Step 9: Make predictions on the testing data

```
[In]: predictions = pipeline_model.transform(testing_data)
```

Step 10: Evaluate the model

```
[In]: evaluator = RegressionEvaluator(labelCol="label",
      predictionCol="prediction", metricName="rmse")
[In]: rmse = evaluator.evaluate(predictions)
[In]: print("Root Mean Squared Error (RMSE) on test data:", rmse)
```

Step 11: Display top five actual vs. predicted

```
[In]: predictions.select("label", "prediction").show(5)
```

Summary

In this chapter, we have explored the topic of pipeline techniques in both Scikit-Learn and PySpark. By harnessing the power of pipelines, data scientists can automate and standardize the steps involved in the modeling workflow. This enables the creation of robust and scalable models, enhances model interpretability, and facilitates the integration of additional preprocessing steps and feature engineering techniques.

We have demonstrated how to construct a pipeline that incorporates a feature scaling step using the StandardScaler and a regression step using the LinearRegression class. We have walked through the process of fitting the pipeline to the training data, making predictions on test data, and evaluating the model's performance.

In the next chapter, which marks the final chapter of this book, we will conclude by providing an in-depth exploration of the practical aspects of deploying machine learning models in production using both Scikit-Learn and PySpark.

CHAPTER 18

Deploying Models in Production with Scikit-Learn and PySpark

In this final chapter of the book, we explore the practical aspects of deploying machine learning models using Scikit-Learn and PySpark. Model deployment is the process of making a machine learning model available for use in a production environment where it can make predictions or perform tasks based on real-world data. It involves taking a trained machine learning model and integrating it into a system or application so that it can provide predictions to end users or other systems.

Model deployment is a critical step in the machine learning life cycle, and it's important for several reasons. First, a machine learning model provides value only when used to make predictions or automate tasks in a real-world setting. Deployment ensures that the model is put to practical use, which can lead to cost savings and efficiency improvements. Moreover, deployment allows an organization to scale the use of its machine learning model to serve a large number of users or handle a high volume of data. In a production environment, the model can process data in real-time or batch mode, making it suitable for various applications.

Subsequently, deployed models can automate decision-making processes that would otherwise require manual intervention or human expertise, leading to faster and more consistent decision-making. More importantly, models can be integrated into existing workflows, making them seamless components of existing processes. This integration can be essential for leveraging the model's capabilities within a specific business context, such as automating the assessment of loan or credit card applications to determine whether a user qualifies based on their creditworthiness and financial history.

A. Testas, *Distributed Machine Learning with PySpark*, https://doi.org/10.1007/978-1-4842-9751-3_18

Additionally, deployment allows for monitoring the model's performance in production, which includes tracking metrics, detecting issues such as model drift, and updating the model as necessary to ensure its continued effectiveness. Organizations that successfully deploy and leverage machine learning models can gain a competitive advantage by making data-driven decisions, improving customer experiences, and staying ahead in their industry.

Another crucial reason for why model deployment is important is that deployed models can be accessed by various stakeholders, including data scientists, analysts, developers, and business users, enabling cross-functional collaboration and knowledge sharing.

This chapter demonstrates model deployment using Scikit-Learn and PySpark. However, the reader should be aware that there are other ways to deploy models in production, even though these methods are out of scope in this chapter. One common approach is to deploy the algorithm as an API service, allowing other applications and systems to make HTTP requests to the model for predictions. Frameworks like Flask, Django, and FastAPI are often used to build API endpoints. Another option is containerization. Docker containers are used to package machine learning models along with their dependencies into portable units. Kubernetes and other container orchestration tools enable scaling and management of containerized models.

In complex workflows, organizations use tools like Apache Airflow to orchestrate the deployment of multiple models and data pipelines in a coordinated manner. Similarly, in customer-facing applications, organizations integrate machine learning models into user interfaces, allowing end users to interact with and benefit from the models' predictions or recommendations. For offline or batch processing, organizations deploy models on distributed computing platforms like Apache Hadoop or Apache Spark, which can handle large-scale data processing and model execution. Additionally, specialized model serving frameworks like TensorFlow Serving and MLflow are designed for deploying and managing machine learning models in production, often including features for versioning, monitoring, and scaling.

Finally, there is the option of cloud-based deployment, as cloud providers offer managed machine learning services (e.g., AWS SageMaker, Azure Machine Learning, Google AI Platform) that simplify the deployment of models on their infrastructure, often including auto-scaling and monitoring capabilities.

To demonstrate model deployment with Scikit-Learn and PySpark, we utilize a Multilayer Perceptron (MLP) and work with the Iris dataset. In the next section, we outline the key steps involved in model deployment. Subsequently, we delve into the process of deploying models in Scikit-Learn. In the final segment of this chapter, our focus will shift to PySpark, enabling data scientists who are considering a transition from Scikit-Learn to leverage the advantages offered by this robust distributed data processing framework.

Steps in Model Deployment

In both Scikit-Learn and PySpark, the deployment of models into production typically involves three main steps: model training and evaluation, model persistence, and deployment.

Step 1: Model training and evaluation

To begin, you train and evaluate your model using the provided algorithms, such as `MLPClassifier` in Scikit-Learn and `MultilayerPerceptronClassifier` in PySpark for neural networks. This involves preprocessing the data using available modules, splitting it into training and testing sets, and employing evaluation metrics like accuracy score to assess model performance.

Step 2: Model persistence

Once the training is complete, it is essential to save the trained model to disk for future deployment. In Scikit-Learn, you can use the `joblib` library (or pickle) to efficiently serialize Python objects, including trained models. By using the `joblib.dump()` function, you can save the model as a file, ready to be loaded later for deployment. Similarly, PySpark offers the ML persistence API, allowing you to save both the model configuration and the trained weights using the `save()` method.

Step 3: Model deployment

The final step involves utilizing the saved models. In Scikit-Learn, you can load the saved model file into memory using the `joblib` library and the `joblib.load()` function. This loads the model, making it ready for making predictions on new data. Similarly, in PySpark, the saved model can be deployed by loading it using the ML persistence API. You can then use the loaded model to transform new data by calling the `transform()` method.

By following these steps and techniques, you can effectively deploy your machine learning models using Scikit-Learn and PySpark, ensuring their readiness for production environments.

Deploying a Multilayer Perceptron (MLP)

In the following sections, we provide practical examples of deploying models in production using Scikit-Learn and PySpark. For demonstration purposes, we train, evaluate, and deploy a Multilayer Perceptron (MLP) using the Iris dataset. We are already familiar with this dataset, as we used it in Chapter 15 to train and evaluate a k-means clustering algorithm. It's important to note that the principles discussed here for MLP classification are applicable to other algorithms, whether they are used for regression or classification tasks.

Deployment with Scikit-Learn

In this subsection, we employ Scikit-Learn to construct, train, evaluate, persist, and deploy an MLP classifier on the Iris dataset. The first step is to import the required libraries.

Step 1: Import necessary libraries

The following functions and classes allow us to access and utilize the necessary functionality from the Scikit-Learn library to train, evaluate, save, and deploy the MLP classifier:

```
[In]: import numpy as np
[In]: from sklearn.datasets import load_iris
[In]: from sklearn.neural_network import MLPClassifier
[In]: from sklearn.model_selection import train_test_split
[In]: from sklearn.metrics import accuracy_score
[In]: import joblib
```

Here is a description of what each library does:

- `numpy`: This library will be used to generate some data for new predictions.
- `load_iris`: This function is used to load the Iris dataset.

- `MLPClassifier`: This is a class for implementing the Multilayer Perceptron (MLP)—a type of artificial neural network used for classification tasks.
- `train_test_split`: This is a utility function that will split our dataset into training and testing subsets for the purpose of evaluating the performance of the machine learning model.
- `accuracy_score`: This is a function to calculate the accuracy metric used to evaluate the accuracy of our classification model's predictions by comparing them to the true labels of the test data.
- `joblib`: This module provides tools for saving and loading Python objects, including trained machine learning models. It is commonly used for model persistence, allowing us to save the trained model to disk and load it later for deployment or further use.

Step 2: Load the Iris data

```
[In]: iris = load_iris()
[In]: X = iris.data
[In]: y = iris.target
```

In this step, we load the Iris dataset using the Scikit-Learn built-in `load_iris()` function. The features are assigned to the variable X while corresponding target values are assigned to the variable y.

Step 3: Split the data into train and test sets

```
[In]: X_train, X_test, y_train, y_test = train_test_split(X, y,
      test_size=0.2, random_state=42)
```

This line of code uses the `train_test_split` function to split the data into training and testing sets. X and y are the feature and label sets, respectively, that we want to split. The test_size=0.2 parameter specifies that 20% of the data will be used for the test set, while 80% will be used for the training set. The random_state=42 parameter ensures reproducibility. By using the same seed value, we will get the same data split each time we run the code. The training and testing sets are assigned to variables X_train, X_test, y_train, and y_test, respectively.

Step 4: Create an MLP classifier

```
[In]: mlp = MLPClassifier(hidden_layer_sizes=(100, 50), max_iter=1000,
      random_state=42)
```

This line of code creates an instance of the `MLPClassifier` class. The hidden_layer_sizes parameter specifies the architecture of the neural network by defining the number of neurons in each hidden layer. In this case, the neural network has two hidden layers, with 100 neurons in the first hidden layer and 50 neurons in the second hidden layer.

The max_iter parameter determines the maximum number of iterations or epochs the neural network will be trained for. During training, the model iteratively updates the weights based on the training data to improve its performance. The max_iter parameter sets the maximum number of iterations to 1000 in this case.

Finally, the random_state parameter allows for reproducibility of the results. By setting a value of 42 for the random_state, the same sequence of random numbers will be generated every time we run the code. This ensures that the initialization of the neural network weights and other random processes within the algorithm are consistent across different runs.

Step 5: Train the classifier

```
[In]: mlp.fit(X_train, y_train)
```

This line of code is used to train the MLP classifier on the training data. The `fit()` method trains the model using the provided input features (X_train) and the corresponding target labels (y_train).

During the training process, the MLP classifier learns the relationships between the input features and the target labels. It iteratively adjusts the weights and biases of the neural network based on the training data, aiming to minimize the error or loss function. The number of iterations was determined by the max_iter parameter specified when we created the `MLPClassifier` object. After the `fit()` method is executed, the MLP classifier is trained and ready to make predictions on new, unseen data.

Step 6: Make predictions on test data

```
[In]: y_pred = mlp.predict(X_test)
```

This line is used to make predictions on the test data using the trained MLP classifier model. The `predict()` method applies the trained model to the input data and returns the predicted class labels. For each sample in the test data, it predicts the corresponding target label based on the learned patterns and associations from the training data. The resulting predicted labels are stored in the y_pred variable.

Step 7: Evaluate accuracy

```
[In]: accuracy = accuracy_score(y_test, y_pred)
[In]: print("Accuracy:", accuracy)
[Out]: Accuracy: 1.0
```

This code is used to evaluate the accuracy of the predictions made by the MLP classifier on the test data. The accuracy_score() is a utility function that calculates the accuracy of the classification model's predictions by comparing the predicted labels (y_pred) with the true labels (y_test). By calling accuracy_score(y_test, y_pred), the function computes the accuracy of the predictions by comparing each predicted label with its corresponding true label. It returns a value between 0 and 1, indicating the proportion of correctly predicted labels.

The output indicates that the accuracy score for the MLP classifier is 1.0, which means that 100% of all predictions were correct. This suggests that the model has accurately classified each species of the Iris flower.

Step 8: Save the trained model for deployment

Having trained and evaluated the model performance, we should now be able to move to the persistence step:

```
[In]: joblib.dump(mlp, 'mlp_model.pkl')
```

This line of code uses the joblib library to save the machine learning model to a file named mlp_model.pkl using the dump() function. The mlp parameter represents the machine learning model object. The mlp_model.pkl parameter is the name of the file where the serialized model will be saved. The .pkl extension is commonly used for files that store serialized Python objects using the pickle protocol.

Step 9: Load the saved model

This step involves utilizing the saved model. We can load the saved model file into memory using the joblib library and the joblib.load() function. This loads the model, making it ready for making predictions on new data:

```
[In]: mlp = joblib.load('mlp_model.pkl')
```

This line of code uses the load() function from the joblib library to load the serialized MLP model from the mlp_model.pkl file and assign it to the variable mlp. After executing this line of code, mlp will contain the deserialized model object, which can be used for prediction.

Step 10: Generate new data (simulated example)

```
[In]: np.random.seed(42)
[In]: X_new = np.random.rand(10, 4)
```

This code generates new data using NumPy's random module. The purpose is to use this new data to demonstrate how the loaded model can be utilized. The first line of code sets the random seed for the random number generator. Setting the seed to 42 ensures that the random numbers generated are reproducible.

The second line creates a NumPy array (X_new) that contains random numbers. The `rand()` function generates random numbers from a uniform distribution between 0 and 1. The arguments (10, 4) specify the shape of the array to be created, where 10 represents the number of rows and 4 represents the number of columns.

Step 11: Make predictions on the new data

```
[In]: y_new_pred = mlp.predict(X_new)
[In]: print("Predictions:", y_new_pred)
[Ou]: Predictions: [1 1 1 0 1 1 0 1 1 1]
```

In this final step, we first use the `predict()` method of the mlp model to make predictions on the new dataset X_new. The `predict()` method takes the input data and returns the predicted values based on the trained model. The predictions are stored in the variable y_new_pred. We then print the predictions generated by the model using the `print()` function.

PySpark

We now switch focus to PySpark to show how to train and evaluate an MLP classifier using Spark ML, persist the model, and then load it to make predictions on new data.

The first step is to import the necessary libraries:

Step 1: Import necessary libraries

```
[In]: from pyspark.sql import SparkSession
[In]: from pyspark.ml.classification import
      MultilayerPerceptronClassificationModel,
      MultilayerPerceptronClassifier
```

```
[In]: from pyspark.ml.feature import VectorAssembler
[In]: from pyspark.ml.evaluation import
      MulticlassClassificationEvaluator
[In]: import numpy as np
```

This code imports necessary classes for building, training, evaluating, and deploying an MLP classification model in PySpark. Here is a description of each of the imported classes:

- `SparkSession`: This is the entry point to Spark functionality; it enables interaction with Spark APIs.
- `MultilayerPerceptronClassifier`: This class provides the implementation of the MLP classifier in Spark.
- `MultilayerPerceptronClassificationModel`: This class represents a trained MLP classification model. It contains the learned weights and biases of the neural network, as well as other model-related information.
- `VectorAssembler`: This class is used for combining multiple feature columns into a single vector column.
- `MulticlassClassificationEvaluator`: This class provides evaluation metrics for multiclass classification models, which is particularly relevant in this case due to the Iris dataset containing three species (setosa, versicolor, and virginica), making it a multiclass dataset.
- `numpy as np`: This library will be used to generate some random input data for new predictions.

Step 2: Create a SparkSession

```
[In]: spark = SparkSession.builder.getOrCreate()
```

This code creates a Spark Session to enable interaction with Spark.

Step 3: Create PySpark DataFrame

```
[In]: spark = SparkSession.builder.getOrCreate()
```

```
[In]: data = [(float(X[i, 0]), float(X[i, 1]), float(X[i, 2]), float
      (X[i, 3]), float(y[i])) for i in range(len(X))]
[In]: spark_df = spark.createDataFrame(data, ["sepal_length", "sepal_
      width", "petal_length", "petal_width", "species"])
```

In this step, we first initialize a SparkSession, which serves as the entry point to leverage the capabilities of Spark. Then, we prepare the data by creating a list where each element corresponds to an Iris flower's characteristics, including sepal length, sepal width, petal length, petal width, and its species label. This data is transformed into a format suitable for PySpark. Finally, using the `createDataFrame()` method, we create a PySpark DataFrame named spark_df.

Step 4: Split the data into train and test sets

```
[In]: train_data, test_data = spark_df.randomSplit([0.8, 0.2],
      seed=42)
```

This line of code splits the Spark DataFrame (spark_df) into two separate DataFrames, namely, train_data and test_data, based on a specified split ratio. The `randomSplit()` method provided by Spark DataFrame does this split randomly. It takes two arguments:

1) The [0.8, 0.2] argument is a list specifying the split ratios for the train and test data, respectively. In this case, 80% of the data will be assigned to train_data, and 20% of the data will be assigned to test_data.

2) The seed=42 argument sets the random seed. It ensures reproducibility of the random splitting process. When the same seed is used, the random split will produce the same train-test split if applied multiple times.

Step 5: Assemble features into a single vector column

```
[In]: feature_cols = ["sepal_length", "sepal_width", "petal_length",
      "petal_width"]
[In]: assembler = VectorAssembler(inputCols=feature_cols,
      outputCol="features")
[In]: train_data = assembler.transform(train_data)
[In]: test_data = assembler.transform(test_data)
```

In this step, the features are combined into a single vector using the `VectorAssembler`. In the first line, feature_cols is defined as a list containing the names of the feature columns we want to include in the machine learning model. These columns represent the characteristics of the Iris flowers: sepal length, sepal width, petal length, and petal width.

The second line creates a `VectorAssembler` object named assembler. A `VectorAssembler` is a PySpark feature transformer that is used to assemble a vector of feature values from multiple columns into a single vector column. In this case, it is configured to take the columns specified in feature_cols as input columns and produce a new column called features that contains a vector representation of the selected features.

The last two commands use the `transform` method of the `VectorAssembler` to apply the transformation to the training and testing data. This transformation takes the specified feature columns from each row in the DataFrames (train_data and test_data) and combines them into a single vector column called features. This transformation is essential for machine learning algorithms in PySpark, as they expect input features to be in vector format.

Step 6: Create MLP classifier

```
[In]: mlp = MultilayerPerceptronClassifier(labelCol="species",
      featuresCol="features", layers=[4, 100, 50, 3], seed=42, maxIter=500)
```

In this line of code, the Multilayer Perceptron (MLP) classification model is created using the `MultilayerPerceptronClassifier` class. This neural network consists of multiple layers of interconnected nodes (neurons).

The labelCol="species" parameter specifies the name of the column in the DataFrame that contains the labels or the target variable for classification. In this case, the column name is species. The featuresCol="features" parameter specifies the name of the column in the DataFrame that contains the input features for the classifier. In this case, the column name is features.

The layers=[4, 100, 50, 3] parameter defines the architecture of the Multilayer Perceptron network. The list specifies the number of nodes in each layer, starting from the input layer and ending with the output layer. In this example, the MLP has 1 input layer with 4 nodes (representing the features), 2 hidden layers with 100 and 50 nodes, respectively, and 1 output layer with 3 nodes (representing the Iris species).

The seed=42 parameter sets the random seed for the MLP classifier. It ensures reproducibility of results by initializing the random number generator to a specific state.

The maxIter=500 parameter determines the maximum number of iterations or training epochs for the MLP classifier. It controls how many times the algorithm will go through the training data to update the weights and biases of the neural network.

Step 7: Train the classifier

```
[In]: mlp_model = mlp.fit(train_data)
```

In this single line of code, the MLP classifier (mlp) is being trained on the training data (train_data). The `fit` method is called to initiate the training process and create a trained model. The mlp is the Multilayer Perceptron classifier instance that was defined earlier. It encapsulates the architecture and settings for the MLP network. The `fit(train_data)` is the method call that starts the training process. It takes the train_data DataFrame as input, which contains the labeled training examples with their corresponding features and labels. The `fit` method performs an iterative optimization process, where it updates the weights and biases of the MLP network using the training examples. The number of iterations is determined by the maxIter parameter specified when creating the MLP classifier.

After the training process completes, the `fit` method returns a trained MLP model (mlp_model). This model represents the learned patterns and relationships in the training data and can be used to make predictions on new, unseen data.

Step 8: Make predictions on test data

```
[In]: predictions = mlp_model.transform(test_data)
```

In this single line of code, the trained Multilayer Perceptron model (mlp_model) is used to make predictions on the test data (test_data). The `transform` method is called on the mlp_model to apply the trained model to the test data and generate predictions.

The predictions DataFrame now contains the predictions made by the Multilayer Perceptron model on the test data. We can use this DataFrame to evaluate the performance of the model.

Step 9: Evaluate accuracy

```
[In]: evaluator = MulticlassClassificationEvaluator(labelCol="species",
      predictionCol="prediction", metricName="accuracy")
[In]: accuracy = evaluator.evaluate(predictions)
[In]: print("Accuracy:", accuracy)
[Out]: Accuracy: 1.0
```

In this code, the accuracy of the predictions made by the Multilayer Perceptron model on the test data is being evaluated. The `MulticlassClassificationEvaluator` is used to calculate the accuracy metric, and the result is printed. The evaluator is an instance of the `MulticlassClassificationEvaluator` class. It is created with the following three parameters:

1. The labelCol="species" parameter specifies the name of the column in the DataFrame that contains the actual labels of the test data.

2. The predictionCol="prediction" parameter specifies the name of the column in the DataFrame that contains the predicted labels generated by the model.

3. The metricName="accuracy" parameter specifies the metric to be calculated, which is accuracy in this case.

In the second line of code, accuracy is a variable that holds the result of evaluating the accuracy metric on the predictions. The `evaluate` method is called on the evaluator instance, passing in the predictions DataFrame. The `evaluate` method compares the predicted labels in the prediction column of the predictions DataFrame with the actual labels in the species column. It calculates the accuracy, which is the fraction of correctly predicted labels out of the total number of instances in the test data.

Finally, the accuracy value is printed using the print statement. This value is 1.0 or 100%, indicating that the model has made accurate predictions for each of the Iris species. This means that the model's predictions match the true labels perfectly, resulting in a classification accuracy of 100%, which is a strong indication of the model's performance on the test data.

Step 10: Save the trained model for deployment

```
[In]: model_dir = "/dbfs/mlp_model"
[In]: mlp_model.write().overwrite().save(model_dir)
```

In this code, the trained Multilayer Perceptron model (mlp_model) is saved for deployment. The `write()` method is called on the mlp_model, followed by the `overwrite()` method and the `save()` method.

The first line defines the directory path (model_dir) where the trained model will be saved. In this example, the directory path is set to "/dbfs/mlp_model" using Databricks.

In the second line, the `write()` method is called on the trained Multilayer Perceptron model (mlp_model) to initiate the saving process. The `overwrite()` method ensures that if a model already exists in the specified directory path, it will be overwritten with the new model being saved. This step guarantees that the saved model is the latest version.

The `save()` method saves the trained model to the specified directory path (model_dir). The trained model is serialized and stored in the specified location for future use.

Step 11: Load the saved model

```
[In]: loaded_model = MultilayerPerceptronClassificationModel.
      load(model_dir)
```

In this code, a saved Multilayer Perceptron model is loaded for further use. The `load()` method is called on the `MultilayerPerceptronClassificationModel` class, and the loaded model is assigned to the variable loaded_model.

The class represents the trained Multilayer Perceptron model in Apache Spark's MLlib. The `load()` method is called on this class to load the model from the specified directory path (model_dir). It reads and deserializes the saved model files and constructs a new instance of the `MultilayerPerceptronClassificationModel` class.

The model_dir variable contains the directory path where the trained model was previously saved. The loaded_model variable is assigned the loaded Multilayer Perceptron model, which can now be used for making predictions on new data.

Step 12: Generate new (simulated) data

```
[In]: np.random.seed(42)
[In]: X_new = np.random.rand(10, 4)
```

To use the loaded Multilayer Perceptron model for making predictions on new data, we need to first generate that data. This is achieved by the code snippet shown earlier, which demonstrates the use of the NumPy library in Python to generate random numbers and create a new array with random features.

The first line of the code sets the random seed for the NumPy random number generator. Setting a seed ensures that the random numbers generated will be the same every time the code is run. The second line generates a new NumPy array named X_new with dimensions 10×4. The `rand` function is used to generate random numbers between 0 and 1. Each element in the array represents a random feature value. The first argument, 10, represents the number of rows or samples in the array. The second argument, 4, represents the number of columns or features in the array. As a result, X_new is a 10×4 array containing random feature values.

Step 13: Convert the NumPy array to a Spark DataFrame

```
[In]: new_data = spark.createDataFrame(X_new.tolist(), ["sepal_length",
      "sepal_width", "petal_length", "petal_width"])
```

This single line of code demonstrates the usage of Spark's `createDataFrame` method to create a Spark DataFrame from a NumPy array. More precisely, the code creates a new Spark DataFrame named new_data from a NumPy array X_new. The `createDataFrame` method is called on the spark object, which is the entry point for working with Spark functionalities.

The `tolist()` method converts the NumPy array X_new into a nested Python list. This conversion is necessary because `createDataFrame` expects an input that is compatible with Python data structures. The ["sepal_length", "sepal_width", "petal_length", "petal_width"] provides the column names for the DataFrame. It specifies the names of the columns in the new_data DataFrame, corresponding to the four features in the X_new array. The resulting DataFrame, new_data, will have four columns: sepal_length, sepal_width, petal_length, and petal_width, with the values populated from the corresponding elements in the NumPy array. The new_data variable holds the newly created Spark DataFrame.

Step 14: Assemble features into a single vector column for new data

```
[In]: new_data = assembler.transform(new_data)
```

In this code, the `transform()` method is called on the assembler object to apply feature transformation to the new_data DataFrame. This method takes the new_data as input and applies the specified transformation operations on its columns. The resulting DataFrame, new_data, will have an additional column named features that contains the assembled feature vectors.

Step 15: Make predictions on the new data

```
[In]: new_predictions = loaded_model.transform(new_data)
[In]: new_predictions.select("prediction").show()
[Out]:
+----------+
|prediction|
```

```
+----------+
|       1.0|
|       1.0|
|       1.0|
|       0.0|
|       0.0|
|       1.0|
|       0.0|
|       1.0|
|       1.0|
|       1.0|
+----------+
```

In step 15, predictions are made on the new dataset using the loaded Multilayer Perceptron model (loaded_model). The resulting predictions are then displayed. More details are provided in the following text.

The first line of code applies the trained Multilayer Perceptron model (loaded_model) to the new_data_scaled DataFrame. The `transform()` method is called on the loaded_model object, passing new_data as the argument. This method applies the model to the input DataFrame, making predictions based on the learned model weights and architecture. The resulting DataFrame, new_predictions, will contain the original columns from new_data along with an additional column named prediction that holds the predicted labels or values.

The second line selects the prediction column from the new_predictions DataFrame and displays its contents. The `select()` method is called on new_predictions with the column name prediction as the argument. The `show()` method is then called to display the selected column.

Bringing It All Together

In this section, we combine all the relevant code from the preceding steps into a single code block. This will enable the reader to execute the code as a coherent entity in both Scikit-Learn and PySpark.

Scikit-Learn

```
# Import necessary libraries
[In]: import numpy as np
[In]: from sklearn.datasets import load_iris
[In]: from sklearn.neural_network import MLPClassifier
[In]: from sklearn.model_selection import train_test_split
[In]: from sklearn.metrics import accuracy_score
[In]: import joblib

# Load the Iris dataset
[In]: iris = load_iris()
[In]: X = iris.data  # Features
[In]: y = iris.target  # Target labels

# Split the data into train and test sets
[In]: X_train, X_test, y_train, y_test = train_test_split(X, y, test_
      size=0.2, random_state=42)

# Create MLP classifier
[In]: mlp = MLPClassifier(hidden_layer_sizes=(100, 50), max_iter=1000,
      random_state=42)

# Train the classifier
[In]: mlp.fit(X_train, y_train)

# Make predictions on test data
[In]: y_pred = mlp.predict(X_test)

# Evaluate accuracy
[In]: accuracy = accuracy_score(y_test, y_pred)
[In]: print("Accuracy:", accuracy)

# Save the trained model for deployment
[In]: joblib.dump(mlp, 'mlp_model.pkl')

# Load the saved model
[In]: mlp = joblib.load('mlp_model.pkl')
```

```
# Generate new (simulated) data
[In]: np.random.seed(42)
[In]: X_new = np.random.rand(10, 4)  # New features (4 features for Iris
      dataset)

# Make predictions on the new data
[In]: y_new_pred = mlp.predict(X_new)
[In]: print("Predictions:", y_new_pred)
```

PySpark

```
[In]: from sklearn.datasets import load_iris
[In]: from pyspark.sql import SparkSession
[In]: from pyspark.ml.classification import
      MultilayerPerceptronClassificationModel,
      MultilayerPerceptronClassifier
[In]: from pyspark.ml.feature import VectorAssembler
[In]: from pyspark.ml.evaluation import MulticlassClassificationEvaluator
[In]: import numpy as np

# Create a SparkSession
[In]: spark = SparkSession.builder.getOrCreate()

# Load the Iris dataset
[In]: iris = load_iris()
[In]: X = iris.data
[In]: y = iris.target

# Create a PySpark DataFrame from the Iris dataset
[In]: data = [(float(X[i, 0]), float(X[i, 1]), float(X[i, 2]), float
      (X[i, 3]), float(y[i])) for i in range(len(X))]
[In]: spark_df = spark.createDataFrame(data, ["sepal_length", "sepal_
      width", "petal_length", "petal_width", "species"])

# Split the data into train and test sets
[In]: train_data, test_data = spark_df.randomSplit([0.8, 0.2], seed=42)

# Assemble features into a single vector column
```

```
[In]: feature_cols = ["sepal_length", "sepal_width", "petal_length",
      "petal_width"]
[In]: assembler = VectorAssembler(inputCols=feature_cols,
      outputCol="features")
[In]: train_data = assembler.transform(train_data)
[In]: test_data = assembler.transform(test_data)

# Create MLP classifier
[In]: mlp = MultilayerPerceptronClassifier(labelCol="species",
      featuresCol="features", layers=[4, 100, 50, 3], seed=42, maxIter=500)

# Train the classifier
[In]: mlp_model = mlp.fit(train_data)

# Make predictions on test data
[In]: predictions = mlp_model.transform(test_data)

# Evaluate accuracy
[In]: evaluator = MulticlassClassificationEvaluator(labelCol="species",
      predictionCol="prediction", metricName="accuracy")
[In]: accuracy = evaluator.evaluate(predictions)
[In]: print("Accuracy:", accuracy)

# Save the trained model for deployment
[In]: model_dir = "/dbfs/mlp_model"
[In]: mlp_model.write().overwrite().save(model_dir)

# Load the saved model
[In]: loaded_model = MultilayerPerceptronClassificationModel.
      load(model_dir)

# Generate new data (mock example)
[In]: np.random.seed(42)
[In]: X_new = np.random.rand(10, 4)  # New features (4 features for Iris
      dataset)

# Create a PySpark DataFrame for the new data
[In]: new_data = spark.createDataFrame(X_new.tolist(), ["sepal_length",
      "sepal_width", "petal_length", "petal_width"])
```

```
# assemble features in a single vector
[In]: new_data = assembler.transform(new_data)

# Make predictions on the new data
[In]: new_predictions = loaded_model.transform(new_data)

# Display the predictions
[In]: new_predictions.select("prediction").show()
```

Summary

In this final chapter of the book, we explored the practical aspects of deploying machine learning algorithms using Scikit-Learn and PySpark. We emphasized the critical role that model deployment plays in the machine learning life cycle. We concluded that organizations that successfully deploy and leverage machine learning models can gain a competitive advantage. This advantage comes from making data-driven decisions, improving customer experiences, and staying ahead in their industry.

We outlined the key steps involved in model deployment (training and evaluation, persistence, and deployment) and applied them to a Multilayer Perceptron (MLP) classification model using the Iris dataset. While this chapter primarily focused on Scikit-Learn and PySpark, we also provided information on other methods of deployment. For example, in complex workflows, organizations use Apache Airflow to orchestrate the deployment of multiple models and data pipelines in a coordinated manner.

Index

A

B

C

D

A. Testas, *Distributed Machine Learning with PySpark*, https://doi.org/10.1007/978-1-4842-9751-3

E

F

G

N

S